ENGINEERING PROBLEM SOLVING: A CLASSICAL PERSPECTIVE

ENGINEERING PROBLEM SOLVING: A CLASSICAL PERSPECTIVE

by

Milton C. Shaw

Arizona State University
Tempe, Arizona

NOYES PUBLICATIONS

WILLIAM ANDREW PUBLISHING
Norwich, New York, U.S.A.

Library of Congress Catalog Card Number: 00-022003
ISBN: 0-8155-1447-6
Printed in the United States

Published in the United States of America by
Noyes Publications / William Andrew Publishing
Norwich, New York, U.S.A.
www.williamandrew.com
www.knovel.com

10 9 8 7 6 5 4 3 2 1

Library of Congress Cataloging-in-Publication Data

Shaw, Milton Clayton, 1915-
Analytical engineering: a classical perspective / Milton C. Shaw.
p. cm.
Includes bibliographical references and index.
ISBN 0-8155-1447-6
1. Engineering mathematics. 2. Engineering--History. I. Title.
TA330 .S47 2001
620'.001'51--dc21 00-022003
CIP

MATERIALS SCIENCE AND PROCESS TECHNOLOGY SERIES

Series Editors

Gary E. McGuire, Microelectronics Center of North Carolina
Stephen M. Rossnagel, IBM Thomas J. Watson Research Center
Rointan F. Bunshah, University of California, Los Angeles (1927–1999), founding editor

Electronic Materials and Process Technology

CHARACTERIZATION OF SEMICONDUCTOR MATERIALS, Volume 1: edited by Gary E. McGuire

CHEMICAL VAPOR DEPOSITION FOR MICROELECTRONICS: by Arthur Sherman

CHEMICAL VAPOR DEPOSITION OF TUNGSTEN AND TUNGSTEN SILICIDES: by John E. J. Schmitz

CHEMISTRY OF SUPERCONDUCTOR MATERIALS: edited by Terrell A. Vanderah

CONTACTS TO SEMICONDUCTORS: edited by Leonard J. Brillson

DIAMOND CHEMICAL VAPOR DEPOSITION: by Huimin Liu and David S. Dandy

DIAMOND FILMS AND COATINGS: edited by Robert F. Davis

DIFFUSION PHENOMENA IN THIN FILMS AND MICROELECTRONIC MATERIALS: edited by Devendra Gupta and Paul S. Ho

ELECTROCHEMISTRY OF SEMICONDUCTORS AND ELECTRONICS: edited by John McHardy and Frank Ludwig

ELECTRODEPOSITION: by Jack W. Dini

HANDBOOK OF CARBON, GRAPHITE, DIAMONDS AND FULLERENES: by Hugh O. Pierson

HANDBOOK OF CHEMICAL VAPOR DEPOSITION, Second Edition: by Hugh O. Pierson

HANDBOOK OF COMPOUND SEMICONDUCTORS: edited by Paul H. Holloway and Gary E. McGuire

HANDBOOK OF CONTAMINATION CONTROL IN MICROELECTRONICS: edited by Donald L. Tolliver

HANDBOOK OF DEPOSITION TECHNOLOGIES FOR FILMS AND COATINGS, *Second Edition:* edited by Rointan F. Bunshah

HANDBOOK OF HARD COATINGS: edited by Rointan F. Bunshah

HANDBOOK OF ION BEAM PROCESSING TECHNOLOGY: edited by Jerome J. Cuomo, Stephen M. Rossnagel, and Harold R. Kaufman

HANDBOOK OF MAGNETO-OPTICAL DATA RECORDING: edited by Terry McDaniel and Randall H. Victora

HANDBOOK OF MULTILEVEL METALLIZATION FOR INTEGRATED CIRCUITS: edited by Syd R. Wilson, Clarence J. Tracy, and John L. Freeman, Jr.

HANDBOOK OF PLASMA PROCESSING TECHNOLOGY: edited by Stephen M. Rossnagel, Jerome J. Cuomo, and William D. Westwood

HANDBOOK OF POLYMER COATINGS FOR ELECTRONICS, *Second Edition*:by James Licari and Laura A. Hughes

HANDBOOK OF REFRACTORY CARBIDES AND NITRIDES: by Hugh O. Pierson

HANDBOOK OF SEMICONDUCTOR SILICON TECHNOLOGY: edited by William C. O'Mara, Robert B. Herring, and Lee P. Hunt

HANDBOOK OF SEMICONDUCTOR WAFER CLEANING TECHNOLOGY: edited by Werner Kern

HANDBOOK OF SPUTTER DEPOSITION TECHNOLOGY: by Kiyotaka Wasa and Shigeru Hayakawa

HANDBOOK OF THIN FILM DEPOSITION PROCESSES AND TECHNIQUES: edited by Klaus K. Schuegraf

HANDBOOK OF VACUUM ARC SCIENCE AND TECHNOLOGY: edited by Raymond L. Boxman, Philip J. Martin, and David M. Sanders

HANDBOOK OF VLSI MICROLITHOGRAPHY: edited by William B. Glendinning and John N. Helbert

HIGH DENSITY PLASMA SOURCES: edited by Oleg A. Popov

HYBRID MICROCIRCUIT TECHNOLOGY HANDBOOK, *Second Edition*: by James J. Licari and Leonard R. Enlow

IONIZED-CLUSTER BEAM DEPOSITION AND EPITAXY: by Toshinori Takagi

MOLECULAR BEAM EPITAXY: edited by Robin F. C. Farrow

SEMICONDUCTOR MATERIALS AND PROCESS TECHNOLOGY HANDBOOK: edited by Gary E. McGuire

ULTRA-FINE PARTICLES: edited by Chikara Hayashi, R. Ueda and A. Tasaki

WIDE BANDGAP SEMICONDUCTORS: edited by Stephen J. Pearton

Ceramic and Other Materials—Processing and Technology

ADVANCED CERAMIC PROCESSING AND TECHNOLOGY, Volume 1: edited by Jon G. P. Binner

CEMENTED TUNGSTEN CARBIDES: by Gopal S. Upadhyaya

CERAMIC CUTTING TOOLS: edited by E. Dow Whitney

CERAMIC FILMS AND COATINGS: edited by John B. Wachtman and Richard A. Haber

CORROSION OF GLASS, CERAMICS AND CERAMIC SUPERCONDUCTORS: edited by David E. Clark and Bruce K. Zoitos

FIBER REINFORCED CERAMIC COMPOSITES: edited by K. S. Mazdiyasni

FRICTION AND WEAR TRANSITIONS OF MATERIALS: by Peter J. Blau

HANDBOOK OF CERAMIC GRINDING AND POLISHING: edited by Ioan D. Marinescu, Hans K. Tonshoff, and Ichiro Inasaki

HANDBOOK OF HYDROTHERMAL TECHNOLOGY: edited by K. Byrappa and Masahiro Yoshimura

HANDBOOK OF INDUSTRIAL REFRACTORIES TECHNOLOGY: by Stephen C. Carniglia and Gordon L. Barna

SHOCK WAVES FOR INDUSTRIAL APPLICATIONS: edited by Lawrence E. Murr

SOL-GEL TECHNOLOGY FOR THIN FILMS, FIBERS, PREFORMS, ELECTRONICS AND SPECIALTY SHAPES: edited by Lisa C. Klein

SOL-GEL SILICA: by Larry L. Hench

SPECIAL MELTING AND PROCESSING TECHNOLOGIES: edited by G. K. Bhat

SUPERCRITICAL FLUID CLEANING: edited by John McHardy and Samuel P. Sawan

Preface

In this day and age of ever-shorter time schedules and increasing expectations of productivity gains in every aspect of engineering activity, creative problem solving has become prominent among the most important skills that engineers possess. Unfortunately, in the day-to-day intensity of the search for new engineering solutions, it is easy to lose touch with the foundations of that creativity. This book and its companion, *Dialogues Concerning Two New Sciences,* by Galileo Galilei, 1638, are intended to provide the working engineer a visit home to the foundations of the engineering profession. (The book *Dialogues Concerning Two New Sciences* by Galileo Galilei is available as a free e-book at the web site: www.williamandrew.com.)

Chapter 1 explains what engineers do and how those activities differ from those of physicists, chemists, and mathematicians. Chapters 2–7 of this book are a guided re-examination of the major topics of the main branches of engineering and how those ideas are involved in the design of manufactured products and the solution of engineering problems. In numerous areas, this re-examination of the core topics of solid mechanics, fluid mechanics, and aerodynamics incorporates and interprets passages from the classic work of Galileo, providing rich examples of how new knowledge is developed and the important role of experiment and dialog in that development. The book bases its visit through the development of engineering problem solving tools by considering the use of dimensional analysis where the heavier-handed solution of differential equations may not be required. The elements of dimensional analysis are presented in Ch. 3 and are employed throughout the remainder of the book. Throughout this book, the often-circuitous route in the development of new ideas is emphasized.

Engineering materials presently play a very important role in engineering applications and will undoubtedly play an even more important role in the future. Therefore, Chs. 8 and 9 are devoted to this area, the first covering some of the more general aspects and the second discussing some of the more practical details. This is followed by concepts in electrical and thermal engineering in Chs. 10 and 11.

Engineering Design is the centerpiece of engineering activity and therefore considerable space is devoted to this topic in Ch. 12.

The importance of economics, statistics, and computers is discussed in Chs. 13 through 15. Numerous examples of applications of these topics are presented.

At the end of most chapters, a number of self-study problems are provided. Answers to most of the problems appear at the end of the book.

While essentially no calculus is involved in the book, a very elementary review is presented in Appendix A. Appendices B and C provide a quick reference to conversion factors and abbreviations.

Mid-career engineers will find this book especially valuable as a refresher for the problem solving mind set. It will serve mechanical engineers first working with electrical or materials science concepts as a concise primer. It will help all engineers prepare for certification exams. It will welcome to the profession everyone who is at all interested in pursuing engineering or in knowing more about how engineers think.

Although this is basically not a textbook, it could be used for that purpose as a first or second year elective course for above-average students just beginning a course of study in one of the engineering disciplines. The book enables beginning engineering students to get the big picture before studying each subject in detail. Later, before pursuing each subject in greater depth, the appropriate chapter can be read with greater attention. This is a book which students will use again and again throughout their schooling and professional careers. An earlier version of the present book was successfully used by the author as a freshman elective at MIT and Carnegie-Mellon University.

I am indebted to the many contributions and valuable discussions with coworkers. These include: Professors C. A. Balanis, D. L. Boyer, D.

Kouris, R. J. McCluskey, J. Shah, W. M. Spurgeon, Dr. B. J. Zitzewitz, and Messers D. Ream and J. Summers. However, none of this group is associated with any errors or omissions that remain. I take full responsibility. In undertaking a work of this magnitude, it is to be expected that some errors and inconsistencies will be uncovered by discerning readers and I will be grateful to be informed of these. I also wish to acknowledge contributions of the publisher, William Andrew Publishing. My wife, Mary Jane, spent many hours editing the text, reading proofs, and generally improving the manuscript for which I am most grateful.

Milton C. Shaw
Tempe, Arizona

November, 2000

NOTICE

Dedicated to my wife,
Mary Jane

Contents

1

What Engineers Do

1.0 INTRODUCTION

Engineering is a profession. Its members work closely with scientists and apply new and old scientific effects to produce products and services that people want. Engineers do creative work. They are skilled in the art of inventing new ways of using the forces of nature to do useful things. Scientists strive to understand nature while engineers aim to produce useful products subject to economic and societal constraints.

Engineers deal with reality and usually have a set of specific problems that must be solved to achieve a goal. If a particular problem is unusually difficult, it may have to be solved approximately within the time and cost limitations under which the engineer operates.

Engineering problems usually have more than one solution. It is the aim of the engineer to obtain the best solution possible with the resources available. A criterion for measuring the degree of success of a solution is usually adopted and an attempt is made to optimize the solution relative to this criterion. The engineer rarely achieves the best solution the first time; a design may have to be iterated several times.

Engineers are professionally responsible for the safety and performance of their designs. The objective is to solve a given problem with the simplest, safest, most efficient design possible, at the lowest cost.

Figure 1.1 shows the major human activities represented as three poles labeled as follows:

- Ideas
- Nature
- People and Things

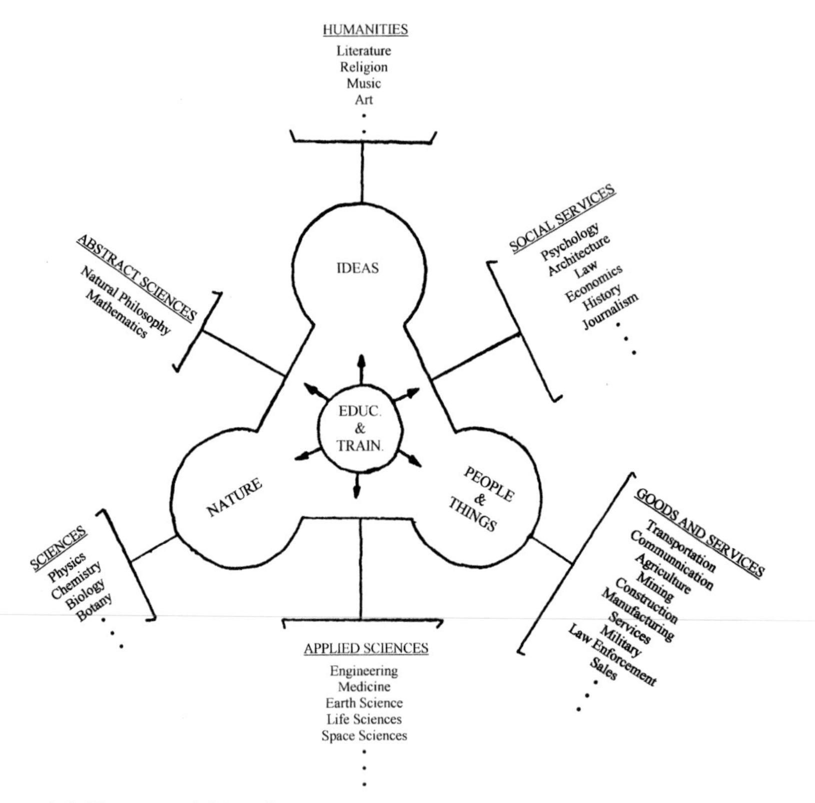

Figure 1.1. Human activities chart.

The major activities of different individuals are grouped around these poles as follows:

- Closest to the ideas pole—Humanities
- Closest to the nature pole—Sciences
- Closest to the people and things pole—Goods and Services
- Between the ideas and nature poles—The Abstract Sciences

- Between the nature and people and things poles—The Applied Sciences
- Between the people and things and the ideas pole—The Social Sciences

Engineering is obviously one of the applied sciences. The specific activities of the engineer cover a wide spectrum (Fig. 1.2). They range from the role of a pure scientist (research), to that of a sales or applications engineer who has more to do with people-oriented subjects such as psychology and economics.

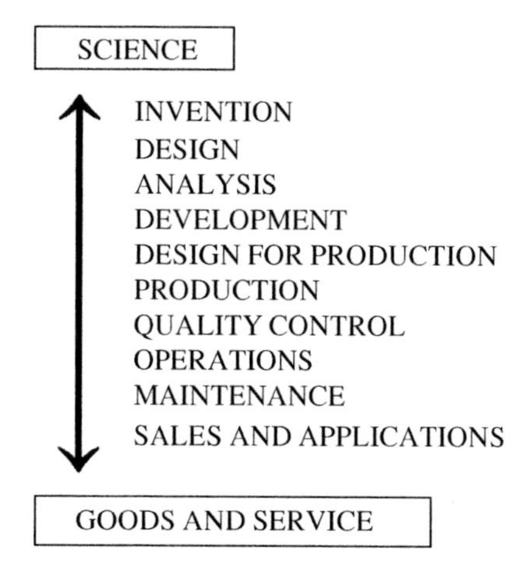

Figure 1.2. Spectrum of engineering activities.

2.0 ENGINEERING EDUCATION

The formal training of engineers, in the modern sense, is only about 125 years old, and the engineering curriculum has gradually evolved until today it contains subjects that may be divided into the following four categories:

- Science (Physics, Chemistry, and Mathematics)
- Engineering Science
- Applied Engineering
- Humanities and Social Sciences

The four years that are normally spent in obtaining an undergraduate degree are about equally divided among these four types of subjects.

The science subjects are normally physics, chemistry, and mathematics. An engineering science subject presents the principles of science in a form well-suited for the solution of a particular class of engineering problem. An example of an important engineering science subject taken by many engineers is Fluid Mechanics. This subject came into being as an integrated subject as recently as 1940.

The applied engineering subjects are concerned with the art of engineering. In these, little new knowledge is presented. Instead, students are trained to solve real problems, preferably under the guidance of an experienced engineer. A typical applied engineering subject is engineering design.

A very important aspect of engineering education is the development of communication skills (written, verbal, and visual). These attributes are usually covered in special subjects as well as in connection with reports associated with experiments.

Humanities and social science subjects are included in the curriculum because an engineer usually must deal with people and problems associated with people.

An important item, usually not treated in a formal way, is the development of a professional sense of responsibility. This includes the habit of getting jobs done on time with a reasonable degree of completeness and with emphasis on precision and a logical approach.

Engineering education is divided into several branches depending on the subject matter of the engineering science and applied engineering subjects. Figure 1.3 lists the most common engineering disciplines. These are arranged with regard to an increasing emphasis on a quantitative approach. Those branches of engineering near the top of the list are most like chemistry while those subjects near the bottom of the list are most like physics. Mechanical engineering is almost as quantitative a discipline as electrical engineering.

The location of the other fields of engineering may be readily placed in Fig. 1.3. For example:

- Nuclear and aeronautical engineering is most like mechanical engineering
- Naval architecture is most like civil engineering
- Petroleum engineering corresponds to materials and chemical engineering

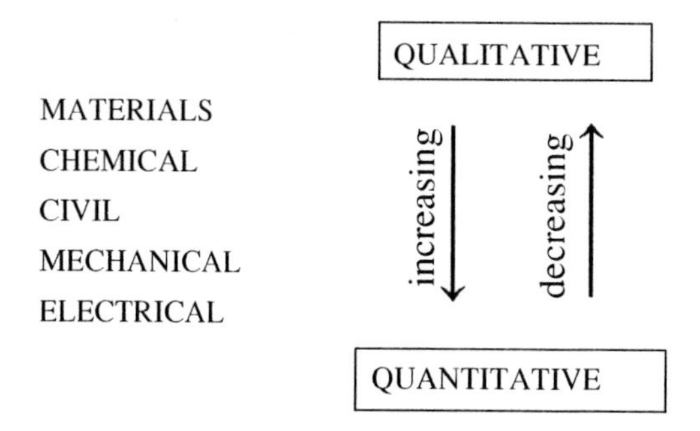

Figure 1.3. Engineering disciplines.

Despite the fact there are now a large number of engineering areas of specialization, the philosophy of approach is the same for all. This involves a strong scientific base and the practice of breaking a large complex problem into smaller manageable units that are represented by an approximate model of the real situation. One of the most important aspects of engineering is this approach to the solution of problems. This is more or less the same for all branches. Only the vehicle for discussion is different.

As with other professionals, education of the engineer does not end with graduation from a four-year curriculum. It is important to keep abreast of new developments in technology that are appearing with ever increasing frequency. In the future, it will be necessary for successful engineers to devote between 10 and 20% of their total effort to learning new analytical procedures and becoming aware of new technology. This will usually be done by self-study, attendance at technical society meetings, or participation in special short courses (one or two week's duration). It may even become common for engineers to return to a university for from six months to a year of full-time study every ten years or so.

Graduate work in engineering is relatively recent. Before World War II, the Ph.D. in Engineering was virtually unknown. This is illustrated by the fact that the first Ph.D. in Mechanical Engineering was awarded at MIT in 1930 and by 1941 only 14 such degrees had been awarded there. Today, this same department annually produces Ph.D.s by the dozens. The growth of graduate training in engineering has occurred very rapidly. About half of the engineers presently continue their formal studies for an advanced degree after graduation.

3.0 OBJECTIVE AND PROCEDURE

The objective of this book is to paint a clear picture of the activities of the engineer including the nature of the special analytical subjects involved and how these are applied to the solution of real world problems. Dimensional analysis is the vehicle that will be used in this book to discuss the analytical side of engineering. This is often the first approach to the solution of a difficult problem. It is particularly useful here in that it enables the highlights of the core engineering subjects to be considered without becoming bogged down with less important details. Subjects to be considered in this way include:

- Solid mechanics
- Fluid mechanics
- Aerodynamics
- Electrical engineering
- Materials science
- Materials engineering
- Energy conversion

Also, since cost plays such a dominant role in the solution of engineering problems, an introduction to engineering economics is included. Because engineering frequently involves experimental measurements, it is appropriate that an introduction to statistics also be included. Computers play such an important role in engineering today that the final chapter is devoted to the history and applications of computers in engineering design, production, and communications. The art of engineering is represented by a discussion of design and creativity.

4.0 GALILEO

Throughout the book, the often circuitous route taken in the development of new ideas is emphasized. In this connection, a classical volume by Galileo will be used by way of illustration. Where possible, Galileo's view of things is related to modern technology; appropriate passages are considered and discussed in most chapters.

Galileo (1564–1642) was born in Pisa and studied at the university there. From 1589 to 1610, he was a member of the faculty of the University of Padua where most of his innovative ideas were formulated. He wrote two important books: "Dialogs Concerning the Two Chief World Orders," (1628) and "Dialogs Concerning Two New Sciences" (1638).

The first of these books discussed experimental observations that supported the Copernican theory that the Earth and all other planets revolve about the sun and that the Earth is not the center of the universe. The second book is concerned mainly with the resistance of materials to fracture and the motion of bodies. Both of these books emphasize the importance of experiments in deriving physical laws as opposed to a reliance on proverbial beliefs, authority, or purely theoretical reasoning.

Selected passages from the second of these two books will be used here to demonstrate:

- The importance of experimental verification.
- How very simple experiments that do not require sophisticated instrumentation may be devised.
- The power of inductive reasoning.
- The importance of explaining experimental observations in fundamental terms.
- The importance of dialog and group action in the solution of complex problems.
- The skillful use of example.
- That even the most famous people may be wrong on occasion.
- The origin of several important physical concepts including those of inertia, buoyancy, surface tension of liquids, the density of air, and its resistance to motion (drag).
- How understanding one physical phenomenon in fundamental terms often provides an explanation for other related ones, or even suggests new inventions.
- To illustrate the value of simulation in the solution of engineering problems.
- That in the approximate solution of engineering problems, relative values rather than absolute numerical values are often sufficient, more easily obtained, and in some cases, preferred.

- The importance of approximation in the solution of real engineering problems and of ignoring second and higher order effects, but at the same time being sure that the effects ignored are small compared with those retained.

It is important that an engineer understand that the solution of new problems rarely involves a direct path of reasoning as textbook explanations usually do. Instead, several false starts are often involved before a solution satisfying all requirements and constraints is obtained.

Galileo's book is written as a rambling dialog between a professor (Salviati representing Galileo) and two students (Sagredo, an A student, and Simplicio, a C student) over a four-day period. To say that the book rambles is an understatement, and a first reading is apt to be confusing when following the order in which the material is presented. Therefore, selected passages concerned with topics being discussed in more modern terms are suggested at appropriate points in this text. It is hoped that this will not only illustrate the working of a great mind, but also serve as an introduction to one of the great pieces of classical scientific literature.

The first segment of the Galileo text to be considered is the first page (passage number 49) followed by the passages numbered 50 through 67. It is suggested that these pages be read after considering the following discussion of their content.

In passage 49, Galileo mentions visiting the arsenal at Venice and interacting with the artisans there involved in the design and construction of various instruments and machines. He considers this useful since many of them by experience have sound explanations for what they do. However, he finds that all their explanations are not true and warns against false ideas that are widely accepted (proverbial concepts).

Galileo's main objective is to explain the resistance to fracture that materials exhibit. He begins by discussing the role that geometrical size plays relative to strength, and concludes that two geometrically similar machines made from identical material will not be proportionately strong.

In support of this view, several examples are discussed:

- The case of similar wooden rods of different size loaded as cantilevers
- The paradox of a brittle column being eventually weaker with three point support than with two point support
- The strength of two geometrically similar nails loaded as cantilevers

In each of these cases, loading is not in simple tension but involves bending. Fracture involves not only the applied load, but also its distance from the point of fracture (called a bending moment—the product of a force and the perpendicular distance from the force to a center of rotation). At this point in the dialog, Galileo does not distinguish between strength in simple tension and strength in bending. However, in the second day, he derives a number of relationships for the relative strengths of beams of different geometries loaded in bending. These will be considered in Ch. 4.

Galileo next tackles the origin of the strength of different materials in simple tension. Fibrous materials, such as rope and wood, loaded parallel to the grain, are considered first. The question of how a long rope consisting of relatively short fibers can be so strong is considered next. The role that helical entwinement of individual fibers plays is qualitatively explained as is the mechanics of a capstan and that of a clever device for controlled descent of a person along a rope (Fig. 3 in Galileo).

The breaking strength of other nonfibrous materials is next considered from two points of view—the force associated with the vacuum generated when two surfaces are rapidly separated, and the possibility of some sort of adhesion existing between minute particles of the material. The role of a vacuum which accounts for the tensile strength of a column of water turns out to be negligible for a solid such as copper. Galileo anticipates the fact that solids consist of extremely small particles (now called atoms) that are held together by some sort of adhesive substance (now called atomic bonds). He further concludes that when a metal is raised to a high temperature, the adhesive substance is reversibly neutralized and the metal melts. These matters are discussed from a modern perspective in Ch. 5 of this book.

In passage 60 of the Galileo text, Sagredo makes the following Yogi Berra-like observation. "...although in my opinion nothing occurs contrary to nature except the impossible, and that never occurs." However, a short time later he makes a more meaningful observation that establishment of a vacuum, as when two smooth flat plates separate, cannot be responsible for the resistance to fracture since generation of the vacuum follows separation by fracture (cause must precede effect).

It is suggested that, at this point, passages 49–67 of the Galileo text be read to gain an appreciation for the way in which a problem is tackled by Galileo and the many interesting discussions that, temporarily, intervene. Galileo, being unusually inquisitive, cannot resist exploring many side issues, but eventually returns to the main problem.

Beginning in passage 68, Galileo considers the nature of infinity and whether infinity can be subdivided. This consists mainly of some clever geometrical considerations designed to better understand the meaning of infinity and whether subdivision of a solid into extremely small particles is feasible. The end result appears to be that infinity is not a number, but an entirely different concept. It, therefore, cannot be treated like a number and subjected to arithmetical operations such as addition, subtraction, multiplication, division, or extraction of a root. It is suggested that the material from passages 68 to 85 of the Galileo text be merely scanned. Material discussed during the remainder of the first day will be covered in subsequent chapters of this book, mainly because important concepts of significance today are clearly identified, and, in many cases, verified by elegantly simple experiments.

2

Rigid Body Mechanics

1.0 HISTORICAL INTRODUCTION

Engineering mechanics involves the effects of force and motion on bodies. It had its beginning with Archimedes (287–212 BC). He is credited with explanations for the lever and for buoyancy. Galileo first demonstrated that falling bodies of different weight would fall at the same speed (if you account for the drag resistance of air). This was contrary to the view of Aristotle who held, apparently without experimental verification, that the velocity of a falling object is proportional to its weight. Galileo said that bodies of similar size and shape, but of different weight, would strike the ground at the same time when released from a great height. Folklore suggests that Galileo demonstrated this by dropping objects from the Tower of Pisa, but there is no mention of this in his writings. Instead, Galileo used a simple pendulum to slow the rate of fall of bodies of different weight and compared times for a large number of complete excursions. This was necessary since the pendulum clock had not yet been invented by Huygens (1657). Short times were expressed in terms of pulse beats and longer times in terms of the weight of water flowing from a large reservoir through an orifice.

It is suggested that the account of Galileo's famous pendulum experiments and the way in which he handled the question of the role of drag resistance of air be read at this point (passages 128–135). Galileo correctly points out that the time for a complete oscillation (period) is independent of the arc of traverse but proportional to the square root of the length of string. Thus, while the decrease of amplitude of swing may vary with drag, the traverse time will not.

Galileo made important contributions to the concept of inertia and clearly distinguished between uniform motion and accelerated motion. This was extensively discussed in the dialogs of the third day, in terms of a geometrical approach reflecting influence of earlier Greek philosophers.

Isaac Newton (1642–1727), who was born the same year that Galileo died, was familiar with his writings and influenced by them. One of Newton's major contributions was the invention of calculus. Others involved contributions in the field of mechanics that were presented in the form of Newton's three basic laws:

1st law: An object remains at rest or in uniform motion (constant velocity) unless acted upon by an unbalanced force.

2nd law: The acceleration (a) of a body is proportional to the resultant force acting on the body. The proportionality constant is mass, m (inertia to change in velocity):

Eq. (2.1) $$F = ma$$

3rd law: Forces of action and reaction between interacting bodies are equal, opposite, and collinear.

Newton also made many other contributions to science including the theory of light and the universal law of gravitation which states that the force of attraction between bodies is:

Eq. (2.2) $$F = (Gm_1m_2)/r^2$$

where m_1 and m_2 are the masses of two interacting bodies separated by a distance (r) and G is a constant equal to $6.673 \times 10^{-11}\ m^3kg^{-1}s^{-2}$.

A body in the vicinity of the earth is attracted to the center of the earth with a force corresponding to the acceleration due to gravity:

$$g = 32.2 \text{ ft.sec}^{-2}\ (9.81 \text{ m.sec}^{-2})$$

This is essentially a constant for all points on the surface of the earth and for altitudes less than about 30 km (18.9 mi.).

The entire field of rigid body mechanics may be divided as follows:

- Statics
- Kinematics
- Kinetics
- Vibrations

2.0 STATICS

Statics involves bodies at rest or with uniform motion (no acceleration) acted on by a system of forces. By Newton's second law, if the resultant external force on a body is zero, forces and moments in all directions must be in equilibrium, i.e.:

Eq. (2.3) $$\Sigma F = 0$$

Eq. (2.4) $$\Sigma M_0 = 0$$

where the subscript 0 represents any axis of rotation perpendicular to the plane of the forces. A force is a vector quantity having both magnitude and direction, and a moment is the product of a force and the perpendicular distance from the force to the axis of rotation (moment is also a vector quantity).

Figure 2.1 (*a*) shows an elementary problem in statics where the problem is to find forces acting as supports at A and B. The beam is assumed to be rigid and the friction force on the roller at A is assumed to be negligible relative to other forces. The first step is to show the isolated beam with all forces acting on it. This is called a *free body diagram* [Fig. 2.1 (*b*)]. In engineering mechanics, it is convenient to resolve forces into components. Since Fig. 2.1 is for a two dimensional problem, only two coordinate directions are involved (x and y). The unknown force at A will be as shown in Fig. 2.1 (*b*) while that at B will have two components. There are three unknowns, hence three equations will be required for a solution. These equations are:

Eq. (2.5) $\Sigma F_x = 0;\ F_{Bx} - 50 = 0$

Eq. (2.6) $\Sigma F_y = 0;\ F_A + F_{By} - 86.6 = 0$

Eq. (2.7) $\Sigma M_A = 0;\ (86.6)(3) + F_{Bx}(0.25) - 50(0.5) - F_{By}(5) = 0$

where ΣF_x is the vector sum of all components of force in the x direction.

Solving these three equations:

$$F_A = 37.14\ \text{N}$$

$$F_{By} = 49.46\ \text{N}$$

$$F_{Bx} = 50\ \text{N}$$

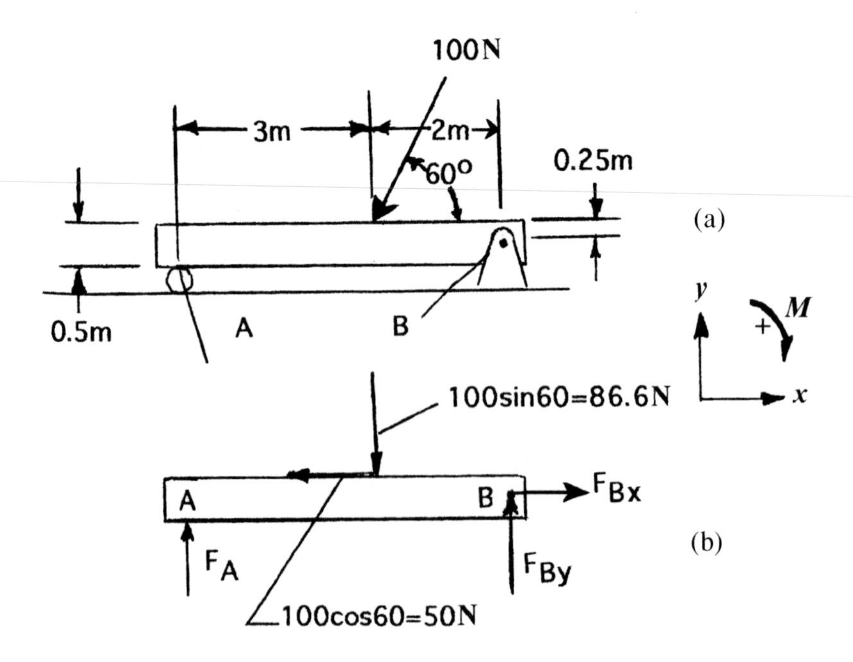

Figure 2.1. (*a*) Simple beam on two supports with load as shown, and (*b*) free body diagrams for (*a*).

Moments could have been taken about any point in the xy plane, but it is convenient to select a point where a force acts.

Figure 2.2 (a) shows a simple cantilever beam with a built-in support at A. This type of support provides a resisting couple as well as horizontal constraint as shown in the free body diagram [Fig. 2.2 (b)]. A couple is a moment consisting of two equal opposing forces (F_c) separated by a perpendicular distance (c) (a pure twisting action).

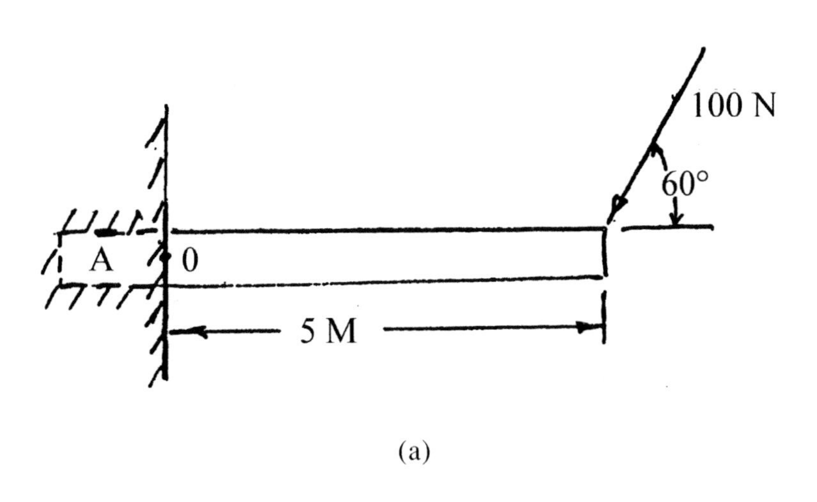

(a)

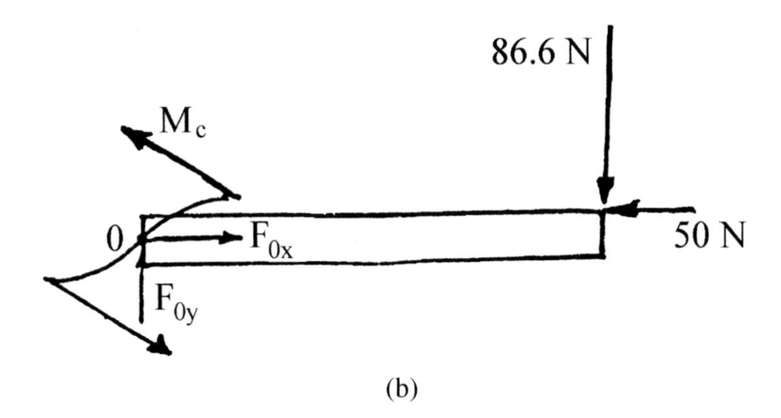

(b)

Figure 2.2. (a) Cantilever beam, and (b) free body diagram for (a).

The moment of a couple is:

Eq. (2.8) $M_c = F_c(c)$

and it may be considered to act at any point in the xy plane. In this case, for static equilibrium:

$$F_{ox} = 50.0 \text{ N}$$
$$F_{oy} = 86.6 \text{ N}$$
$$M_c = (86.6)(5) = 433 \text{ N}$$

3.0 TRUSSES

Figure 2.3 (*a*) shows a structural member called a plane truss. This particular design is called a Howe truss commonly used for roofs. This truss has pin connections at each end, and if loaded by a rope as shown, there are four unknowns (F_{Ax}, F_{Ay}, F_{Bx}, and F_{By}) as shown in the free body diagram [Fig. 2.3 (*b*)], but only three equations ($\Sigma F_x = 0$, $\Sigma F_y = 0$, and $\Sigma M = 0$). Such a structure is indeterminate. However, if the truss would be symmetrically loaded as by a snow load, it would be structurally determinate ($\Sigma F_{Ax} = 0$, $\Sigma F_{Bx} = 0$, and $\Sigma M = 0$). It would also be structurally determinate if the support at either A or B was a roller.

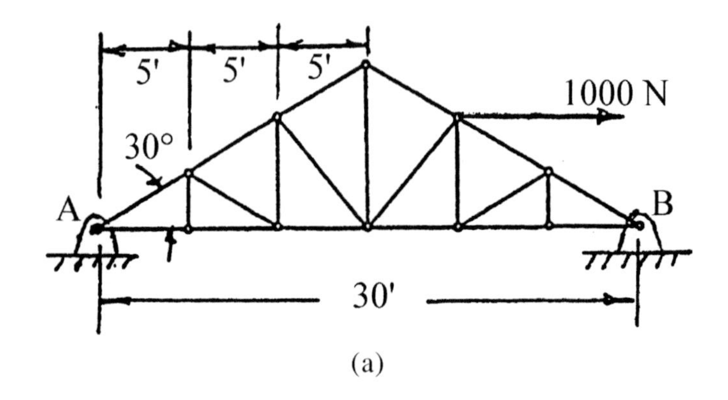

Figure 2.3. *(a)* Howe roof truss with pin connections at *A* and *B*, and *(b)* free body diagram for *(a)*.

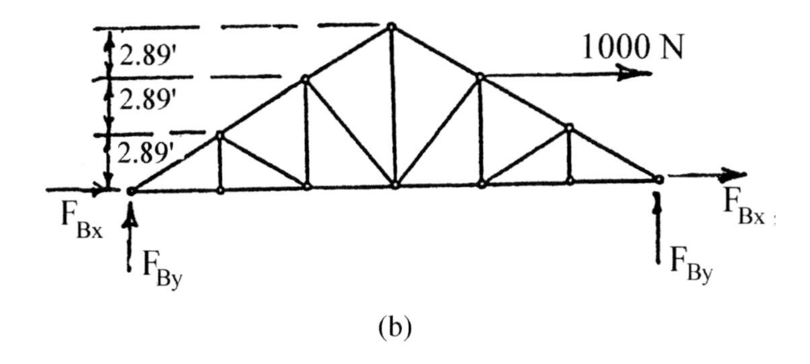

Figure 2.3. *(Cont'd.)*

Since the load is applied by a rope, the load must be a tensile load. Flexible members (rope, cable, chain, etc.) have no load capacity in compression. Normally, the weight of the elements of a truss are ignored and the individual struts are considered to be loaded, in either tension or compression, and external loads are considered to be applied only at pins.

The load in internal members of a truss may be determined by sectioning the truss and taking either half as a free body showing all forces acting including those in the sectioned member. Figure 2.4 (*a*) shows Fig. 2.3 (*a*) with a roller support at *B* and with the components of force at *A* and *B* for static equilibrium of the entire truss. Consider the truss to be sectioned along *EF* in Fig. 2.4 (*a*). The left hand section is shown as a free body in Fig. 2.4 (*b*). For static equilibrium, the loads in elements *AC* and *AD* may be found by solving the following three equations:

Eq. (2.9) $\Sigma F_x = 0; F_{Dx} + F_{Cx} - 1000 = 0$

Eq. (2.10) $\Sigma F_y = 0; F_{Cy} - 193 = 0$

Eq. (2.11) $F_{Cy}/F_{Cx} = \tan 30$ (resultant force F_C is in direction *AC*)

Solving these three equations:

$$F_{Cy} = F_{Ay} = 193 \text{ N}$$
$$F_{Cx} = 334 \text{ N}$$
$$F_D = F_x = 666 \text{ N}$$

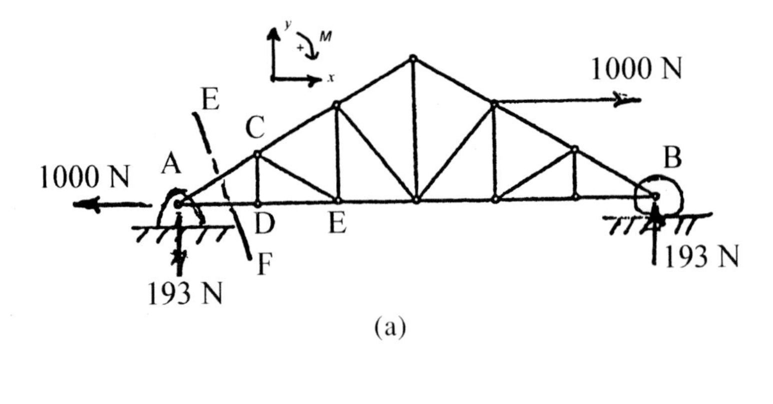

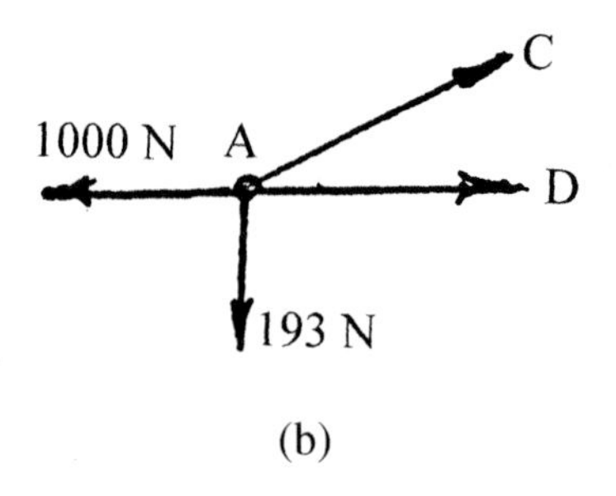

Figure 2.4. *(a)* Howe roof truss with pin connection at A and rollers at B, shown sectioned along EF, and *(b)* free body diagram for left hand section.

The few truss problems considered here are extremely easy ones, but collectively illustrate all of the principles involved in the solution of the more complex problems met in practice. It should be noted that values are given to only three significant figures. This is all that is justified when one considers the number of approximations made in this sort of analysis. When angles are small, sin θ = tan θ = θ (radians) which corresponds to only an error of 0.5% for θ = 10°. Distributed forces are approximated by replacing a distributed load by a number of concentrated loads acting at the center of mass of each element. Friction forces are handled by assuming a coefficient of friction f_s = tangential force/normal force for static surfaces and f_k for moving surfaces (usually $f_k < f_s$). A convenient way of estimating f_s or f_k is by measuring the angle of repose (Fig. 2.5). A weighted specimen is placed on a flat surface and the surface is rotated upward until it is about to slide (for f_s) or slides at constant velocity (for f_k).

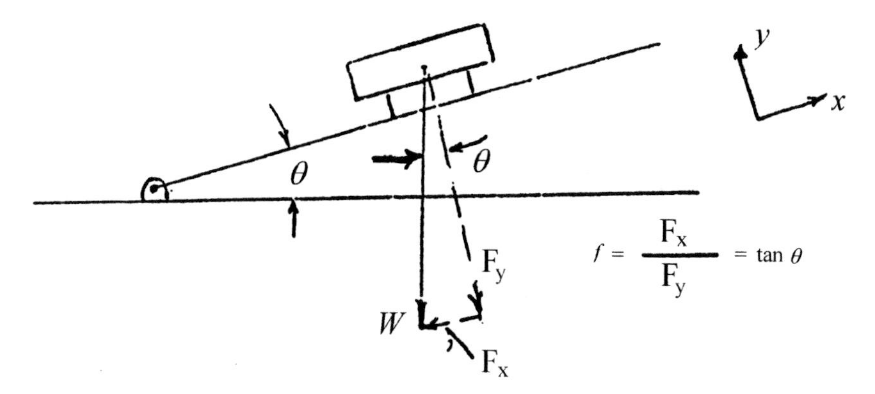

Figure 2.5. Determination of friction coefficient (f) by measurement of angle of repose θ.

4.0 FRICTION

Figure 2.6 illustrates an important statics problem involving friction. This shows a belt driving a pulley at constant speed (V) without slip. T_1 and T_2 are tensile forces in the belt leaving the pulley and entering the pulley respectively. If the belt drives the pulley $T_1 > T_2$ and the work done per unit time (power $= P$) will be:

Eq. (2.12) $$P = (T_1 - T_2)V$$

The value of the static friction force to prevent slip may be estimated by a simple application of integral calculus as follows: Fig. 2.6 (b) shows an incremental section of the belt as a free body. The tensile force directed to the left is $T + dT$ while that directed to the right is T. The static friction force acting to the right when slip is impending will be $f_s 2T\sin d\alpha/2 \cong f_s T d\alpha$, since $d\alpha$ is an infinitesimally small angle. For static equilibrium:

Eq. (2.13) $$\Sigma F_x = 0; \quad (T + dT - T)\cos\frac{d\alpha}{2}$$

but since $d\alpha$ is very small, $\cos d\alpha/2 \cong 1$, and

Eq. (2.14) $$dT/T = f_s d\alpha$$

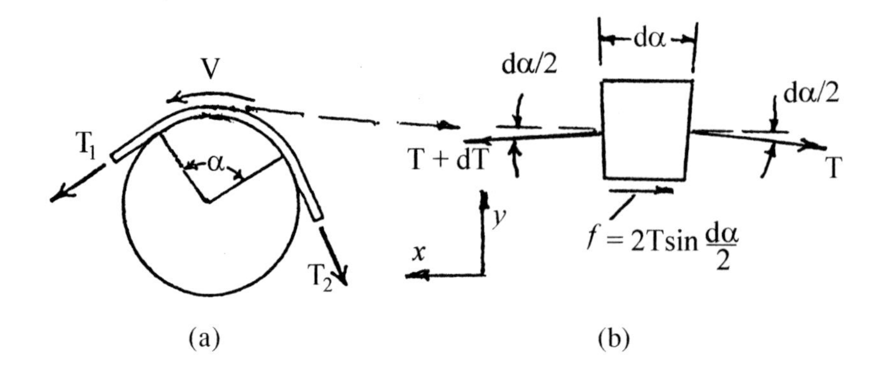

Figure 2.6. *(a)* Belt driven pulley with arc of contact α, and *(b)* free body diagram for element of belt.

Integrating both sides from 1 to 2 and considering the static friction coefficient f_s to be constant:

Eq. (2.15) $\quad$ $\ln T_1/T_2 = f_s(\alpha_2 - \alpha_1) = f_s\alpha$

(A brief review of elementary calculus is given in App. A.)

Hence,

Eq. (2.16) $\quad$ $T_1/T_2 = e^{f_s\alpha}$ (where e = 2.72 and α is in radians)

This nondimensional equation is very useful and has many applications. It should be noted the force ratio is independent of the pulley diameter, but increases exponentially with the product ($f_s\alpha$).

For the problem of Fig. 2.6, if f_s (impending slip) = 0.3 (leather on cast iron) and $\alpha = \pi/2$ radians (90°), then the value of T_1/T_2 when slip occurs will be 1.60. If the value of T_1 for the above case is not to exceed 1,000 lbs (4,448 MN), then the maximum horsepower (1 hp = 33,000 ft. lb/min.) that may be transmitted at a belt speed of 1,000 fpm (305 meters/minute) without slip may be calculated from Eq. (2.12).

The maximum hp without slip will, therefore, be:

$$hp_{max} = [(1{,}000 - 625)/33{,}000](1{,}000) = 11.4 \ (8.5 \text{ kW})$$

Another application is to estimate the mechanical advantage of a capstan. The static coefficient of friction when slip is impending for hemp rope on steel is about 0.25. For four complete turns about the capstan ($\alpha = 25.1$ radians), the value of T_1/T_2 from Eq. (2.16) is:

$$T_1/T_2 = 2.72^{(0.25)(25.1)} = 533$$

Thus, a force of 10 pounds (44.5 N) applied at T_2 would give a force of 5,330 lbs (23.7 kN) at T_1.

Figure 2.7 shows a simple dynamometer for loading a rotating shaft. From Eq. (2.16), $T/W = e^{f\alpha}$ where f is the kinetic coefficient of friction. At a constant speed, V, the horsepower dissipated will be:

Eq. (2.17) $$\text{hp} = \frac{(T - W)V}{33{,}000} = \frac{(e^{f\alpha} - 1)WV}{33{,}000}$$

for $f_k = 0.40$
$\alpha = \pi$
$W = 100$ lbs (44.8 N)
$V = 1{,}000$ fpm (305 meters/minute):

$$\text{hp} = [(3.52 - 1)(100)(1{,}000)]/33{,}000 = 7.63 \text{ (5.69 kW)}$$

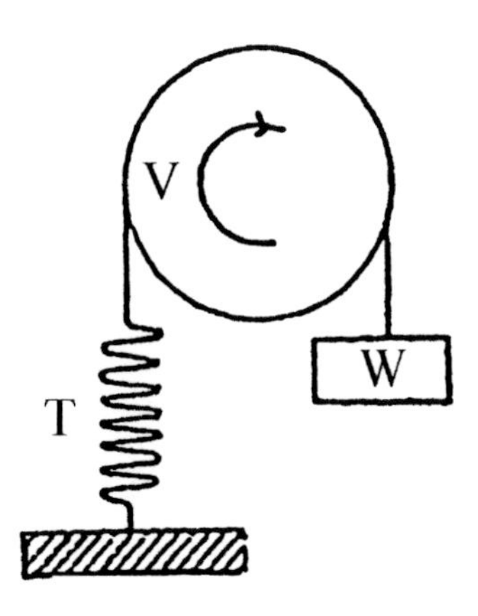

Figure 2.7. Dynamometer for measuring load applied to rotating shaft.

5.0 GALILEO REGARDING ROPE

In discussing the strength of a rope made of short fibers, Galileo correctly suggests that both twist and compression play a role. The latter concept is suggested by Galileo in passage 57 of the Galileo text: "… the very force which pulls it in order to snatch it away compresses it more and more as the pull increases."

A simple three strand rope is shown in Fig. 2.8. Here, three preliminary ropes of diameter (d_2) are twisted together in close-packed helical fashion to form the final rope of diameter (d_1). These three preliminary ropes A, B, and C are laid together to form right hand helixes, in this case. Each of the preliminary ropes consists of secondary three strand, closely-packed helical ropes of diameter (d_3) that have a left hand helical lay. The preliminary rope (A) is made up of secondary ropes D, E, and F. Other layers of closely-packed helixes may be added to form ropes of greater diameter.

When a tensile force (T) is applied to the rope, it may be resolved into components along and normal to the axis of the preliminary ropes [as shown in Fig. 2.8 (a)]. The normal component (T_n) of one preliminary rope is constrained by its neighbor resulting in a closed circle of internal compressive forces proportional to the applied load T.

Force component T_{a1} will be transferred to a preliminary rope and this may, in turn, be resolved into normal and axial components as shown at (b) in Fig. 2.8. The normal components will, in turn, give rise to a closed circle of internal compressive forces as before. This is the source of the compressive component of strength of a rope to which Galileo refers. The helical twist gives rise to an amplification of the static coefficient of friction by the angle of twist while the high internal compression developed gives rise to very high friction forces. This combination of high compression and amplification of static friction due to twist results in rope being a very stable structure when loaded in tension. Of course, the ratios d_1: d_2: d_3 and the helix angle should be carefully selected in designing a rope so that the load is uniformly distributed with a minimum of internal adjustment by slip when load is applied.

The novelty always popular at children's birthday parties called the "Chinese Finger Trap" (Fig. 2.9) represents a dual application of the principle of rope. Eight thin strips of material are woven into a hollow tube, four strips with a right hand lay and four with a left hand lay. When fingers are inserted into the ends of the tube (at A and B in Fig. 2.9) and pulled outward, the fingers are very tightly gripped and cannot be removed.

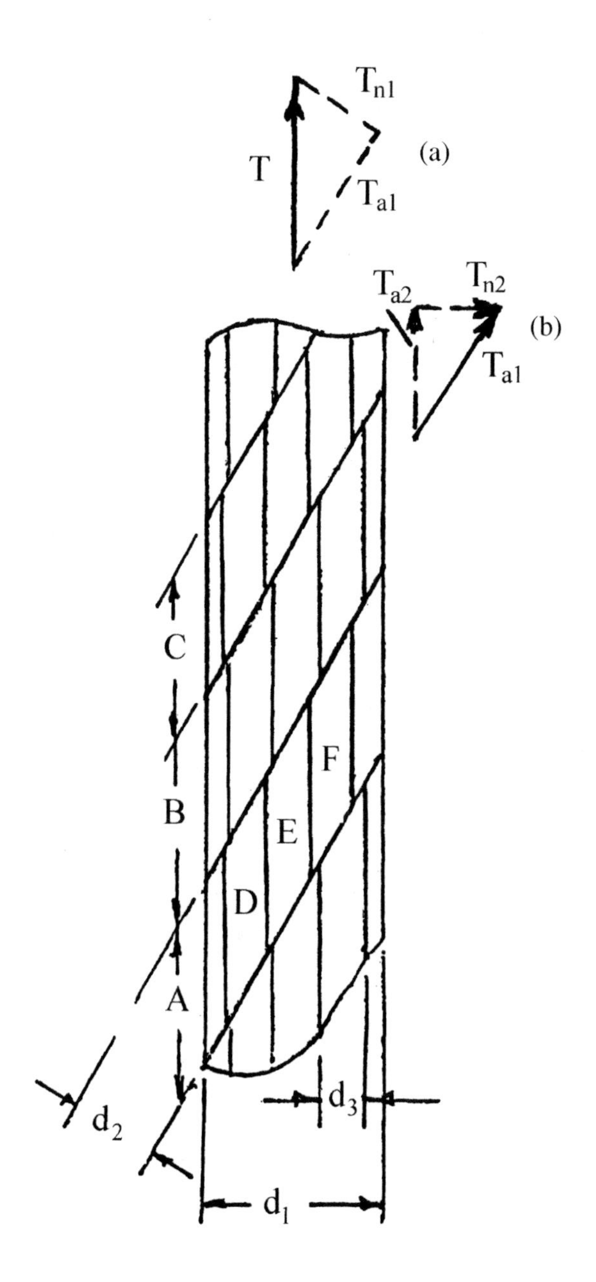

Figure 2.8. Schematic structure of rope.

Figure 2.9. Chinese finger trap.

Braided rope is now available that resembles the finger trap even more closely than conventional rope. This rope has more than four helical threads spiraling left and right around a central core. The advantages of this rope are that it will not twist when loaded in tension, it runs more smoothly over pulleys (due to symmetry), and gripping of the core will be more uniform.

6.0 KINEMATICS

Kinematics concerns the motion of bodies without reference to forces. It is involved with relationships between displacements (s), velocities ($v = ds/dt$), and accelerations ($a = dv/dt = d^2s/dt^2$) for a wide variety of machine elements such as cams, gears, and mechanical linkages.

Problems of this sort are frequently solved by writing and solving differential equations using appropriate initial conditions to evaluate constants of integration. The following problem is a relatively simple one that illustrates the procedure.

6.1 Projectiles

A projectile is set free from point (0) in Fig. 2.10 having an initial velocity V_0 inclined θ to the horizontal at time $t = 0$. Air drag is ignored and the acceleration due to gravity is considered constant at 9.81 m/s^2 (32.2 ft/sec^2). The projectile does not interact with any forces and acceleration in the x direction is zero. The maximum elevation y_m above $y = 0$ is to be found. The following analysis is involved (see App. A if needed):

Eq. (2.18) $d^2y/dt^2 = -g$

Eq. (2.19) $dy/dt = -gt + C_1$

Eq. (2.20) $y = -gt^2/2 + C_1t + C_2$

Eq. (2.21) $d^2x/dt^2 = 0$

Eq. (2.22) $dx/dt = C_3$

Eq. (2.23) $x = C_3t + C_4$

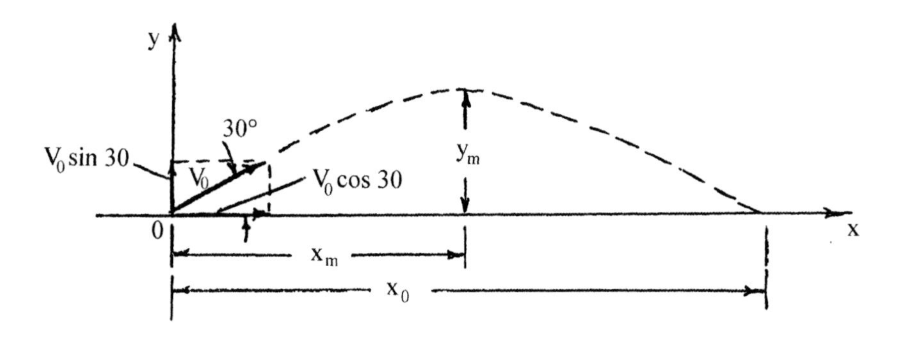

Figure 2.10. Path of projectile released from ground at initial velocity (V_0) at an angle of 30°.

Initial conditions are as follows:

$$\text{when } t = 0\text{: } x = 0, y = 0; C_2 = C_4 = 0$$

$$dx/dt = V_0 \cos\theta = C_3,\ dy/dt = V_0 \sin\theta = C_1$$

Hence,

Eq. (2.24) $y = -gt^2/2 + V_0t \sin\theta$

Eq. (2.25) $x = V_0t \cos\theta$

Eliminating t:

$$y = -\frac{g}{2V_0^2 \cos^2 \theta} x^2 + (\tan\theta)x \qquad \text{(a parabola)}$$

The maximum elevation above $y = 0$ will occur when $dy/dt = 0$. Substituting this into Eq. (2.19) and solving for t_m:

Eq. (2.26) $\quad t_m = (v_0 \sin\theta)/g$

and from Eq. (2.24):

Eq. (2.27) $\quad y_m = (V_0 \sin\theta)^2/2g$

As an example, let $\theta = 30°$ and $V_0 = 60$ m•s^{-1}:

$$t_m = [60\ (\sin 30)]/9.81 = 3.06 \text{ s}$$

$$y_m = (60 \sin 30)^2/[2(9.81)] = 45.9 \text{ m (150 ft)}$$

The corresponding horizontal distance is:

$$x_m = V_0 t_m \cos\theta = (60)(3.06)(\cos 30) = 159 \text{ m (522 ft)}$$

The horizontal distance traveled before striking the ground will be:

$$2x_m = 2(159) = 318 \text{ m (1040 ft).}$$

6.2 Crank Mechanism

Possibly the most widely used linkage requiring kinematic analysis is the basic crank mechanism of the reciprocating internal combustion engine shown diagrammatically in Fig. 2.11, where:

R = the radius of the crankshaft in inches

$2R$ = the stroke in inches

L_R = the connecting rod length in inches

N = the rpm of the engine (here considered a constant)

A = the big end bearing linking the connecting rod to the crank shaft

B = the wrist pin that attaches the connecting rod to the piston

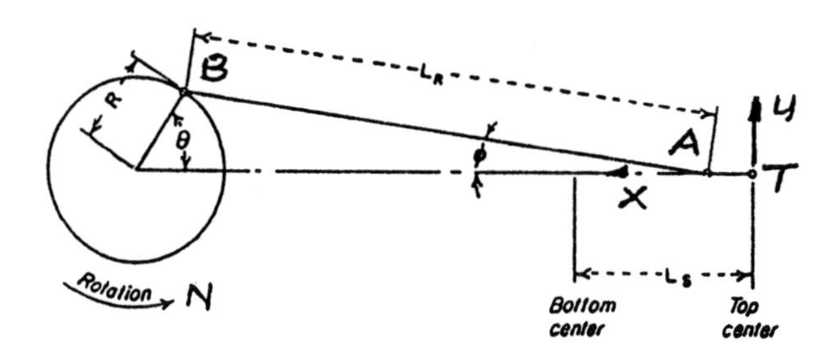

Figure 2.11. Internal combustion engine crank mechanism.

Figure 2.12 shows a typical connecting rod, with the location of its center of mass at (C). The displacement (x), velocity (dx/dt), and acceleration (d^2x/dt^2) for different points on the linkage are of interest in engine design. These values at (A) (the wrist pin) are of particular interest and the analysis given below illustrates just how complex pertinant such calculations can be.

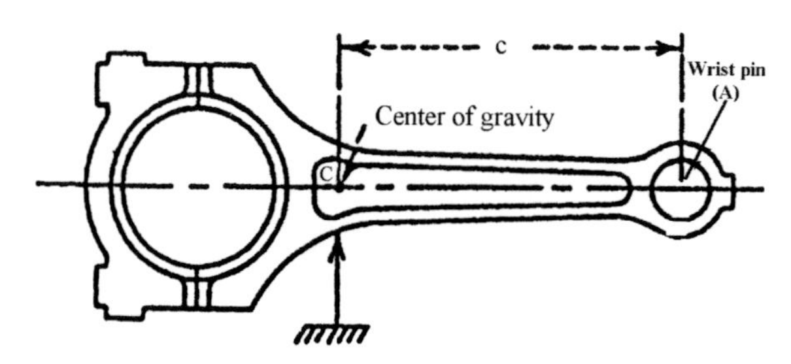

Figure 2.12. Connecting rod.

The displacement x from the top center position T ($\theta = 0$) will be:

Eq. (2.28) $$x = L_R + R - L_R \cos \phi - R \cos \theta$$

The relation between ϕ and θ will be:

Eq. (2.29) $$L_R \sin \phi = R \sin \theta$$

When ϕ is eliminated between Eqs. (2.28) and (2.29):

Eq. (2.30) $$x = R\left[1 - \cos\theta + \frac{L_R}{R} - \sqrt{\left(\frac{L_R}{R}\right)^2 - \sin^2\theta}\right]$$

The displacement of the piston from top center thus may be obtained for any crank angle θ if R and L_R are known.

The velocity of the piston may be found as follows:

Eq. (2.31) $$\frac{dx}{dt} = \frac{dx}{d\theta}\frac{d\theta}{dt}$$

where $d\theta/dt = (2\pi N)/60$ [radians per sec.], and

Eq. (2.32) $$\frac{dx}{dt} = \left(\frac{2\pi}{60}\right)\left(\frac{R}{12}\right)\left[\sin\theta + \frac{\sin 2\theta}{2\sqrt{\left(\frac{L_R}{R}\right)^2 - \sin^2\theta}}N\right]$$

The acceleration of the piston may be similarly found to be

Eq. (2.33)

$$\frac{d^2x}{dt^2} = \left(\frac{2\pi}{60}\right)^2\left(\frac{R}{12}\right)\left\{\cos\theta + \frac{\left[\left(\frac{L_R}{R}\right)^2 - 1\right]\cos 2\theta + \cos^4\theta}{\left[\left(\frac{L_R}{R}\right)^2 - \sin^2\theta\right]^{1.5}}\right\}N^2$$

When these three equations are evaluated for given values of R, L_R, and N at a number of values of crank angle θ, a plot of piston displacement, velocity, and acceleration such as that shown in Fig. 2.13 is obtained. These results are for the R-1820 that was the most widely used aircraft engine in World War II. This engine had a stroke ($2R$) of 6.875 in. (175 mm) and a connecting rod length of 13.75 in. (349 mm). Figure 2.13 is for a constant engine speed of N of 2500 rpm.

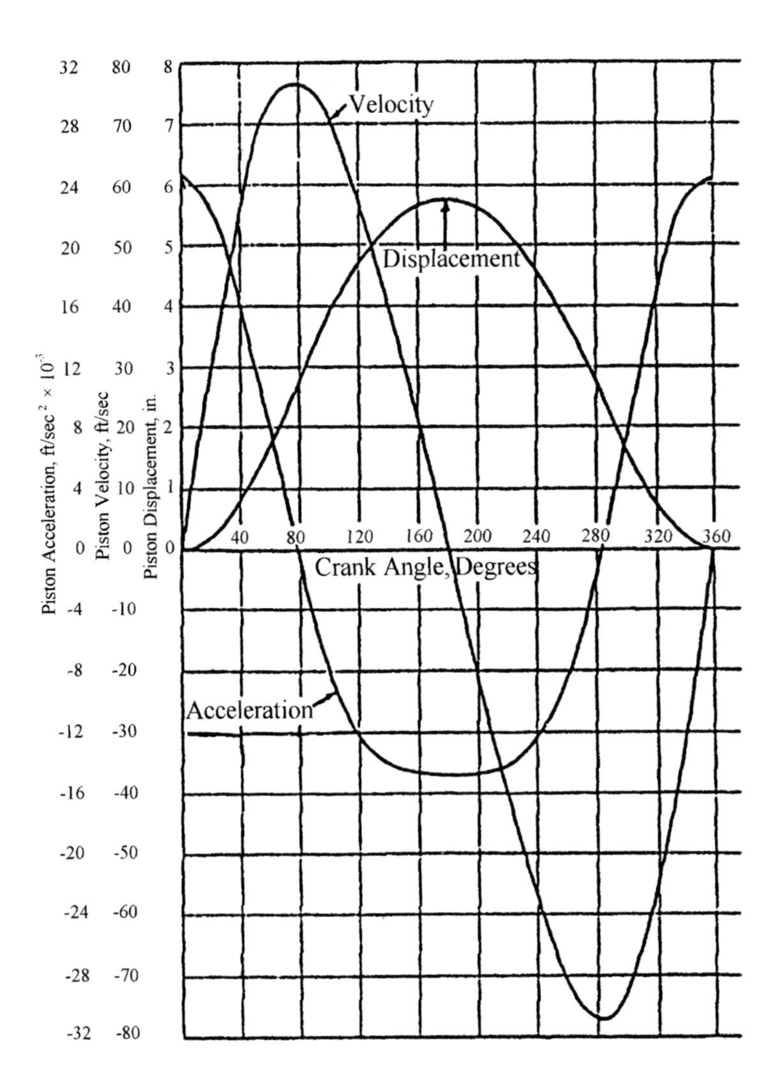

Figure 2.13. Typical plot of piston displacement, velocity, and acceleration versus crank angle.

Since the digital computer had not yet been invented during World War II, extremely complex calculations such as this had to be performed manually on the equivalent of a hand held calculator (which at the time of WW II was a mechanical device more than an order of magnitude greater in size and weight than present day hand held electronic calculators). A very large number of person days went into the production of diagrams such as Fig. 2.13.

Today such plots could be generated very quickly with a personal programmable computer. All of this is for just one point (wrist pin bearing at (*A*) on the linkage of Fig. 2.11. Several other points are of interest in engine design.

6.3 Pulleys

Figure 2.14 illustrates still another type of kinematics problem. A force (F_1) and displacement (y_1) at (A) will displace a force (F_3), a distance (y_3) at (B). Velocities of the rope at different points are $\dot{y}_1$, $\dot{y}_2$, and $\dot{y}_3$ as indicated. The free body diagram for the center pulley is shown in Fig. 2.13 (b), and it is evident that $\dot{y}_2 = \dot{y}_1/2$.

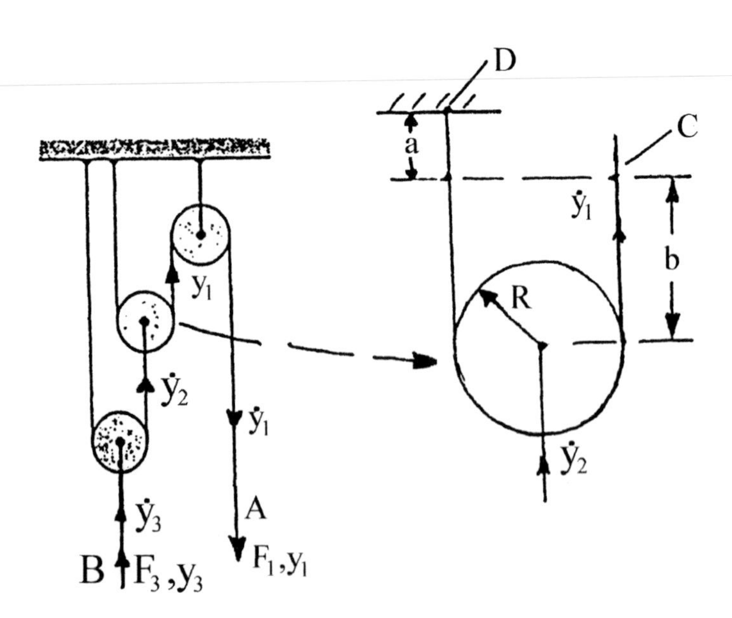

Figure 2.14. System of pulleys having mechanical advantage of 4.

Analysis for the left hand pulley similarly gives: $\dot{y}_3 = \dot{y}_2/2 = \dot{y}_1/4$. Hence, ignoring friction losses, the input power ($F_1 y_1$) will equal the output power ($F_3\dot{y}_3$) and the mechanical advantage is:

$$F_3/F_1 = \dot{y}_1/\dot{y}_3 = 4$$

With mechanical losses the input power must be greater than the output power, and the mechanical efficiency (e) will be:

Eq. (2.34) $$e = P_{out}/P_{in}$$

7.0 KINETICS

This area of engineering mechanics involves application of Newton's Second Law to problems where acceleration is involved. In such cases, the vector sum of all forces must equal the mass of the body times the acceleration of its center of mass. In addition, the vector sum of all moments about the center of mass must equal the rate of change of angular momentum about the center of mass of the body.

Kinetics problems may involve linear motion, angular motion, or a combination of the two. Since accelerations are involved, a kinematic analysis is often required before proceeding with a kinetic analysis. In general, kinetics problems are more difficult than statics problems, but the basic approach to be followed is the same.

8.0 VIBRATION

An important area of dynamics has to do with problems where bodies oscillate in response to external disturbances. The exciting disturbances may be due to imbalance of rotating machinery; a sudden shock, as in an earthquake; or a periodic series of forces, as in an internal combustion engine. Self-excited vibrations exist when a system, once disturbed, vibrates indefinitely. This occurs when the energy stored exceeds that which is released during a cycle of vibration. In such cases, the amplitude of vibration continues to increase until the increase in energy lost per cycle in friction (called damping) equals the excess of energy stored. The flutter of aircraft wings, which occurs under certain flight conditions, is an example of a self-excited vibration. The chatter of a metal cutting tool is another, and the sound emitted by a wire vibrating in a brisk wind is still another.

Ways of preventing unwanted vibration include the following:

- Increasing the stiffness of the structure
- Avoiding operation at certain unstable speeds or frequencies
- Increased damping, i.e., converting vibrational energy irreversibly into heat by solid or fluid friction
- Improved balancing
- Isolation of machinery and instrumentation from adjacent vibrating equipment

Figure 2.15 shows a very simple vibrating system consisting of a mass (m) supported by a spring having a spring constant (k) (force per unit length of deflection). In Fig. 2.15, the origin is taken at the equilibrium position of the mass. If the mass is extended downward to (A) and released, it will oscillate as shown.

The frequency of this free vibration is called the natural frequency of the system. When an unstable system is excited by a fluctuating force at a frequency close to the natural frequency of the system, a large amplitude of vibration is apt to occur.

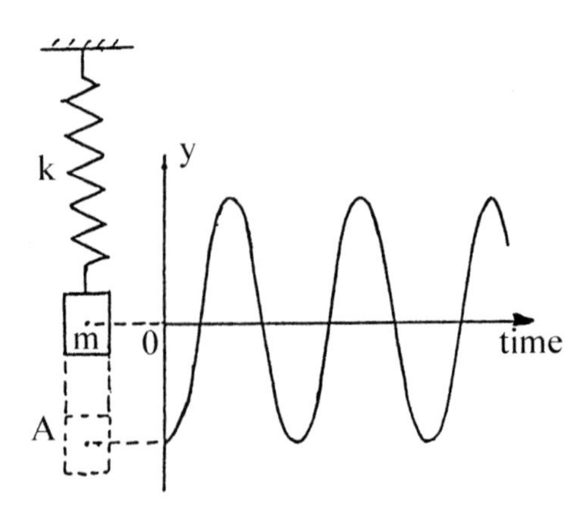

Figure 2.15. Simple one dimensional spring-mass system.

9.0 SUMMARY

In this chapter, the important aspects of rigid body mechanics have been presented. This discourse gives an idea of the scope of the subject, and how this area of engineering is related to other core subjects to be discussed later. Where possible, the material presented has been discussed in terms of the Galileo dialogs. It is not expected that it will be possible to solve many real problems after reading this chapter. However, it is hoped that a better understanding will be gained of what engineering mechanics is about, and that some useful information will have been absorbed.

PROBLEMS

2.1 If the mass of the earth is 6×10^{24} kg, its radius is 6.39×10^{6} m (~ 4,000 miles) estimate the mean acceleration due to gravity (ft/s^2) by use of Eqs. (2.1) and (2.2):

a) At the surface of the earth

b) At an elevation of 50 km (31 miles)

c) At an elevation of 500 km (310 miles)

d) At an elevation of 5,000 km (3,100 miles)

2.2 A weight of 10 lbs is supported by two ropes as shown in Fig. P2.2. Draw a free body diagram and determine the tensile force in each of the ropes.

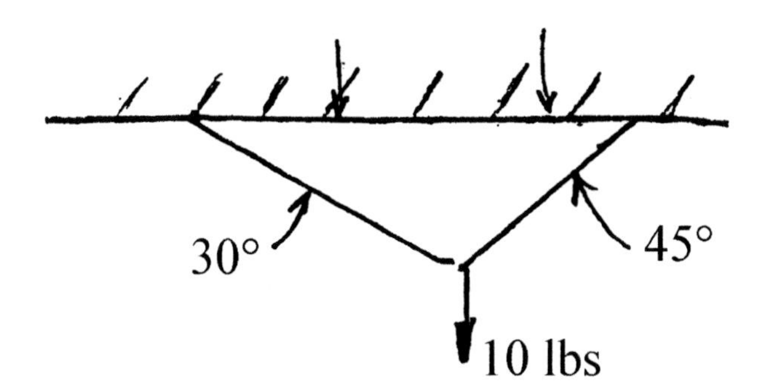

Figure P2.2.

2.3 For the cantilever shown in Fig. P2.3:

a) What is the resisting moment at 0 (M_0) if the weight of the beam is neglected?

b) What is the vertical resistance at 0 if the weight of the beam is neglected?

c) Repeat a) if the weight of the beam is 50 lbs/ft.

d) Repeat b) if the weight of the beam is 50 lbs/ft.

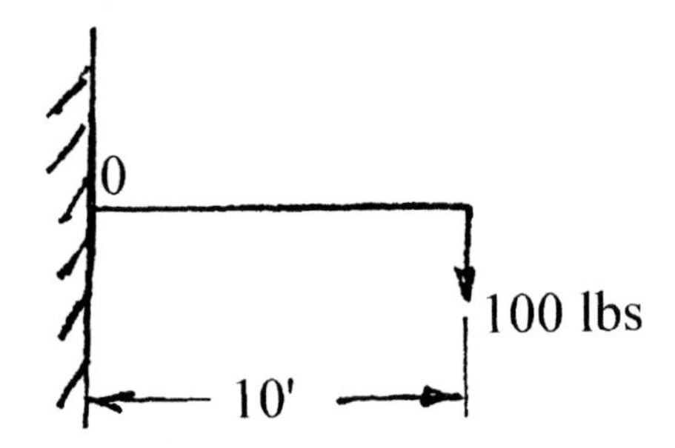

Figure P2.3.

2.4 For the beam shown in Fig. P2.4, draw a free body diagram and find the forces at (*A*) and (*B*).

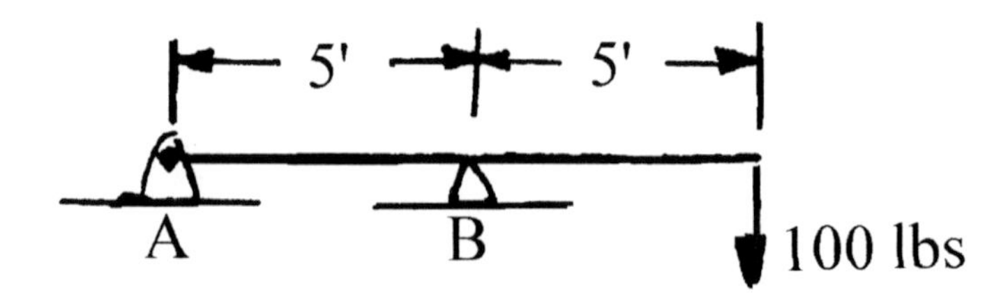

Figure P2.4.

2.5 For the Howe truss shown in Fig. P2.5, find the force in member (CD). Is this a tensile or compressive force?

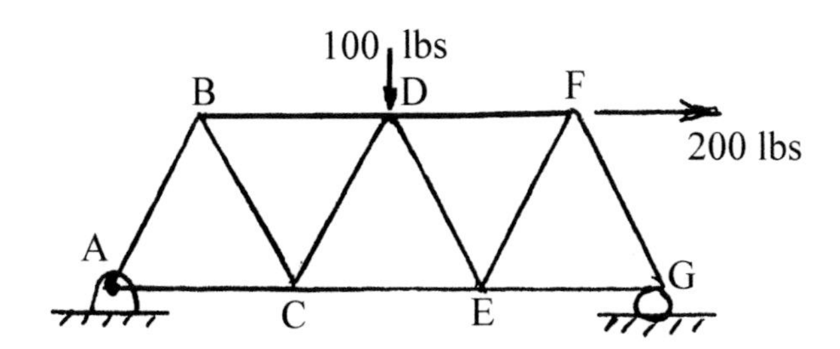

Figure P2.5.

2.6 For the Howe truss shown in Fig. P2.5, find the force in member (DE). Is this a tensile or compressive force?

2.7 Find the force in member (BC) for the truss in Fig. P2.5 if all members have the same length and all angles are the same.

2.8 A weight of 100 lbs is being lifted by pulling a rope passing over a pulley (Fig. P2.8). The 1 in. diameter bearing at the center of the 10 in. diameter pulley has a coefficient of friction of 0.10. What is the minimum static coefficient of friction at the face of the pulley so that the rope does not slip over the surface of the pulley?

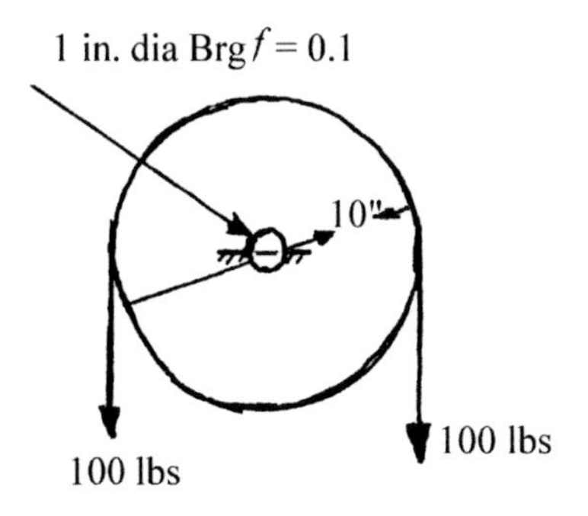

Figure P2.8.

2.9 For the example of reciprocating engine kinematics given in Fig. 2.13,

a) What is the maximum piston velocity in mph?

b) What is the maximum acceleration expressed as a multiple of (g) (gravitational acceleration)?

2.10 For the problem of Fig. 2.10, consider the case where the terrain is not level but the projectile lands on a table mesa as shown in Fig. P2.10 (with all other conditions the same). Find:

a) The distance x_0'

b) The total time of flight

c) The resultant velocity of impact and

d) The angle of impact θ'

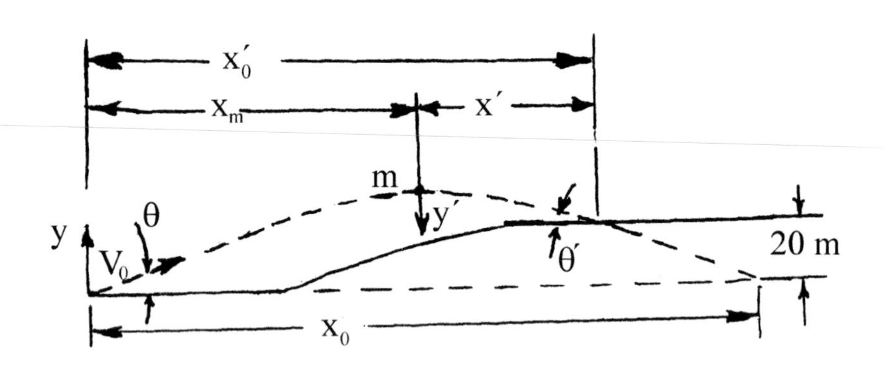

Figure P2.10.

2.11 The free body diagram of Fig. 2.14(*b*) relative to forces will be as shown in Fig. P2.11 for static equilibrium.

a) Calculate the rate of work going into this free body and the rate of work leaving.

b) Assuming no mechanical losses, what does an energy balance reveal?

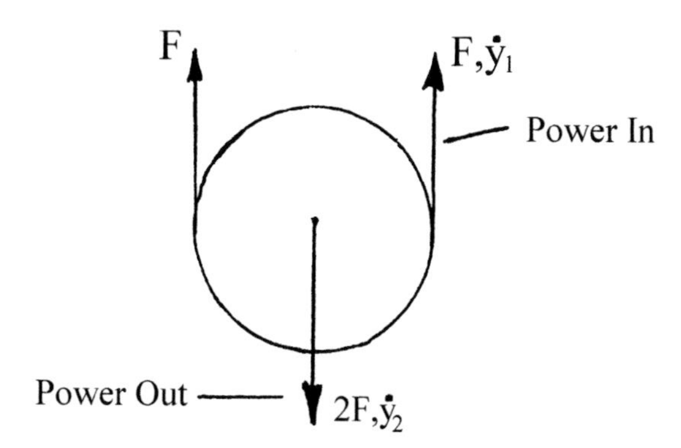

Figure P2.11.

2.12 What is the mechanical advantage (W/F) for the block and tackle shown in Fig. P2.12?

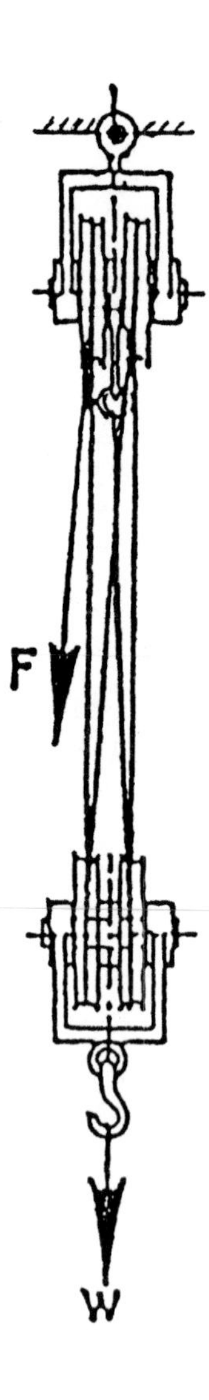

Figure P2.12.

2.13 The cantilever in Fig. P2.13(*a*) weighs 100 pounds, and a weight of 100 pounds acts at the end of the beam. The beam at (*b*) has the same weight per unit length with no load at the end. Find ℓ such that beams (*a*) and (*b*) have the same strength. (This problem is discussed by Galileo in the Second Day.)

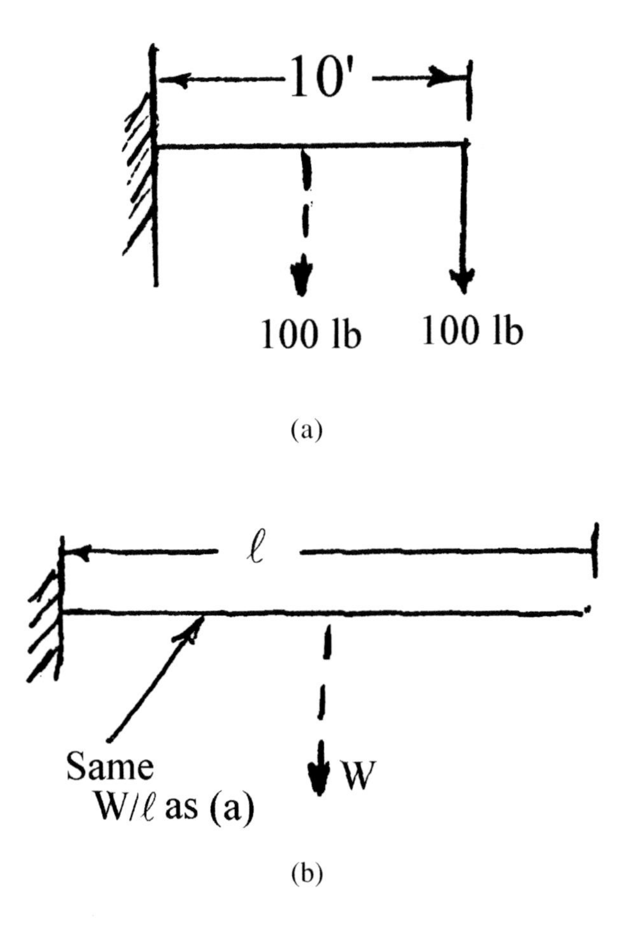

Figure P2.13.

3

Dimensional Analysis

1.0 INTRODUCTION

Many engineering problems are too complex to solve in closed mathematical form. In such cases, a type of analysis, which involves the dimensions of the quantities entering the problem, may be useful. This is called dimensional analysis. Uses for dimensional analysis include the following:

- To reduce the number of variables to be studied or plotted
- In planning experiments
- In designing engineering models to be built and studied and in interpreting model data
- To emphasize the relative importance of parameters entering a problem
- To enable units of measurement to be changed from one system to another

The last of these is a common, although relatively trivial, application.

Before discussing the details of dimensional analysis, it is useful to consider the nature of the end result. This will be done in terms of the projectile problem considered in the previous chapter (Fig. 2.10) where it was found that the elevation of the projectile would be as follows [see Eq. (2.24)]:

$$y = -gt^2/2 + v_0 t \sin \theta$$

where t is the time elapsed after launching the projectile with an initial velocity, v_0, inclined at an angle (θ) to the horizontal.

In considering this problem by the method of dimensions, it would first be observed that the main dependent variable (y) will be some function (ψ_1) of v_0, t, and g, i.e.,

Eq. (3.1) $$y = \psi_1(v_o, t, g)$$

After performing the dimensional analysis, the following result would be obtained using the procedure outlined later in this chapter:

Eq. (3.2) $$y/gt^2 = \psi_2(v_0/gt)$$

where ψ_2 is some function of the quantity in parenthesis. The number of variables has been reduced from four (y,v_0, t, and g) to two (y/gt^2 and v_0/gt), but ψ_2 must still be evaluated either experimentally or by further analysis.

Dimensional analysis thus represents a partial solution to the problem but one that is very useful and easily obtained. In the solution of complex problems, it can play a valuable role.

2.0 DEFINITIONS

Before considering the method of proceeding from Eq.(2.24) to Eq. (3.2) in this chapter, a few quantities will be defined. Fundamental or primary dimensions are properties of a system under study that may be considered independent of the other properties of interest. For example, there is one fundamental dimension in any geometry problem and this is length (L). The fundamental dimensions involved in different classes of mechanical problems are listed in Table 3.1 where L stands for length, F for force, and T for time. Dimensions other than F, L, and T, which are considered fundamental in areas other than mechanics, include: temperature

(θ), heat (H), electrical charge (Q), magnetic pole strength (P), chemical yield (Y), and unit cost (¢).

Table 3.1. Fundamental Dimensions for Different Types of Problems

Type of Problem	Fundamental Dimensions
Geometry	L
Statics	F
*Temporal	T
Kinematics	L, T
Work or Energy	F, L
Momentum or Impulse	F, T
Dynamics	F, L, T

*Temporal problems are those involved in time study, frequency analysis, time tables, etc.

Dimensional equations relate the dimensions of the fundamental quantities entering a problem to nonfundamental or secondary quantities. For example, acceleration (a) is a quantity of importance in most kinematic problems and the dimensions of acceleration are related to fundamental dimensions L and T by the following dimensional equation:

$$a = [LT^{-2}]$$

It is customary to enclose the combination of fundamental dimensions in brackets when writing a dimensional equation.

Dimensional units are the basic magnitudes used to specify the size of a fundamental quantity. In engineering problems, lengths are frequently measured in feet and times in seconds. In this system of dimensional units, acceleration would be expressed in units of ft/sec^2.

A nondimensional quantity is one whose dimensional equation has unity or $F^0L^0T^0$ on the right side. The quantities (y/gt^2) and (v_0/gt) in Eq. (3.2) are nondimensional. A group of variables that cannot be combined to form a nondimensional group is said to be dimensionally independent.

3.0 FUNDAMENTAL QUANTITIES

The choice of fundamental quantities is somewhat arbitrary, as the following discussion will reveal. In Table 3.1, the fundamental quantities for the general dynamics problem were stated as F, L, and T. The quantities M (mass), L, and T could also have been used. Then F would be a secondary variable related to fundamental variables M, L, and T by a dimensional equation based on Newton's second law:

$$F = ma = [MLT^{-2}]$$

F, L, and T will be used here since most engineering quantities are expressed in units of force and not in units of mass.

The principle of dimensional homogeneity first expressed by Fourier (1822) states that the dimensions of each term of any physically correct equation must be the same. For example, Eq. (2.24) is readily found to meet this test:

$$[L] = [(L/T^2)(T^2)] - [(L/T)(T)]$$

4.0 PROCEDURE

We are now ready to return to the projectile problem and consider the steps involved between Eq. (2.24) and Eq. (3.2). Since this is a kinematics problem there are two fundamental dimensions (L, T). Table 3.2 gives the fundamental dimensions for all quantities entering the problem.

The quantities g and t are dimensionally independent (may not be combined by raising to powers and multiplying together to form a nondimensional group). If a function such as Eq. (3.2) exists, then the following may also be written:

Eq. (3.3) $$y/gt^2 = \psi_2(g, t, v_0/gt)$$

The left side of this equation is nondimensional, and by the principle of dimensional homogeneity, all terms of ψ_2 must also be nondimensional.

The quantities g and t, therefore, cannot appear in ψ_2 except where combined with other quantities to form a nondimensional group. Thus,

Eq. (3.2) $$y/gt^2 = \psi_2(v_0/gt)$$

Table 3.2. Dimensional Equations for Projectile Problem

Quantity	Dimensions
Vertical displacement, y	$[L]$
Acceleration due to gravity, g	$[LT^{-2}]$
Time, t	$[T]$
Initial velocity, v_0	$[LT^{-1}]$

Exponents a and b required to make yg^at^b nondimensional may be found by writing simultaneous equations as follows (although these exponents may usually be written by inspection):

$$yg^at^b = \underbrace{[L(LT^{-2})^aT^b]}_{\text{I}} = \underbrace{[L^0T^0]}_{\text{II}}$$

Equating exponents of L in I and II yields:

$$1 + a = 0$$

While equating the exponents for T in I and II gives:

$$-2a + b = 0$$

When these equations are solved simultaneously, a and b are found to be -1 and -2 respectively, in agreement with the left-hand side of Eq. (3.2).

The result of a dimensional analysis is sometimes written in symbolic form in terms of nondimensional Pi quantities as follows:

Eq. (3.4) $\pi_1 = \psi(\pi_2, \pi_3, \text{etc.})$

where π_1 is a nondimensional group involving the main dependent variable (y in the projectile problem) and the other Pi values represent the remaining nondimensional quantities entering the problem.

Buckingham (1914) formulated a theorem which states that the number of Pi quantities remaining after performing a dimensional analysis is equal to the difference between the number of quantities entering the problem and the maximum number of these that are dimensionally independent. The maximum number of dimensionally independent quantities will always be equal to or less than the number of fundamental dimensions needed to write all dimensional equations. In the projectile problem, two fundamental dimensions (L, T) are involved and the maximum number of dimensionally independent quantities should, therefore, be two (g and t).

Applying Buckingham's Pi Theorem to this case:

Total quant. - dimensionally independent quant. = Pi quant.

i.e., $4 - 2 = 2$

The dimensionally independent quantities can usually be chosen in more than one way. There will often be several correct answers to a dimensional analysis. One of these answers may prove to be more convenient for a given purpose than the others. In the projectile problem, all of the pairs of dimensionally independent quantities listed in Table 3.3 could have been used and the dimensionless equations listed opposite each pair of dimensionally independent quantities would have been obtained.

Table 3.3. Possible Dimensional Analyses for Projectile Problem

Dimensionally Independent Quantities	Resulting Equation
g, t	$y/gt^2 = \psi[v_0/gt]$
g, v_0	$(yg)/v_0^2 = \psi[(gt)/v_0]$
t, v_0	$y/v_0t = \psi[(gt)/v_0]$

While there are no hard and fast rules concerning choice of dimensionally independent quantities, the following considerations serve as a useful guide:

- The main dependent variable should not be chosen
- Variables having the greatest significance should be chosen
- The variables chosen should represent as many physically different aspects of the problem as possible

In performing a dimensional analysis, it is important to include all quantities of importance to the problem. Otherwise, an incorrect and misleading result will be obtained. It is less important to include a variable about which doubt exists than to omit one which proves to be significant. Frequently, combinations of variables that are known to appear in a given class of problems in a unique association can be treated as a single quantity. This will result in fewer Pi quantities in the final result. Another method of reducing the resulting Pi quantities to a minimum is to use the maximum number of dimensionally independent quantities possible. Results of auxiliary or approximate analysis can sometimes be combined with conventional dimensional reasoning to greatly increase the power of dimensional analysis. Such auxiliary analysis includes:

- Arguments involving symmetry
- Assumptions of linearity or known behavior
- Special solutions for large or small values of variables

The meaning of some of these abstract statements will become clear after examples considered later have been considered.

5.0 CHANGE OF UNITS

Dimensional equations are useful where it is desired to change from one system of units to another. Let the dimensional equation for a given quantity (Q) be

Eq. (3.5) $$Q = [\alpha^a \beta^b]$$

If the new unit for fundamental dimension (α) is smaller than the old unit by a factor (f) while the new unit for (β) is smaller by a factor (g), then the number of units of Q in the new system of measurement per unit of Q in the old system will be $(f^a g^b)$.

As an example, consider the conversion of an acceleration (a) from 32.2 ft/sec^2 into the equivalent number of in/hr^2. The appropriate dimensional equation is:

Eq. (3.6) $\quad a = [LT^{-2}]$

The new unit for L is smaller than the old one by a factor 12 while the new unit for T is larger by a factor of 3,600. Thus, each ft/sec^2 will correspond to $(12)(1/3,600)^{-2}$ in/hr^2 and the answer to the problem is $(32.2)(12)(3,600)^2 = 50 \times 10^8$ in/hr^2.

6.0 GALILEO REGARDING MOTION OF A PROJECTILE

In dialogs of the Fourth Day, Galileo discusses the motion of projectiles in considerable detail. It is suggested that the reading of Galileo's discourse concerning projectiles be postponed until after reading the following discussion, since the material in Galileo is scattered and difficult to follow at some points.

Galileo first proves that a projectile traveling in air will follow the path of a parabola. This is based on the properties of a parabola as a conic section given by Appolonius (~200 B.C.). After this, it is verified that assumptions that air drag and variations of gravitational attraction due to changes in distance from the center of the earth are negligible. However, it is acknowledged that a weight falling from a great height will assume a terminal velocity due to drag, and that the distance traveled to achieve terminal velocity will depend upon the shape of the body and vary inversely with its density.

In passage 281-283 of the Galileo text, an unusual approach to the velocity of a projectile is taken in which the missile is released from rest and allowed to fall freely to the ground (having the velocity attained part way down added as a uniform velocity in the horizontal direction). The point at which the horizontal velocity is introduced corresponds to the highest point reached by a projectile. Galileo refers to the elevation of the initial point of release as the sublimity, the elevation at the point where uniform horizontal velocity is introduced as the *altitude*, and twice the horizontal distance traveled before the body strikes the ground as the *amplitude*.

Since accurate means of measuring distance, time, and velocity were not available, a scheme was adopted where distances on a geometrical diagram are proportional to all three of these quantities. This procedure is described in passage 281 of the Galileo text in terms of Fig. 110.

The reason for inverting the problem, starting with free fall from the *serenity* followed by the addition of a uniform horizontal velocity at the *altitude* is that this is consistent with the very clever experiment Galileo devised to record the paths of projectiles. The experimental technique is illustrated in Fig. 3.1. This involves two plane surfaces having maximum slopes at right angles to each other. A smooth sphere is released at (*a*) and rolls down the first plane to (*b*). At this point, it has a velocity corresponding to free fall through the vertical distance between (*a*) and (*b*). At (*b*), the ball enters the second plane and moves down this with a constant horizontal velocity corresponding to that attained at (*b*), plus a vertical component of velocity corresponding to a vertical free fall from (*b*) to (*c*). The resultant velocity at (*c*) corresponds to the launching velocity at angle θ.

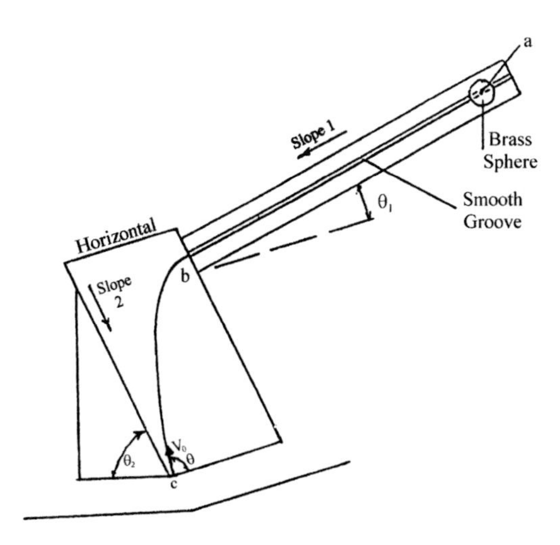

Figure 3.1. Arrangement used by Galileo in recording the path of a body subjected to uniform velocity in one direction and uniform acceleration in the orthogonal direction. This corresponds to the path taken by a projectile leaving the ground with a velocity (v_0) at an angle (θ) to the horizontal.

The method of recording the path taken from (*b*) to (*c*) is not given in the fourth day dialog, but is in passage 185 of the Galileo text. Here, it is mentioned that a parabola will be traced by a brass ball about the size of a walnut rolling down an inclined plane if the second plane in Fig. 3.1 is a metallic mirror. This is, presumably, a front surface mirror from which the silver is removed as the ball rolls over the surface. Use of a soot covered plane surface would serve equally well to record the parabolic path of the sphere moving horizontally with uniform speed, but vertically with gravitational acceleration.

The behavior of the sphere on plane 1 is described in passages 212 and 213 of the Galileo text. Here Salviati also describes how the mean time of descent was measured for different values of angle θ and distances of fall along the incline, and by repeating identical experiments many times.

It is suggested that passages 185, 212, and 213 now be read following passages 267-293 of the Galileo text.

7.0 SIMPLE PENDULUM

A pendulum consisting of a weight (W) suspended by a cord of length (ℓ) is shown in Fig. 3.2. An expression for the period of oscillation (P) is desired when friction at 0 and air resistance to motion are negligible. Consideration of this problem reveals the quantities listed in Table 3.4 to be of possible importance.

Table 3.4. Dimensions Considered in Pendulum Problem

Quantity	Symbol	Dimensions
Period	P	$[T]$
Length of cord	ℓ	$[L]$
Gravity	g	$[LT^{-2}]$
Weight of bob	W	$[F]$

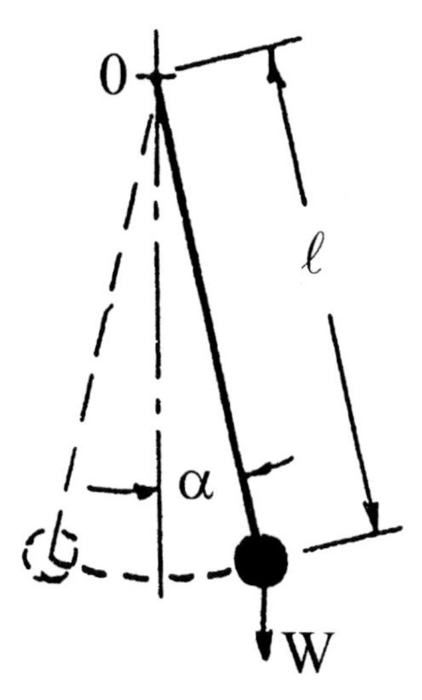

Figure 3.2. Simple pendulum.

The quantities(ℓ), (g), and (W) are dimensionally independent, hence there will be one π expression in this problem (by Buckingham's Pi Theorem: 4 - 3 = 1). This single Pi quantity will be found to be $P/(\ell/g)^{0.5}$.

The general dimensionless result expressed symbolically in Eq. (3.4) assumes a special form in this case since π_2, π_3, etc., are all unity. In this instance:

Eq. (3.7) π_1 = a nondimensional constant

and the answer to this problem is, therefore:

Eq. (3.8) $P = C\,(\ell/g)^{0.5}$

where C is the constant that may be found to be 2π by a single experiment or by the conventional application of Newton's second law.

It should be noted that the weight of the bob which was incorrectly assumed to be of importance, disappeared during the course of the dimensional analysis. This is not always the case when superfluous variables are included in the analysis. For example, if we had considered the maximum angle of swing (α) to be significant, then the end result would have been:

Eq. (3.9) $P/(\ell/g)^{0.5} = \psi(\alpha)$

This is as far as we may proceed by dimensional analysis alone. Experiments will reveal that $\psi(\alpha)$ is a constant (2π).

PROBLEMS

3.1 Explain why angle is a nondimensional quantity regardless of size or method of measurement (radians or degrees).

3.2 Combine the following variables into a nondimensional quantity or show they are dimensionally independent:

a) Force (F), mass density ($\rho = [FL^{-4}T^{2}]$), velocity (V), and a diameter (D)

b) Fluid flow rate ($Q = [FT^{-1}]$), a length ℓ, and viscosity ($\mu = [FTL^{-2}]$)

c) Surface tension ($T_e = [FL^{-1}]$, viscosity (μ), and mass density (ρ)

3.3 Are the following quantities dimensionally independent? If not, form them into one or more nondimensional groups.

a) Acceleration (a), surface tension ($T_e = [FL^{-1}]$), and area (A)

b) A volume (B), an angle (α), a velocity (V), and acceleration (a)

c) Mass density ($\rho = [ML^{-3}]$, viscosity ($\mu = [FTL^{-2}]$), and bulk modulus ($K = [FL^{-2}]$)

d) Acceleration (a), surface tension ($T_e = [FL^{-1}]$), velocity (V), and mass ($M = [FL^{-1}T^{2}]$)

3.4 Are the following quantities dimensionally independent? If not, form them into one or more nondimensional groups.

a) Mass density ($\rho = [ML^{-3}]$), velocity (V), diameter (D), and surface tension (T_e)

b) Mass density ($\rho = [ML^{-3}]$), surface tension ($T_e = [FL^{-1}]$), and viscosity ($\mu = [FTL^{-2}]$)

c) Mach number ($N_M = [F^0, L^0, T^0]$, kinematic viscosity ($\nu = [L^2T^{-1}]$), and surface tension ($T_e = [FL^{-1}]$)

3.5 When a capillary tube of small inside diameter is immersed in a liquid of surface tension $T_e = [FL^{-1}]$ and specific weight (γ), the liquid will rise to a height (h) as shown in Fig. P3.5 and $h = \Psi(T_e, d, \gamma)$. Perform a dimensional analysis and, thus, reduce the number of variables.

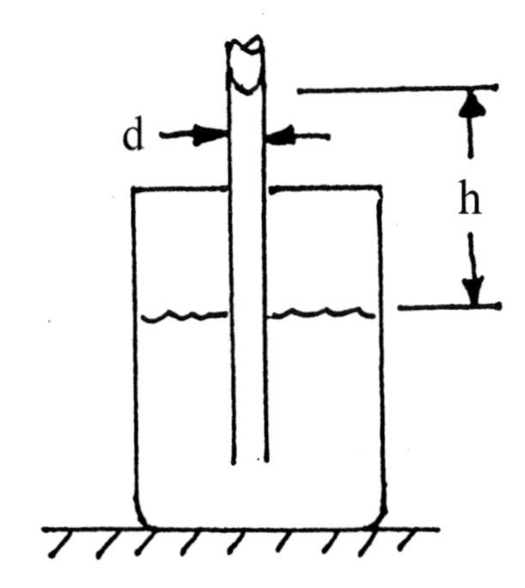

Figure P3.5.

3.6 It is reasonable to assume that (h) will be proportional to (γ). What result is obtained when this is incorporated into the solution of Problem 3.5?

3.7 The drag force (F_D) on a body of a given shape (such as a sphere) will be influenced by the variables listed below having dimensions indicated.

Quantity	Dimensions
Drag force, F_D	$[F]$
Size of body, S	$[L]$
Density of fluid, ρ	$[FL^{-4}T^{2}]$
Relative velocity, V	$[LT^{-1}]$
Viscosity of fluid, μ	$[FTL^{-2}]$
Acceleration due to gravity, g	$[LT^{-2}]$
Velocity of sound in fluid, C	$[LT^{-1}]$

The drag force is the main dependent variable in this case, and there will be three dimensionally independent quantities involving F, L, and T. These may be taken to be S, ρ, and V.

a) Show that S, ρ, and V are dimensionally independent.

b) How many nondimensional quantities ($\pi_1, \pi_2, \pi_3 \ldots$) will there be, in this case?

c) Perform a dimensional analysis and express π_1 involving F_D as a function of the other π quantities; i.e., find $\pi_1 = \psi(\pi_2, \ldots)$.

3.8 The velocity of a wave in deep sea water (V) is a function of the wavelength (λ) and the acceleration due to gravity (g) (Fig. P3.8). By dimensional analysis, derive an expression for the wave velocity in deep water. Will the wave length be smaller in a high wind?

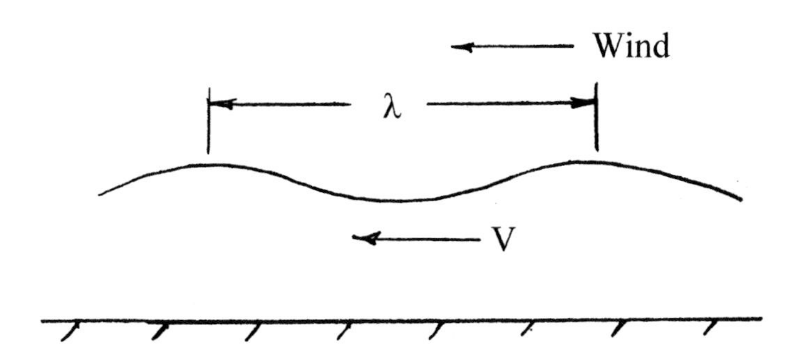

Figure P3.8.

3.9 Water has a surface tension of 73 dynes/cm. To how many (in.lb/in^2) does this correspond?

3.10 A small spherical drop of liquid (diameter d) such as one of those found on a cabbage leaf by Galileo (passage 115) oscillates between a sphere and an ellipse under the action of surface tension when the leaf is disturbed (Fig. P3.10). If the drop is very small, gravitational forces will be negligible and the period of the oscillation (P) will be:

$$P = \psi(T_e, d, \rho)$$

where ρ is the density of the liquid ($FL^{-4}T^{2}$). Perform a dimensional analysis for this problem. Will the frequency of oscillation of the drop when disturbed increase or decrease as the drop gets smaller?

Figure P3.10.

3.11 Surface tension (T_e) is sometimes determined by noting the volume (B) of a drop delivered from a narrow dropping tube of diameter (d) at its tip (Fig. P3.11) (Tate's pendant drop method). The formation of the pendant drop and its subsequent detachment are complex phenomena but the following variables are of importance:

$$B = \psi(d, \gamma, T_e)$$

where γ is the specific weight of the fluid. Perform a dimensional analysis for this problem with T_e the main dependent variable.

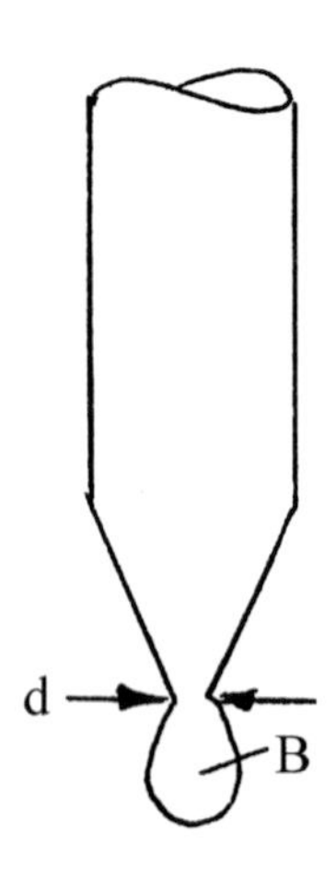

Figure P3.11.

3.12 If the velocity of water (V) flowing through the orifice in the tank (Fig. P3.12) is a function of gravity (g) and the height of the fluid (h):

$$V = \psi(g, h)$$

a) Perform a dimensional analysis.

b) By what factor will the velocity change when (h) is doubled?

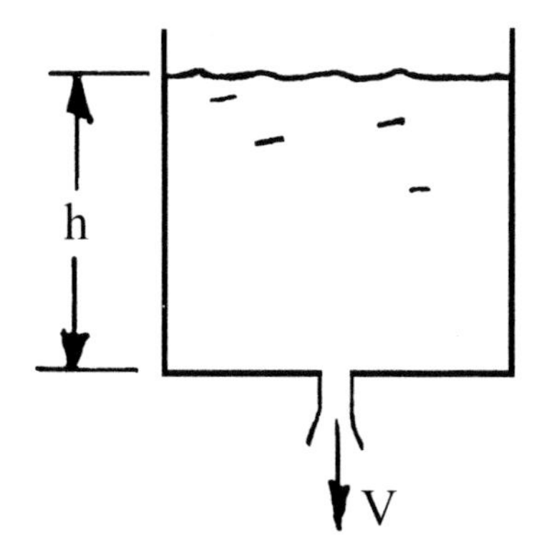

Figure P3.12.

3.13 Repeat Problem 3.7, taking M, L, and T as the fundamental set of dimensions instead of F, L, and T.

3.14 Repeat Problem 3.12 taking M, L, and T as the fundamental set of dimensions instead of F, L, and T.

4

Deformable Body Mechanics

1.0 INTRODUCTION

All bodies undergo deformation when subjected to a force. In some cases, deformation may be ignored and these are problems of rigid body mechanics (Ch. 2). In other cases, deflections are of importance. Such problems are considered in this chapter, including:

- Problems involving the strength of internal members
- Problems where deformation is important in design
- Problems involving the stability of structures
- Problems involving impact

Deformable body mechanics is sometimes referred to as mechanics of materials and strength of materials.

2.0 STRESS AND STRAIN

If a bar is subjected to tension by force (W), it will elongate an amount $\Delta \ell$ over a given gage length ℓ_0 (Fig. 4.1). Instead of studying the relation between W and $\Delta \ell$, it is preferable to use the following intensive quantities:

Eq. (4.1) $S = \text{tensile stress} = W/A_0$

Eq. (4.2) $e = \text{tensile strain} = \Delta \ell / \ell_0$

where ℓ_0 and A_0 are initial unloaded values and where force and deformation are perpendicular to area (A). The quantities S and e are referred to as normal stress and normal strain. For the elastic region of stress (Fig. 4.2):

Eq. (4.3) $S = Ee$

where the proportionality constant is called the modulus of elasticity or Young's modulus and Eq. (4.3) is known as Hooke's law.

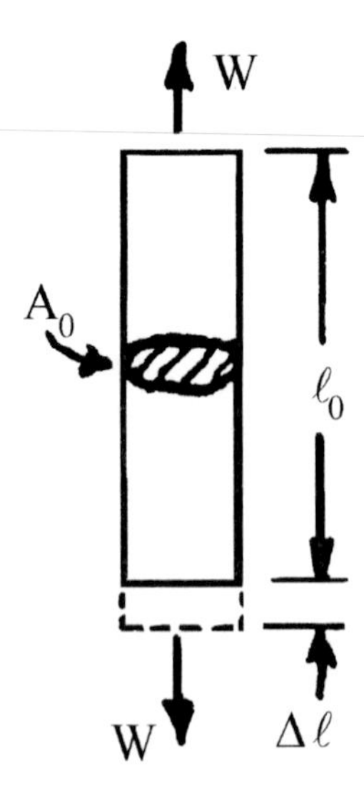

Figure 4.1. Tensile test.

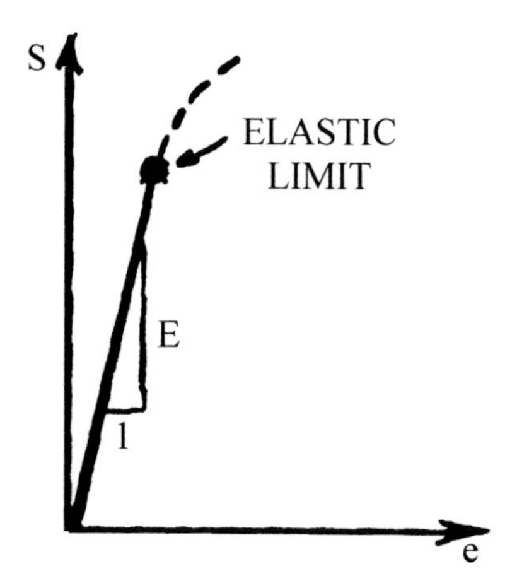

Figure 4.2. Stress-strain curve for tensile test in elastic region.

For shear stress (τ) and shear strain (γ) the force and associated deformation are in the plane of the area (Fig. 4.3) and:

Eq. (4.4) $\tau = V/A$

Eq. (4.5) $\gamma = \Delta x/\Delta y$ (for small elastic strains)

where V is a shear force in the plane of A. Below, the elastic limit (τ) is proportional to γ(Fig. 4.4) and:

Eq. (4.6) $\tau = G\gamma$

Here, the proportionality constant (G) is called the shear modulus. The dimensions of τ and G are $[FL^{-2}]$ and γ is nondimensional.

The stress strain curve for a ductile metal will be as shown in Fig. 4.5. The stress where S vs e ceases to be linear is called the yield point (S_y), the maximum value is called the ultimate stress (S_u), while the stress where gross fracture occurs is called the fracture stress (S_f).

Initially, deformation is uniformly distributed along gage length (ℓ_0) in Fig. 4.1, but in the vicinity of the ultimate stress, a localized neck begins to form and A_0 begins to decrease more rapidly than W increases as strain progresses. This gives rise to a maximum stress.

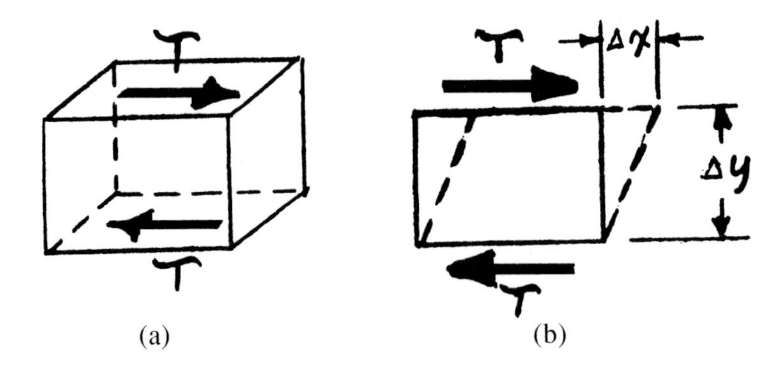

Figure 4.3. *(a)* Shear stress and *(b)* shear strain.

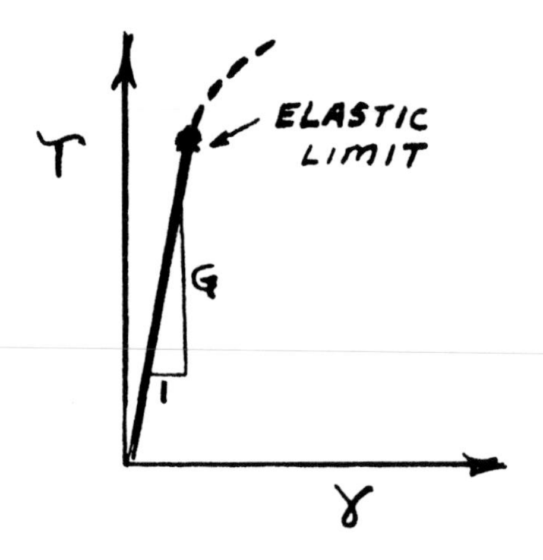

Figure 4.4. Shear stress (τ) vs shear strain (γ) in the elastic region.

For large plastic strains, elastic action is so small it is ignored. In the plastic region, it is necessary to redefine stress and strain in terms of instantaneous values of ℓ and A instead of initial values of ℓ_0 and A_0. Then,

Eq. (4.7) $\sigma = F/A$

Eq. (4.7a) $d\varepsilon = d\ell/\ell$

where A and ℓ are instantaneous values that change with strain. Integrating both sides of the equation for $d\varepsilon$ (see App. A if needed):

Eq. (4.8) $\varepsilon = \ln \ell/\ell_0$

The symbols for stress and strain in the elastic region are S and e, but in the plastic region, where A and ℓ are no longer constant, they are σ and ε. However, this is not of concern in this chapter since stress and strain in the vicinity of the yield point are involved where Eqs. (4.1) and (4.2) are satisfactory.

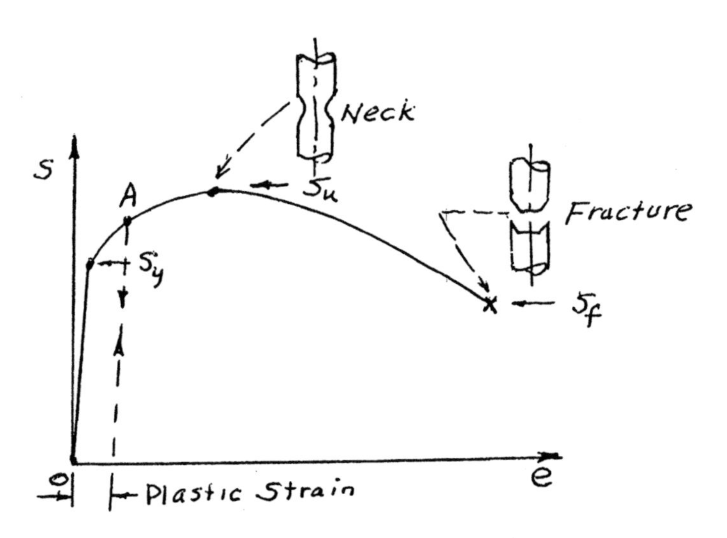

Figure 4.5. Complete S-e tensile curve for a ductile metal.

If a tensile specimen is loaded to a point below the yield point, it will return to its original shape on unloading. If, however, a specimen is loaded beyond the yield point and then unloaded, it will return along the dashed line in Fig. 4.5 parallel to the original elastic curve. The dashed curve intersects the e axis with a permanent strain. Strains greater than that at the yield point are called plastic strains while those below the yield point are elastic strains.

If a specimen that has undergone plastic strain is unloaded and then reloaded, it will follow the dashed curve in Fig. 4.5 until point A is reached, and then follow the original curve with further loading. Thus, loading a specimen into the plastic region results in an increase in yield stress (S_y), but no change in Young's modulus (E).

In engineering mechanics, the reason for the shift from elastic to plastic behavior at the yield point is not a concern. Why it occurs is a matter for materials science, and this will be discussed in Ch. 8.

An important experimental observation is that in the plastic region there is almost no change in the volume of a specimen regardless of the change in shape involved. This was demonstrated experimentally by Galileo in passages 97–99 of the Galileo text where the dimensions of a silver rod were reduced by drawing it through a succession of dies. The volume of silver remained constant even with a drawing ratio (initial dia./final dia.) of 200. In this passage of Galileo, the extreme ductility of pure gold is also demonstrated by noting that 10 layers of very thin gold leaf applied to a silver rod are reduced in thickness by a factor of 200 without fracture. It should be noted that a cubit is about 1.5 ft and that "mean proportional" equals mean value.

In this chapter, we are primarily concerned with the elastic behavior of materials. Deformation of materials involving large plastic strains is more a concern of materials processing. Brittle materials tend to have a fracture strain that is not far beyond the yield strain while a perfectly brittle material (glass) has a fracture strain below the yield strain.

The strains corresponding to the yield, ultimate, and fracture conditions for a ductile metal will be approximately as follows:

$$e_y \cong 0.001 \text{ (elastic)}$$

$$e_u \cong 0.200 \text{ (elastic)}$$

$$\varepsilon_f \cong 1.00 \text{ (plastic)}$$

3.0 BEAM STRENGTH

There are many structural members that are long, relatively thin, beams. The cantilever with built-in end of Fig. 4.6 (a) is an example. The bending moment at A due to external load (W) is $M_A = W\ell$. Figure 4.6 (b) shows the free body diagram for the beam section at A. This free body diagram must satisfy the following equilibrium equations:

Eq. (4.9) $\Sigma F_x = 0$

Eq. (4.10) $\Sigma F_y = 0$

Eq. (4.11) $\Sigma M_A = 0$

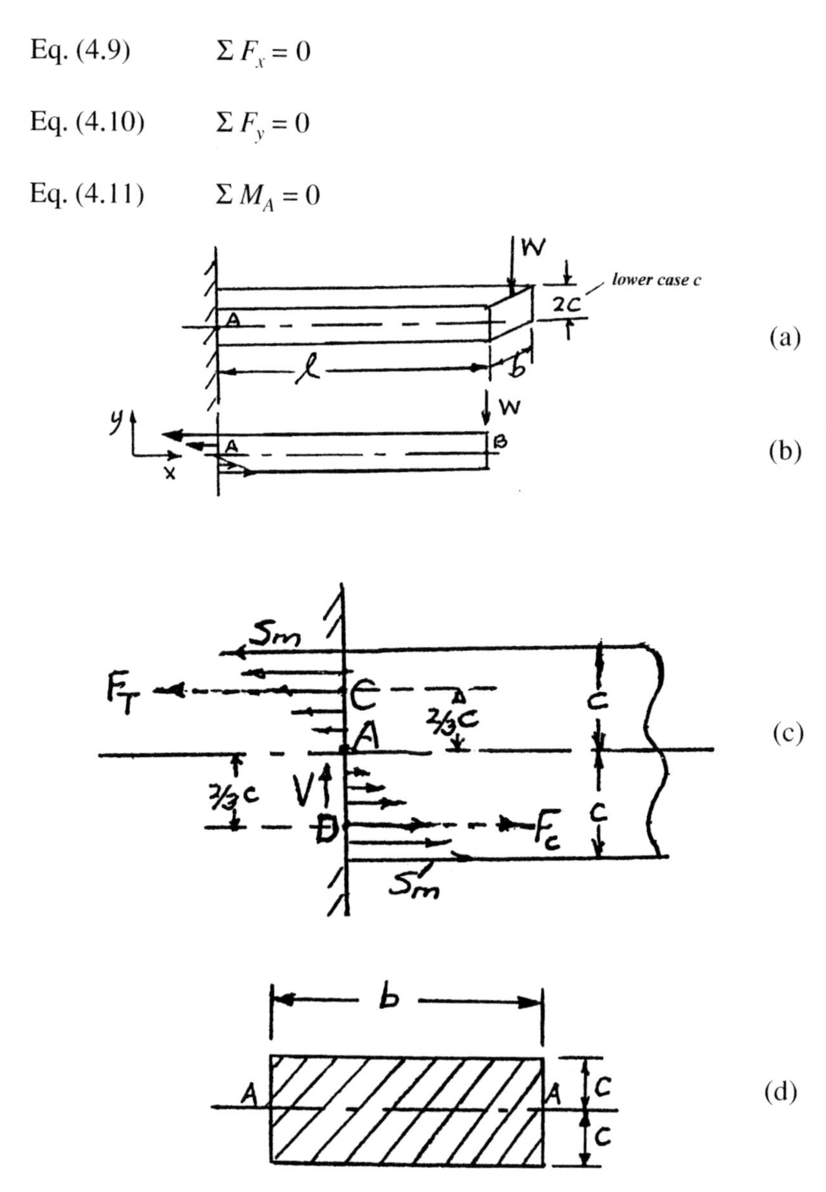

Figure 4.6. Simple end-loaded cantilever of rectangular cross section *(a)* schematic, *(b)* free body diagram, *(c)* enlargement of section at built-in end, and *(d)* cross section of beam.

The stress distribution at A in Fig. 4.6 (b) satisfies these equations. Line AB is called the neutral axis since the stress and strain remains zero along this line. Points above AB are in tension (stretched) while those below are in compression. Figure 4.6 (c) shows the section at A enlarged. The distributed tensile stress above A may be replaced by force (F_T) at C while the distributed compressive stress below A may be replaced by F_C at D, where

Eq. (4.12) $F_T = (S_m/2)(bc)$

Eq. (4.13) $F_C = (S_m/2)(bc)$

Forces (F_T) and (F_C) constitute what is called a couple having a moment.

Eq. (4.14) $M_C = (S_m/2)(bc)(2)(2/3c) = 2/3\ S_m bc^2$

Since $F_T = -F_C$, Eq. (4.9) is satisfied. Equation (4.10) is satisfied if the shear force (V) at $A = W$. Equation (4.11) is satisfied by equating M_C and M_A:

Eq. (4.15) $(2/3)S_m bc^2 = M_A$

or

Eq. (4.16) $S_m = M_A/(2/3\ bc^2) = 3/2W\ell/(bc^2)$

This is often written:

Eq. (4.17) $S_m = M_A c_m/I_N$

where I_N is called the moment of inertia of the cross section of the beam about A–A, the neutral axis of the beam [Fig. 4.6 (d)], and for the rectangular beam of Fig. 4.6 (d):

Eq. (4.18) $I_N = (I/12)\, b(2c_m)^3$

The quantity (c_m) is the distance from the neutral axis (A–A) to the outer most point on the beam.

The moment of inertia is introduced since handbooks give values of I_N for different cross sectional shapes. For example, if the beam of Fig. 4.6 (*a*) was a rod of radius (r) instead of a rectangle, then I_N would be $(\pi/4)r^4$ and from Eq. (4.18):

Eq. (4.19) $\quad S_m = (M_A r)/(\pi 4 r^4) = 4W\ell/\pi r^3$

In solving beam strength problems, it is helpful to begin by plotting diagrams of shear force (V), and bending moment (M) along the length of the beam. Figure 4.7 shows such diagrams for a cantilever at (*a*) and a centrally loaded beam on two supports at (*b*). The point where the maximum tensile stress (S_m) is located will be at A in both cases where M is a maximum.

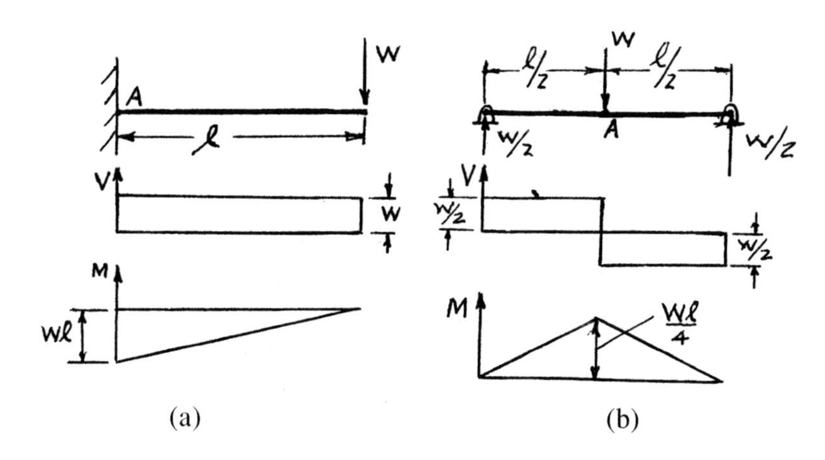

Figure 4.7. Shear and bending moment diagrams for beams of constant cross section *(a)* cantilever beam, and *(b)* simple centrally loaded beam.

In considering the strength of a beam, it is customary to ignore the shear stress ($V/2cb$) unless the beam has a relatively low value of ℓ/c. For the cantilever of Fig. 4.6, $S_m/\tau = [(1.5W\ell)/(bc^2)][(2bc/W)] = 3\ell/c$. It seems reasonable to ignore τ if it is less than about 5% of S_m. This corresponds to $\ell/c > 7$. For the case of Fig. 4.6, ℓ/c is more likely to be about 100.

4.0 GALILEO REGARDING BEAM STRENGTH

It is suggested that passages 151–165 of the Galileo text be read at this point to see what he has to say about the strength of beams when the weight of the beam is neglected. After discussing the law of the lever and introducing the concept of moment of a force, Galileo considers the strength of a cantilever. He employs the free body diagram shown in Fig. 4.8. This does not satisfy equilibrium Eqs. (4.9) and (4.10), but does satisfy Eq. (4.11). From Eq. (4.11):

Eq. (4.20) $$M_A = W\ell = S_m(2cb)(c/2)$$

or

Eq. (4.21) $$S_m = (W\ell)/(bc^2)$$

Comparing this with Eq. (4.16), it is evident it gives a maximum stress that is only two-thirds of the correct value.

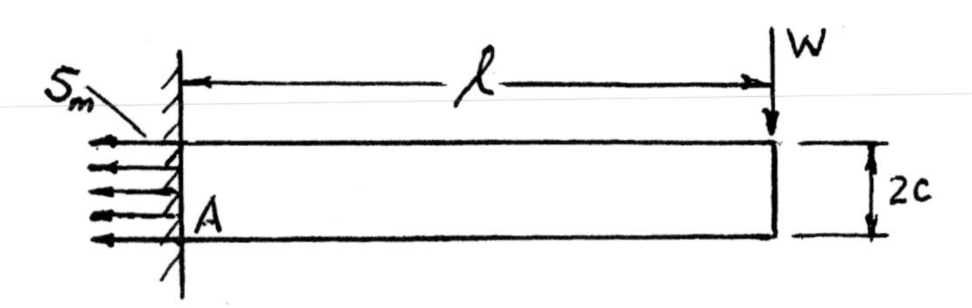

Figure 4.8. Free body diagram used by Galileo that does not satisfy equilibrium equations $\Sigma F_x = 0$, and $\Sigma M_A = 0$.

However, Galileo does not seek or use absolute values of strength but compares the strength of one beam with that of another. What is used is the proportionality:

Eq. (4.22) $$S_m \sim (W\ell)/(bc^2)$$

which follows from either Eq. (4.16) or (4.21), and states that maximum stress (S_m) is proportional to $(W\ell)/(bc^2)$.

In passage 158, Galileo compares the relative strength of a rectangular cantilever beam oriented with the long side horizontal with that when the long side is vertical. For the beam of Fig. 4.6, if $b/2c = 2$, then, with the long side oriented parallel to the load, the strength will be twice as great as for the other orientation. Galileo then includes the weight of a beam, correctly suggesting that this involves an additional moment equal to (wt. of beam)($\ell/2$).

The rest of the second day is devoted primarily to discussion of the relative strengths of prisms and cylinders that fail under their own weight. By prism, he means a column having a square cross section.

Figure 4.9 shows a cantilever of circular cross section subjected to a uniformly distributed load (w), which is its own weight per unit length.

Eq. (4.23) $w = (\gamma \pi d^2)/4$

where γ is the specific weight of the material $[FL^{-3}]$. The dotted line is the equivalent force acting through the center of mass of the cantilever which gives a bending moment at A of $(w\ell^2)/2$. The maximum stress at B will be

Eq. (4.24)

$$S_m = (Mc_m)/I_N = [(wl^2)/2][(d/2)]/[(\pi/4)(d/2)^4] = (16wl^2)/(\pi d^3)$$

substituting from Eq. (4.23):

Eq. (4.25) $S_m = (4\gamma\ell^2)/d$

Thus, for the same material (same γ and same S_m), when ℓ is doubled in length, d must increase by a factor of 4. This is illustrated in Fig. 4.10, which shows the relative shapes of bodies of the same material about to fail under their own weights.

Galileo suggests that this explains the basic shapes of tree branches and the appendages of animals and people. As limbs get longer, their diameter must increase as the square of their length, and there is a limiting size that cannot be exceeded for any given material. Galileo also says this explains why small animals are relatively stronger than large ones (a dog may carry three other dogs on his back but a horse cannot support even one other horse).

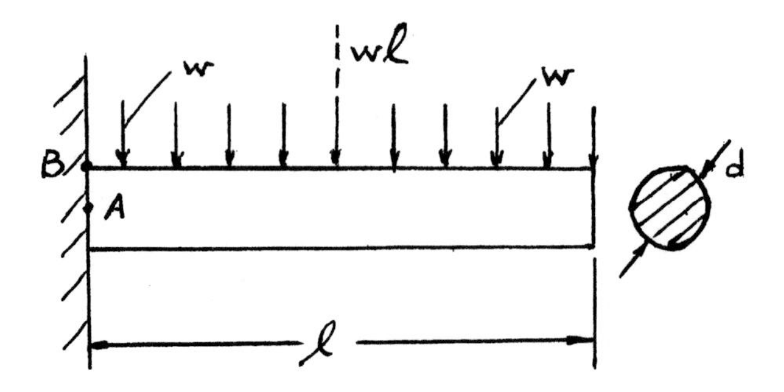

Figure 4.9. Cantilever of circular cross section subjected to uniform distributed load (w), $[FL^{-1}]$.

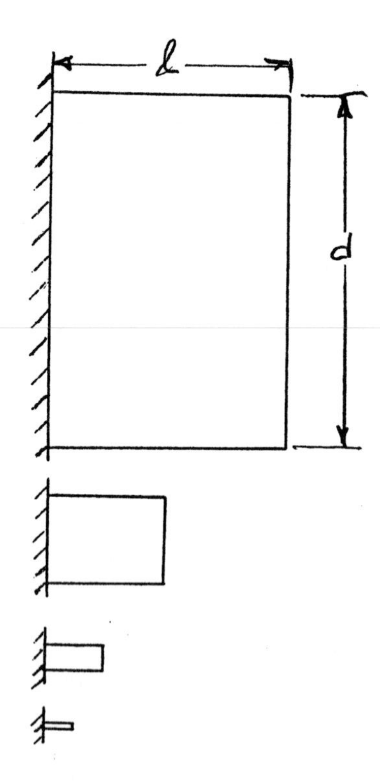

Figure 4.10. Relative shapes of cylindrical cantilevers of different diameter that will just fail under their own weight.

In passage 173, Galileo observes that the total force required to break the beam of Fig. 4.11 $(A + B)$ depends on the location of the fulcrum (0). Since the breaking moment at the fulcrum (M) will always be the same for a given beam, then:

Eq. (4.26)

$$A + B = M/a + M/b = (Mb + Ma)/ab = (M\ell)/(ab) = 1/ab$$

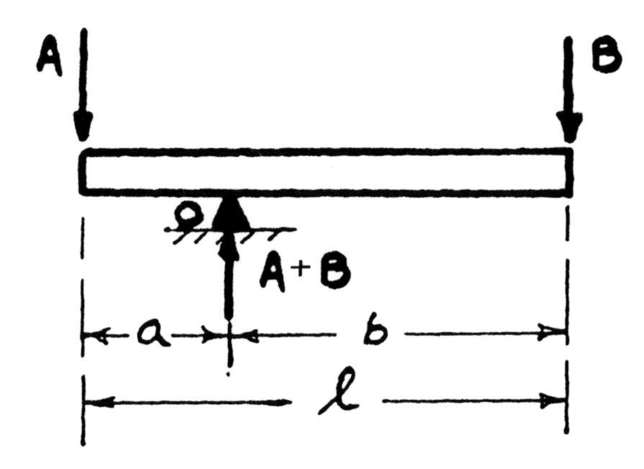

Figure 4.11. Rod being broken by three point loading.

Galileo then introduces the important concept of an optimum and shows that the force at the fulcrum $(A + B)$ will be a minimum when $a = b$. This may be seen by writing Eq. (4.26) as follows:

Eq. (4.27) $$A + B = (M\ell)/[a(\ell - a)]$$

where ℓ is the total length of the beam $(a + b)$.

The quantity $(A + B)$ will be a minimum when $[a(\ell - a)]$ is a maximum and the corresponding value of a may be found in Table 4.1 as follows (ℓ is a constant):

Table 4.1.

Let a equal	Then $a(\ell - a)$ equals
0.1ℓ	$(0.1\ell)(0.9\ell) = 0.09\ell^2$
0.2ℓ	$(0.2\ell)(0.8\ell) = 0.16\ell^2$
0.3ℓ	$(0.3\ell)(0.7\ell) = 0.21\ell^2$
0.4ℓ	$(0.4\ell)(0.6\ell) = 0.24\ell^2$
0.5ℓ	$(0.5\ell)(0.5\ell) = 0.25\ell^2 = \text{MAX}$
0.6ℓ	$(0.6\ell)(0.4\ell) = 0.24\ell^2$
0.7ℓ	$(0.7\ell)(0.3\ell) = 0.21\ell^2$
etc.	etc.

The same result may be obtained less laboriously by calculus (see App.A) noting that $(\ell - a)$ will be a maximum when:

Eq. (4.28) $$d[a(\ell - a)]/da = \ell - 2a = 0$$

or when

$$a = \ell/2 = b$$

Beginning in passage 173 of the Galileo text, several circular beams of the same length, diameter, and material that are supported in different ways are compared relative to strength. In all cases, the beams are subjected to a uniformly distributed load (w = weight/length) corresponding to the weight of the beam. Figure 4.12 shows four such cases with corresponding moment diagrams. The tendency for fracture to occur will be proportional to the maximum bending moment in each case from which it follows that:

- Cases (*a*) and (*b*) have equal tendencies to fracture
- Cases (*c*) and (*d*) have twice the tendency to fracture as either (*a*) or (*b*)

A comparison of cases (*b*) and (*d*) explains the paradox that a stone column resting on three equally spaced supports has a greater tendency to fracture under its own weight than when only two supports at the ends are used. If the ground under either end support settles more than under the other two, then case (*d*) pertains for which the tendency for rupture at the center of the beam is twice that for case (*c*).

It is suggested that passages 165–177 of the Galileo text be scanned now in order to obtain a feeling for how Galileo explains the failure of beams under their own weight.

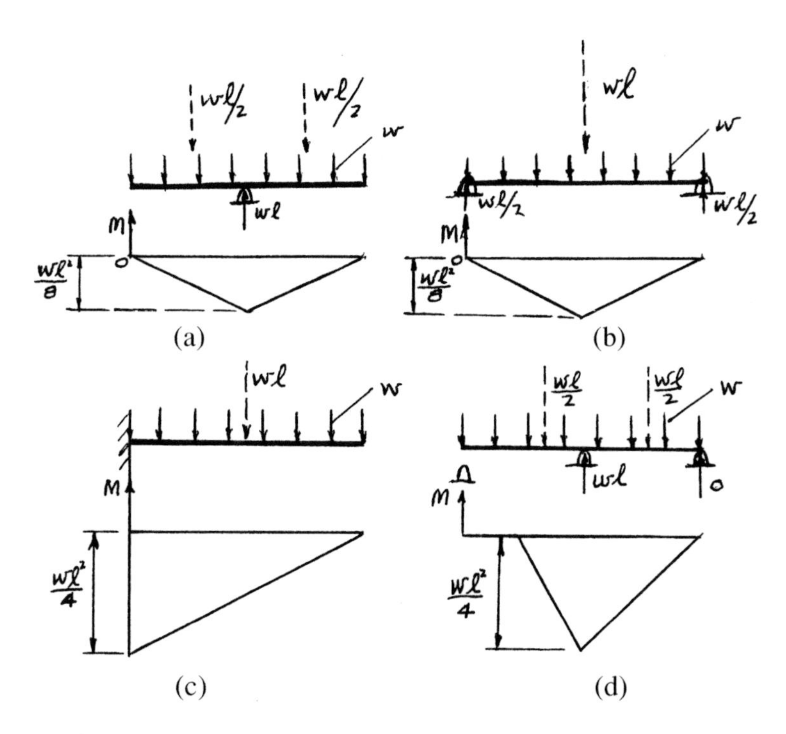

Figure 4.12. Uniformly loaded beams of the same length and material supported in different ways with moment diagram shown beneath each beam *(a)* single central support, *(b)* beam supported at two ends, *(c)* cantilever, and *(d)* beam initially on three equally spaced supports after one end support has been removed by settling of the ground.

5.0 STRENGTH-TO-WEIGHT RATIO

Beginning in passage 178, Galileo introduces the concept of strength-to-weight ratio first with regard to the optimum shape of a cantilever beam, and then by substituting tubing for a solid rod. He proves that a cantilever beam [Fig. 4.13 (*a*)] that has a lower surface of parabolic shape will weigh only two-thirds as much as a comparable beam that is of constant depth along the entire length of the beam [Fig. 4.13 (*b*)]. The depth of the beam (y) in Fig. 4.13 (*a*) at any value of x is:

Eq. (4.29) $$y = (2c/\ell^{0.5})x^{0.5}$$

From Eqs. (4.16), (4.17), and (4.29), it is found that the maximum tensile stress at any section located a distance (x) from the origin in Fig. 4.13 (*a*) will be:

Eq. (4.30) $$S_{mx} = [M_x(y/2)]/I_x = (6Wx)/by^2 = 3W\ell/2bc^2 \text{ (a constant)}$$

Thus, for the parabolic cantilever of Fig. 4.13 (*a*), the maximum stress will be the same at all points along the length of the beam. The maximum stress for the beam of Fig. 4.13 (*b*) will be the same as that for Fig. 4.13 (*a*) but will occur at A. The strength-to-weight ratio for the beam of Fig. 4.13 (*a*) will be 1.5 times that for the beam of Fig. 4.13 (*b*).

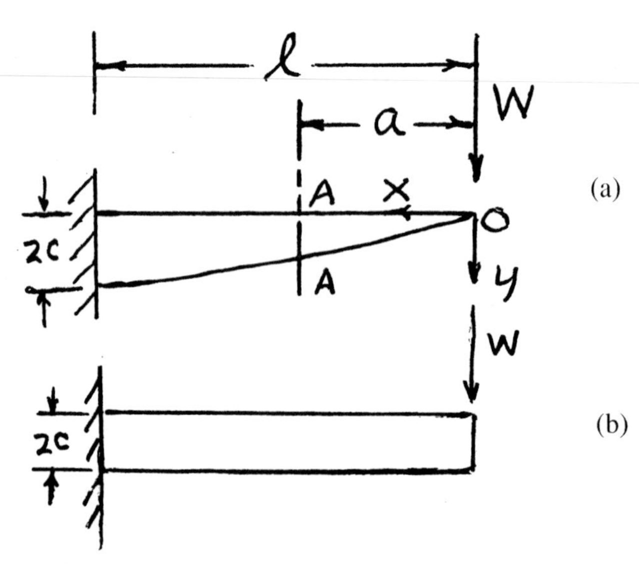

Figure 4.13. *(a)* Cantilever of parabolic shape and constant width, *b*, that gives constant maximum tensile strength at all points along its length and *(b)* cantilever of constant width, *b*, and constant height, 2*c*, along its length.

Continuing the discussion of cantilevers of high strength-to-weight ratio, Galileo compares solid [Fig. 4.14 (*a*)] and hollow [Fig. 4.14 (*b*)] beams of the same length, material, and weight.

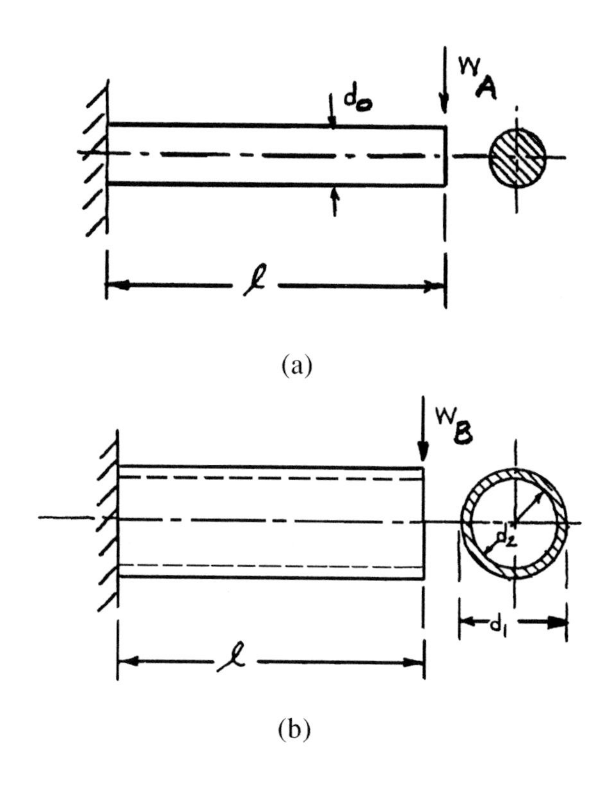

Figure 4.14. End loaded cantilever beams of same length *(a)* circular cross section and *(b)* annular cross section of same area as *(a)*.

From Eq. (4.17) for the solid cylindrical cantilever of Fig. 4.14 (*a*):

Eq. (4.31) $$S_{ma} = [W\ell(d_0/2)]/\pi/4(d_0/2)^4 = (32/\pi)(W\ell/d_0^3)$$

For the hollow cantilever of Fig. 4.14 (*b*):

Eq. (4.32) $$S_{mb} = [W\ell(d_1/2)]/\{\pi/4[(d_1/2)^4 - (d_2/2)^4]\}$$
$$= (32/\pi)(W\ell)(d_1/[d_1^4 - d_2^4])$$

For the same W and ℓ:

Eq. (4.33) $S_{ma}/S_{mb} = (d_1^4 - d_2^4)/(d^1 d_0^3)$

If, in addition, the weights of the two beams are the same:

Eq. (4.34) $(\pi d_0^2)/4 = (\pi/4)(d_1^2 - d_2^2)$

Eliminating d_2 from Eqs. (4.33) and (4.34):

Eq. (4.35) $S_{ma}/S_{mb} = (d_1^4 - 9d_0^4)/(d_1 d_0^3) = 7d_0^4/d_1 d_0^3$

For example, if $d_1 = 2d_0$, then $S_{ma}/S_{mb} = 3.5$

Thus, in general, the ratio of strengths-to-weight for the beams of Figs. 4.14 (*a*) and 4.14 (*b*) will be given by Eq. (4.35) which will give 3.5 for the example where $d_1 = 2d_0$.

It is suggested that passages 178–189 of the Galileo text be read at this point to observe what Galileo writes about the importance of reducing the weight of structures.

6.0 BEAM DEFLECTION

The deflection of a beam is often important in design. While a general discussion of beam deflection is too complex for present purposes, dimensional analysis of the deflection of an elastically end-loaded cantilever is useful. Figure 4.15 shows a cantilever of constant cross section along its length with an end deflection (δ). It is reasonable to assume that δ will be a function of the variables indicated below.

Eq. (4.36) $\delta = \psi(W, \ \ell, \ E, \ I_N)$
$[L] \quad [F] \ [L] \ [FL^{-2}] \ [L^4]$

The deflection should be expected to vary with the strain at all points which in turn will be a function of the bending moment ($W\ell$), the distribution of area about the neutral axis measured by I_N, and the elastic stiffness of the material (E). Since this is a statics problem only two fundamental dimensions are involved $[F]$ and $[L]$. The dimensions of each of the five variables are indicated beneath each variable in Eq. (4.36).

Variables W and ℓ are dimensionally independent and may be combined with each of the remaining three variables to form nondimensional groups. To satisfy dimensional homogeneity, Eq. (4.36) then becomes:

Eq. (4.37) $$\delta/\ell = \psi_1\,[(E\ell^2)/W],\,(I_N/\ell^4)$$

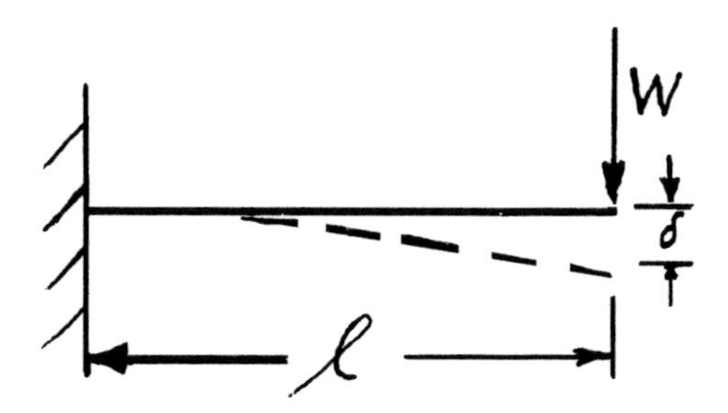

Figure 4.15. Deflection of end-loaded cantilever.

For a linearly elastic beam, δ will vary inversely with the stiffness of the material and the moment of inertia of the section about the neutral axis. Hence, Eq. (4.37) may be written:

Eq. (4.38) $$\delta/\ell = (\text{const.})(W/E\ell^2)(\ell^4/I_N)$$

from which it follows that:

Eq. (4.39) $$\delta \sim (W\ell^3)/(EI_N)$$

It is important to note that E and I_N appear only as the product EI_N. This will be the case for all elastic deflection beam problems, and represents an important observation for dimensional analysis. Equation (4.36) could have been written

Eq. (4.40) $$\delta = \psi(W, \ell, EI_N)$$

which would also lead to Eq. (4.39) after dimensional analysis.

The proportionality constant in Eq. (4.39) cannot be obtained by dimensional analysis, but must be obtained by performing a few simple experiments or a complete analysis involving calculus. Either approach would reveal that the constant is one-third. In this case, the experiments could consist of a wooden yard stick and/or a circular dowel being clamped in a vice with different unsupported lengths as δ is measured for a range of values of W. For each measurement, the constant in Eq. (4.39) would be calculated and the average for all measurements taken as the assumed value.

7.0 COLUMNS

For long slender columns that are end-loaded, stability, may be a problem. Figure 4.16 shows such a column with pin connected ends. The dashed line shows the deflected column. As the load is increased, the deflection δ will gradually increase as will the maximum stress. At a critical elastic stress, a long slender beam may buckle. For a beam of a given length (ℓ), moment of inertia (I_N), and Young's modulus (E), there will be a critical load W_C at which buckling will occur. As in other beam deflection problems, E and I_N will appear only as the product (EI_N). Therefore:

Eq. (4.41) $$W_C = \psi(\ell,\ EI_N)$$
$$[F] \qquad [L]\ [LF^2]$$

In this case, there will be only one nondimensional group $(W_C\ell^2)/(EI_N)$, and this must equal a constant. Thus:

Eq. (4.42) $$W_C = (\text{const.})(EI_N/\ell^2)$$

A more complete analysis reveals the constant to be π^2 for the pin connected column. This is known as the Euler (1757) equation for the column shown in Fig. 4.16. In general:

Eq. (4.43) $$W_C = k\pi^2(EI_N/\ell^2)$$

where k depends upon the degree of restraint at the ends of the column. Figure 4.17 shows values of k for several types of end restraint.

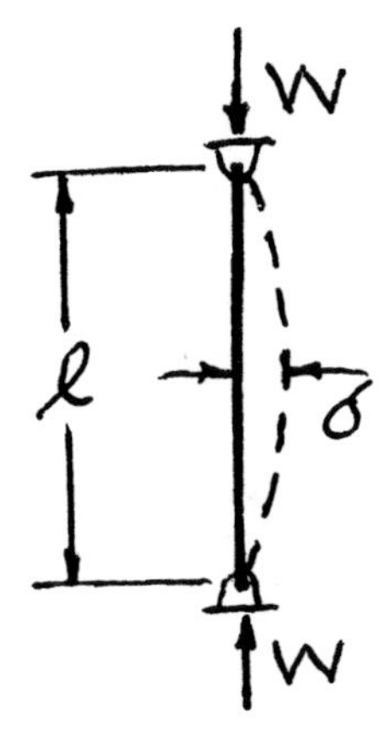

Figure 4.16. End-loaded slender column with pinned ends.

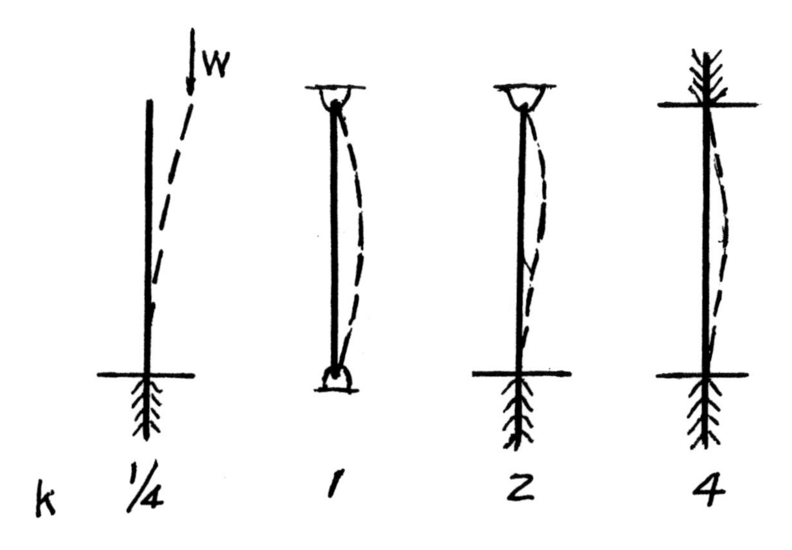

Figure 4.17. End-loaded column with different degrees of constraint at the ends.

8.0 IMPACT

In passages 291–293, Galileo discusses the impact of one body on another in general terms. He mentions the roles of relative motion, resilience of target, and angle of impact. He also indicates that he has spent much time attempting to explain the unusual destructiveness of a suddenly arrested projectile compared with that of slowly applied loads. Near the end of the fourth day, he promises to give his views on the mysteries of impact as soon as consideration of the paths of projectiles is completed. This, presumably, would have been covered on the fifth day, but unfortunately there was no fifth day. It would have been interesting to learn how his geometrical approach would have been adapted to the problem of impact. This is a problem best considered in terms of work (energy).

When a body is lifted against gravity to a high elevation, work is done that is equal to the product of the weight and the vertical distance involved. When a body falls through a vertical distance, its velocity increases and potential energy is converted to kinetic energy which is available to do work on another body as the first body comes to rest. If a moving body strikes a deformable body and is brought to rest as stresses and strains develop, the energy of the body on impact is converted to internal strain energy. If this occurs without any loss of energy in the form of heat, then the energy released by the falling body as it is brought to rest must equal the internal energy absorbed by the target.

Figure 4.18 shows a weight (W) falling through a distance (h) before impacting a deformable body (B). If B deforms elastically an amount Δ_D as weight (W) is brought to rest, then equating external and internal energies:

Eq. (4.44) $$W(h + \Delta_D) = \tfrac{1}{2}(k\,\Delta_D)(\Delta_D)$$

where k is the spring constant of body (B) when slowly loaded $[k = (W/\Delta_S)]$ and Δ_D and Δ_S are dynamic and static deformations respectively.

Equation (4.44) may be rewritten as follows:

Eq. (4.45) $$\Delta_D^2 - (2W/k)\,\Delta_D - (2W/k)\,h = 0$$

This is a quadratic equation whose maximum root is:

Eq. (4.46) $\Delta_D = W/k + [(W/k)^2 + 2(W/k)h]^{0.5}$

Substituting for $k = W/\Delta_S$:

Eq. (4.47) $\Delta_D/\Delta_S = 1 + [1+(2h/\Delta_S)]^{0.5} = S_D/S_S = n$

where S_D and S_S are dynamic and static stresses respectively, and n is the factor by which S_D exceeds S_S and is called the impact factor. For example, if W just touches D and then is suddenly released, $h = 0$ and $n = 2$. If, however, $h = 10\Delta_S$, then $n = 5.58$.

The above, very approximate, analysis assumes D remains elastic, that there is no lost energy during impact, and that W does not rebound from D before the maximum dynamic stress is obtained. Actually, all of these assumptions are unlikely to pertain and values of n as computed above will always be greater than experimental values. However, the analysis gives a useful qualitative explanation as to why n is always greater than one for impact loading.

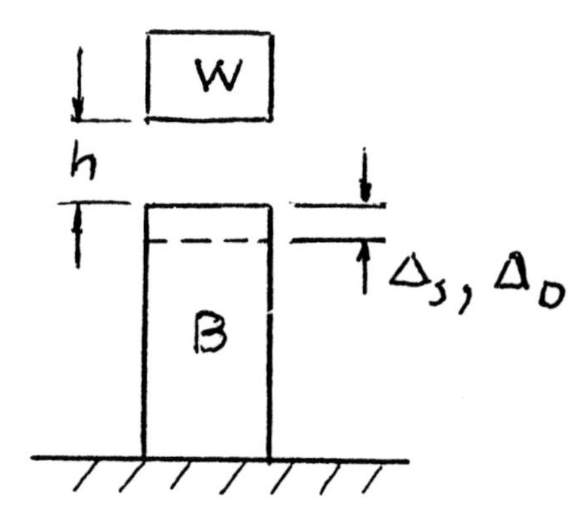

Figure 4.18. Body *(B)* subjected to weight *(W)* dropping through distance *(h)*. For static loading, Δ_S equals maximum deflection of *B* and for impact loading, Δ_D equals Δ.

9.0 COMPOSITE BEAMS

In some cases, beams may consist of layers of materials having different properties. In such cases, the resultant stress may be obtained by superposition. The most important example of a composite member is a reinforced concrete beam. Figure 4.19 shows the cross section of such a beam.

While concrete is so brittle it is considered to have negligible tensile strength, its strength in compression is significant. Therefore, bars of steel that have been produced with many surface protrusions to prevent slip are cast into the beam on the tensile side. The concrete on the compressive side of the neutral axis provides the compressive component of the resisting couple, while only the embedded steel is assumed to provide the tensile component. The moment of inertia of this composite structure will be the sum of that for the concrete rectangle and that for the steel. The steel contribution to the moment of inertia would be $A_S a^2$ where A_S is the area of the steel and a is the distance of the steel bars from the neutral axis.

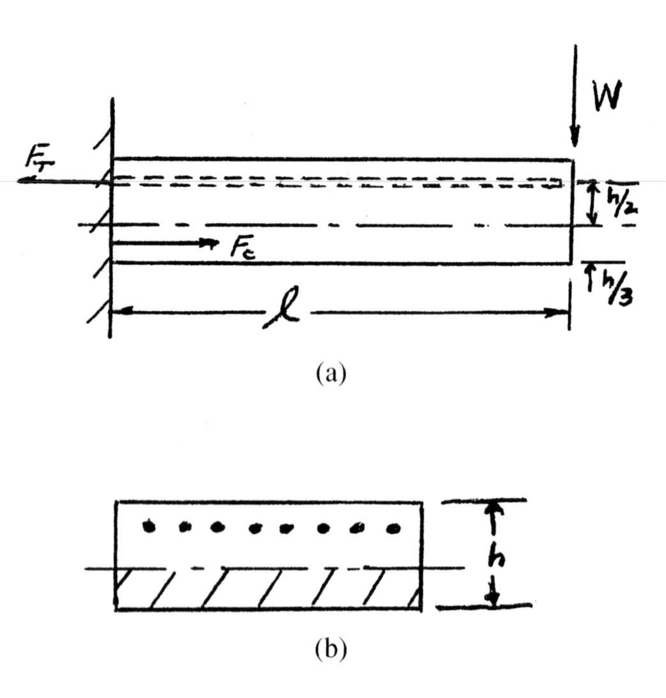

Figure 4.19. Steel reinforced concrete beam *(a)* side view and *(b)* end view.

10.0 VIBRATIONS

A system consisting of a mass (m) and a helical spring (Fig. 4.20) of stiffness, (k lbs/in.) (often called the spring constant), will exhibit a natural frequency of vibration (f) when displaced downward and released. The units of f are cycles per second (or Hertz) and, therefore, the dimension of f is $[T^{-1}]$. This natural frequency may be subjected to dimensional analysis. If the mass of the spring and air friction are ignored, then the variables in Table 4.2 are involved.

The quantities g, k, and m are dimensionally independent. However, when an attempt is made to combine these three quantities with f to form a nondimensional group, it is found that g is not required. This means that g need not be considered in this problem. Thus, there are two dimensionally independent quantities and one Pi quantity which leads to:

Eq. (4.48) $f = C_1(k/m)$

where C_1 is a constant that may be shown to be ½π either experimentally or analytically.

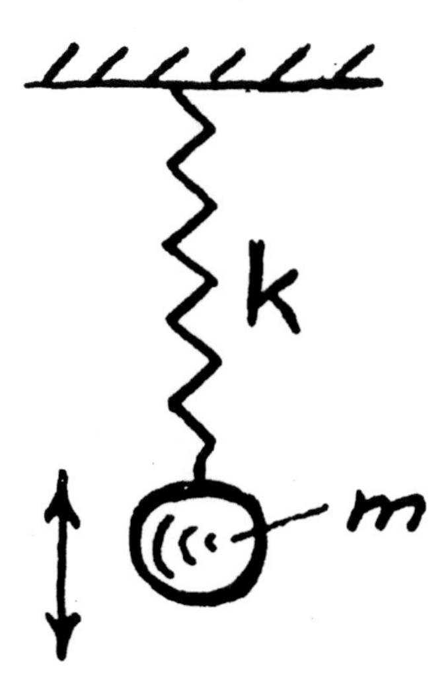

Figure 4.20. Simple spring-mass system.

Table 4.2. Quantities Involved in Dimensional Analysis of Spring-Mass System

Quantity	Symbol	Dimensions
Natural frequency	f	$[T^{-1}]$
Spring constant	k	$[FL^{-1}]$
Mass	m	$[FL^{-1}T^{2}]$
Acceleration due to gravity	g	$[LT^{-2}]$

The similarity between this solution and that for the simple pendulum should be noted. The quantity (g) does not enter this problem since the weight of the body (mg) merely determines the mean vertical position of the lower end of the spring about which it vibrates, but has nothing to do with the time it takes for a complete excursion above and below this mean position.

If the mass of the spring is not small compared with the external mass (m), then it can be shown that mass (m) should be increased by one-third of the mass of the spring when using Eq. (4.48).

Equation (4.48) is useful in correcting difficulties due to vibration. If a motor rests upon a base that has some flexibility which may be expressed in terms of a spring constant (k) and the mass of the motor and base (m), then the natural frequency of the system will be given by Eq. (4.48). When the rotational frequency of the motor corresponds to the natural frequency of the mass-spring system, then the amplitude of vibration will quickly rise to a high level. Not only will the motor move violently, but also the vibration will be transmitted to the floor. There are several remedies for this situation:

- Operate the motor at either a lower or higher speed
- Change the spring constant (k) of the support
- Increase the mass (m) of the system
- Introduce damping as described below

The first three of these items are obvious, but the fourth requires discussion. If a dashpot consisting of a close fitting piston that shears a viscous oil film is introduced as shown in Fig. 4.21, this will impart a resisting force to the mass (m) that is proportional to velocity (V). This will remove energy from the system and limit the vibrational amplitude. The viscous damping force is then:

Eq. (4.49) $$F_v = CV$$

where the proportionality constant (C) is called the damping constant [$FL^{-1}T$]. If the quantity (C) is included in the foregoing dimensional analysis, we obtain:

Eq. (4.50) $$f(m/k)^{0.5} = \psi[C/(mk)^{0.5}]$$

A complete analysis reveals that:

Eq. (4.51) $$\psi = \tfrac{1}{2}\pi[1 - C^2/(4km)]^{0.5}$$

which is in agreement with Eq. (4.50) and reduces to Eq. (4.48) when there is no damping ($C = 0$). Damping is seen to decrease the natural frequency of an undamped system.

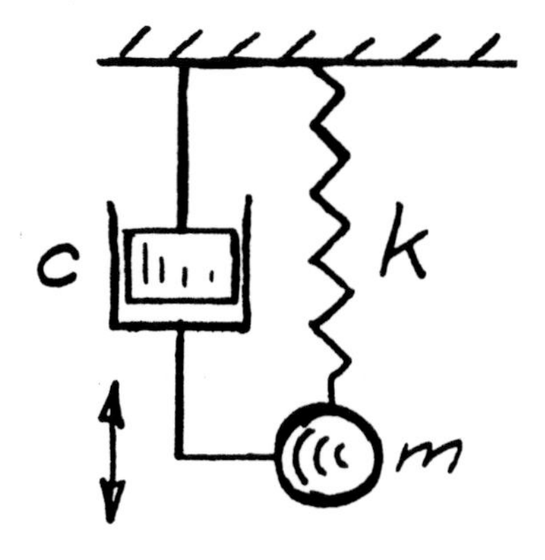

Figure 4.21. Spring-mass system with damping.

Sometimes it is desirable to mount a piece of machinery on a base, which isolates it from the floor, so that other machinery is not disturbed. This may be achieved by using a base of large mass (m) mounted on relatively soft springs (Fig. 4.22) to provide a system of low natural frequency. If the operating frequency of the machine is at least four times that of the natural frequency, good results are obtained. It is then unnecessary to have damping. Damping is beneficial only when the operating frequency must lie below the natural frequency of the system. In such cases, the support should be as stiff as possible (high k).

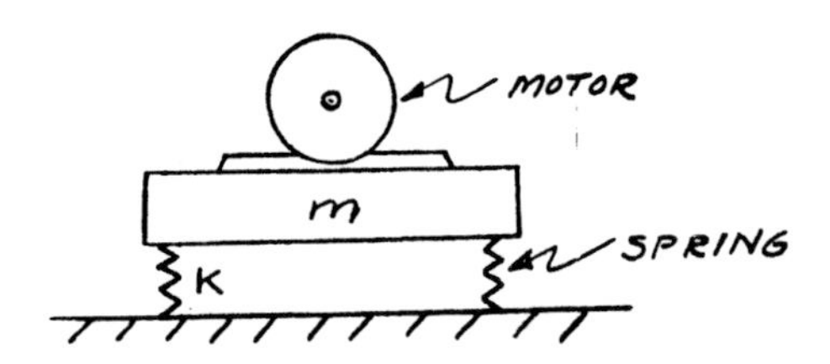

Figure 4.22. Motor on vibration-free mount.

In modeling an undamped structure for natural frequency, we may arbitrarily fix two scale factors, since there were two dimensionally independent quantities in connection with Eq. (4.48). It will usually be convenient to use a geometrically similar model, which fixes one scale factor, and to specify identical materials, which fix the other. With these restrictions, we will have a perfect model if:

Eq. (4.52) $$f(m/k)^{0.5}{}_p = f(m/k)^{0.5}{}_{model}$$

and for such a case, the natural frequency of the scale model will vary inversely with size, if the model is made from the same material as the prototype. That is:

Eq. (4.53) $$f_p/f_m = \ell_m/\ell_p$$

PROBLEMS

4.1 For a metal, the strain at the yield point (e_y) is approximately 0.001 (nondimensional). If the modulus of elasticity for this metal is 10^6 pounds per square inch (psi), what is the stress (S_y) at the yield point?

4.2 If a bar, 1 inch in diameter, is subjected to an axial load of 1,000 lbs, what is the mean tensile stress?

4.3 If the modulus of elasticity for the material of Problem 4.1 is 30×10^6 psi:

a) What is the strain?

b) What is the length of the bar under load, if its initial length (ℓ_0) is 5 inches?

4.4 For a cantilever beam such as that of Fig. 4.6 (*a*), estimate the maximum tensile stress (S_m) if $w = 100$ lbs, $\ell = 10$ ft, $b = 6$ in., and $c = 1$ inch.

4.5 Find S_m if, in Problem 4.4, $c = 3$ in. and $b = 2$ in. (same area $= 2cb = 12$ in.^{2}).

4.6 Find the value of shear stress (τ) in Problem 4.4.

4.7 If the cantilever in Fig. 4.7 (*a*) is a 1 in. diameter rod of length $\ell = 10$ in. and $w = 100$ lbs, find:

a) The maximum tensile stress at *A* (S_A).

b) The shear stress at *A*.

c) The shear stress at $\ell/2$.

4.8 For Problem 4.7, find the maximum tensile stress at $\ell/2$.

4.9 For the beam of Problem 4.7, what is the ratio of the maximum tensile stress at A and the shear stress? (Find S_m/τ at A.)

4.10 For the beam of Fig. 4.7 (b), find the maximum tensile stress (S_m), if $\ell = 5$ ft, $W = 100$ lbs, and the cross section of the beam is 1 in. square throughout its length.

4.11 What is the maximum shear stress for Problem 4.10?

4.12 Two long slender beams have rectangular cross sections with different orientations relative to the load. For beam A, the width = b and the depth = $2c$. For beam B, the width = $2c$ and the depth = b. Find the ratio of S_m for beam A to S_m for beam B.

4.13 What will S_{mA}/S_{mB} be if, in Problem 4.12, $b/2c = n$?

4.14 Two long, slender, rectangular, cantilever beams have the same length and are subjected to the same end load (W). If beam A has a width of 4 in. and a height of 2 in., while beam B has a width of 2 in. and a height of 1 in., find the ratio of the two maximum tensile stresses, S_{mA}/S_{mB}.

4.15 If the circular beam of Fig. 4.9 has a weight of 1 lb/ft, a length of 2 ft, and a diameter of 1 in., find the maximum tensile stress at B due to the weight of the beam.

4.16 For the rectangular beam of constant width (b) shown in Fig. 4.13 where the lower surface has a parabolic shape corresponding to height $y = (2c/\ell^{0.5})x^{0.5}$, show that the weight of the beam will be only two-thirds as great as the weight of a beam of constant height ($2c$) made from the same material.

4.17 Find the maximum tensile stress in the beam of Fig. 4.13 (*a*), if $W =$ 20 lbs, $\ell = 2$ ft, $b = 2$ in., and $c = 1$ inch.

4.18 What is the maximum tensile stress for the beam of Fig. 4.13 (*b*)?

4.19 For the example comparing the strength-to-weight ratios of the beams of Figs. 4.14 (*b*) and 4.14 (*a*), where $d_1 = 2d_0$ find the wall thickness for Fig. 4.14 (*b*) as a fraction of d_0.

4.20 Determine the deflection for the beam of Fig. 4.15, if $W = 100$ lbs, $\ell = 10$ in., the beam is a cylinder having a constant diameter of 1 in., and $E = 30 \times 10^{-6}$ psi (steel).

4.21 Repeat Problem 4.20 if the material is aluminum ($E = 10 \times 10^6$ psi) and the cross section is 1 in.^{2}.

4.22 Find the maximum stress and maximum strain for Problem 4.20.

4.23 Find the maximum stress and maximum strain for Problem 4.21.

4.24 For the beam shown in Fig. 4.16, find the buckling load (W_C), if $\ell =$ 100 in., $E = 30 \times 10^6$ psi (steel), and the moment of inertia (I_N) $= 0.049$ in.^{4} (1 in. diameter rod).

4.25 Find the buckling load for the beam of Problem 4.24, if both ends are fixed and all other conditions are the same.

4.26 Find the buckling load for an end-loaded 2×4 in. wooden beam with both ends pin loaded and $\ell = 8$ ft, $E = 1.2 \times 10^6$ psi and actual dimensions of the 2×4 are $1^5/_8$ inch. by $3^3/_8$ inch.

4.27 For the beam of Fig. 4.19, $W = 500$ lbs, $\ell = 10$ ft, and $h = 6$ inches. Find the area of steel and of concrete required if the maximum stress in the steel = 18,000 psi and the maximum compressive stress in the concrete = 650 psi.

4.28 A clock is regulated by a torsion pendulum which consists of a disc supported by a thin wire of diameter (d) and shear modulus (G) (Fig. P4.28). The disc of mass (m) oscillates periodically to and fro with a period of P sec. Before performing a dimensional analysis:

$$P = \psi_1(D, H, d, \ell, m, G)$$

a) Perform a dimensional analysis.

b) If, in one instance, the disc is made of aluminum and in a second it is made of steel, find p_{Al}/p_{ST}, if all other dimensions remain the same. The ratio:

(specific weight)$_{Al}$/(specific weight)$_{ST}$ = 1/3

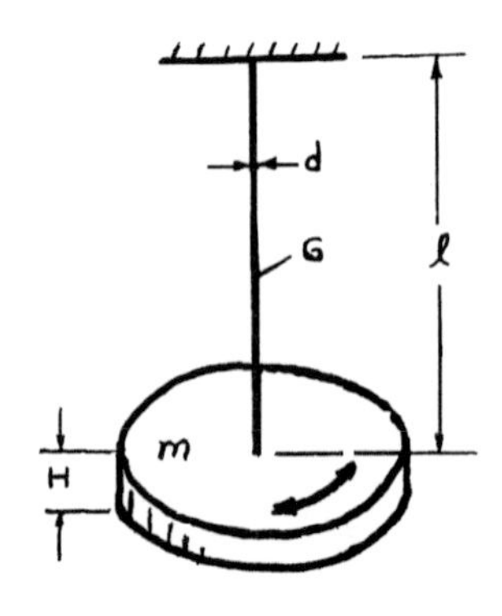

Figure P4.28.

4.29 The deflection of a coiled spring ($\Delta\ell$) depends on:

D = mean coil diameter
d = wire diameter
G = shear modulus of wire
W = load
n = number of turns of wire

Before dimensional analysis:

$$\Delta\ell = \psi_1(D, d, G, W, n)$$

a) Perform a dimensional analysis for $\Delta\ell$.

b) From experience you know that $\Delta\ell \sim W$. Use this fact to simplify part a) above.

c) If the shear modulus of the wire is doubled, how much must the load be increased for the deflection to remain the same? (D, d, and n remain unchanged.)

d) If D and d are both doubled, but G and n remain the same, how much must W change for $\Delta\ell$ to remain the same?

4.30 When a flexible steel rule is clamped to the edge of a table [Fig. P4. 30 (*a*)], the overhanging cantilever will execute a natural frequency (f) when plucked. Before dimensional analysis:

$$f = \psi_1(\ell, W, I_N, E)$$

where

W is the weight of the material per unit length
I_N is the moment of inertia about the neutral axis ($I_N = 1/12\ bh^3$)
E is the Young's modulus of elasticity of the material

a) Perform a dimensional analysis.

b) Has an important variable been omitted or the wrong one used here?

c) Make the necessary correction and use an obvious observation to simplify the result.

d) The "vibrotak" [Fig. P4.30 (*b*)] is an inexpensive instrument for measuring the speed of an engine or a motor. It consists of a small steel wire that may be extended from a chuck as illustrated in Fig. P4.30 (*b*). The extended length of wire is adjusted until it vibrates with maximum amplitude. The natural frequency of the wire then corresponds to the frequency of the device with which the chuck is in contact. If the wire extends 5.05 inches when the amplitude is a maximum, and the frequency of the motor on which it is resting is 2,000 rpm, how far will the wire extend when its amplitude is a maximum at 10,000 rpm? at 20,000 rpm?

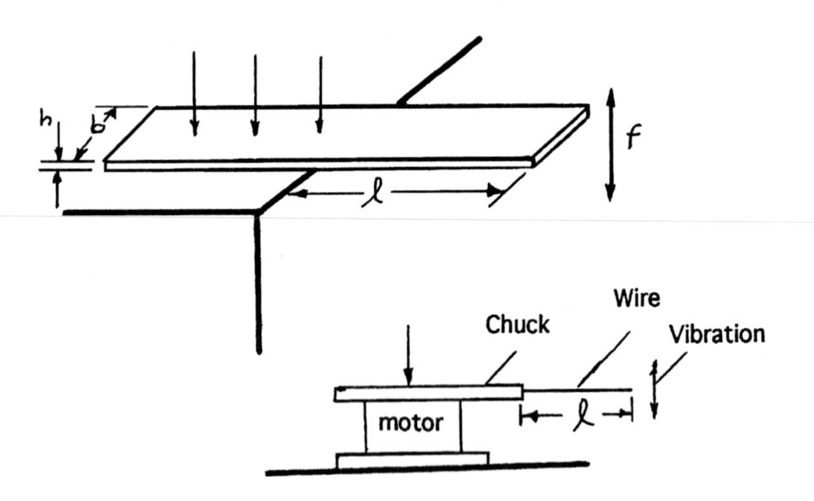

Figure P4.30.

5

Fluid Mechanics

1.0 INTRODUCTION

In this chapter, dimensional analysis is applied to several problems associated with the flow of fluids. In studying this material, attention should be directed not to the fine details but to the underlying philosophy and the manner in which dimensional analysis can be used to knit together a vast number of problems which, at first, appear to be independent. What Galileo has to say concerning fluid statics is also considered.

2.0 FLUID PROPERTIES

A fluid is either a liquid or a gas. While a gas completely fills its container, a liquid will fill a vessel only to a given level. The specific weight (γ) of a fluid is its weight per unit volume $[FL^{-3}]$, while density (ρ) is the mass per unit volume $[FL^{-4}T^{2}]$.

Eq. (5.1) $\rho = \gamma / g$

Unlike compressive stress in a solid that has different values in different directions, pressure (p) is the same in all directions in a fluid. Pressure is the compressive force acting upon a unit area of fluid $[FL^2]$. The pressure gradient ($\Delta p/\ell$), the change in pressure per unit length, is an important variable in fluid flow problems. Other important variables are the volume rate of flow, Q $[L^3T^{-1}]$ and the drag force (D) $[F]$ which acts on a moving body immersed in a fluid.

Viscosity (μ) is a very important fluid property that is responsible for the resistance fluids offer to flow. When a fluid particle is subjected to a shear stress, it undergoes a change in shape as shown in Fig. 4.3 (b). The ratio (dx/dy) is a measure of this change in shape and is called shear strain. The rate of shear strain (R') is

Eq. (5.2) $$R' = [d(dx/dy)]/dt = dV/dy$$

where dV is the difference in velocity of two planes a distance (dy) apart. The rate of shear strain (R') has the dimension $[T^{-1}]$. The quantity dV/dy is the velocity gradient of a particle of fluid as shown in Fig. 5.1.

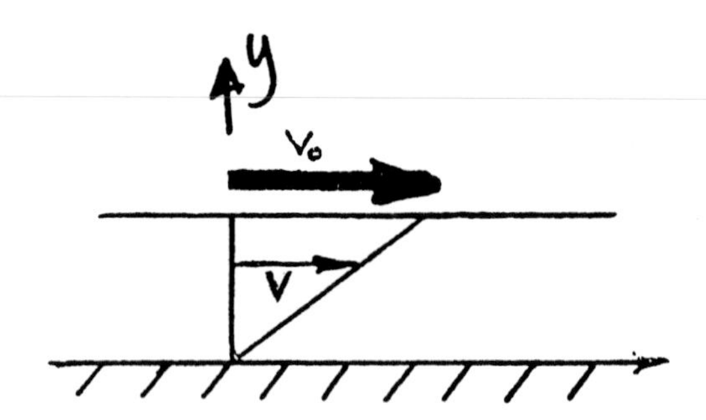

Figure 5.1. Velocity gradient across fluid film under constant pressure.

Newton's law of viscous flow relates the shear stress (τ) and rate of shear (dV/dy) for a fluid particle as follows:

Eq. (5.3) $$\tau = \mu dV/dy$$

The constant of proportionality (μ) is called viscosity. The dimensions of viscosity can be found from Eq. (5.3) to be [FTL^{-2}]. In the English system of units, the unit of viscosity is called the Reyn (after Reynolds) and equals one lb sec/in^2. In the metric system, the unit of viscosity is the Poise (after Poiseuille) and equals one dyne sec/cm^2. One Reyn equals 68,950 Poise. Representative values of viscosity (μ) and specific weight (γ) are given in Table 5.1.

Table 5.1. Representative Values of Viscosity (μ) and Specific Weight (γ)

	Viscosity, μ		Specific Weight, γ	
	Reyn	**Poise**	**lb/in³**	**N/cm³**
Air	2.6×10^{-9}	1.8×10^{-4}	44.3×10^{-6}	12.0×10^{-6}
Gasoline	7.3×10^{-8}	0.5×10^{-2}	0.031	8.41×10^{-3}
Water	1.5×10^{-7}	10^{-2}	0.036	9.77×10^{-3}
Mercury	2.3×10^{-7}	1.6×10^{-2}	0.462	0.125
Kerosene	2.9×10^{-7}	2×10^{-2}	0.031	8.41×10^{-3}
Lubricating Oil	10^{-6}	1	0.030	8.14×10^{-3}
Glycerin	9×10^{-6}	9	0.043	1.17×10^{-2}
Pitch	1	10^{6}		
Glass	10^{9}	10^{15}	0.088	2.39×10^{-2}

The ratio γ for water to γ for air = 813.

Viscosity is usually measured by the time it takes a fluid to flow through a tube under standard conditions. Figure 5.2 is a schematic of a glass Ostwald viscometer. The fluid to be measured is poured into the left tube. The unit is immersed in a constant temperature water bath and when a constant temperature is reached, the fluid is drawn up into the right tube. The time required for the level to fall from A to B is measured with a stopwatch The quantity $\mu/\rho = \nu$ of the liquid, called kinematic viscosity [L^2T^{-1}], is proportional to the time required for outflow. Such a device would be calibrated using a fluid of known viscosity, usually water. The density of the fluid must also be measured at the same temperature as the viscosity, and ν is multiplied by ρ to obtain the absolute velocity (μ).

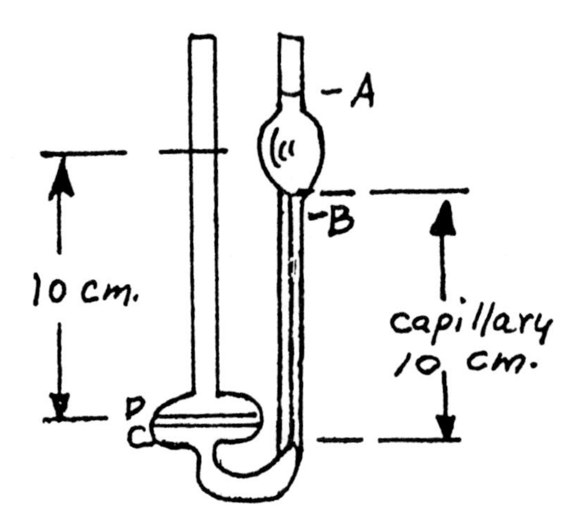

Figure 5.2. Ostwald viscometer.

The viscosity of a fluid varies very rapidly with the temperature, as indicated in Fig. 5.3. This figure shows absolute viscosity μ (to be distinguished from previously discussed kinematic viscosity, $\mu/\rho = \upsilon$) vs temperature for several motor oils. Society of Automotive Engineers (SAE) numbers are used to classify the viscosity of motor oils in the USA.

Two important observations concerning fluids subjected to viscous flow are:

1. There is never any slip between a fluid and a solid boundary. In Fig. 5.1, the fluid particles in contact with the moving and stationary surfaces have the velocities of these surfaces (V and zero respectively).
2. In the absence of a pressure gradient, the fluid velocity varies linearly across a film contained between parallel plates, one that is moving, the other being stationary (as in Fig. 5.1).

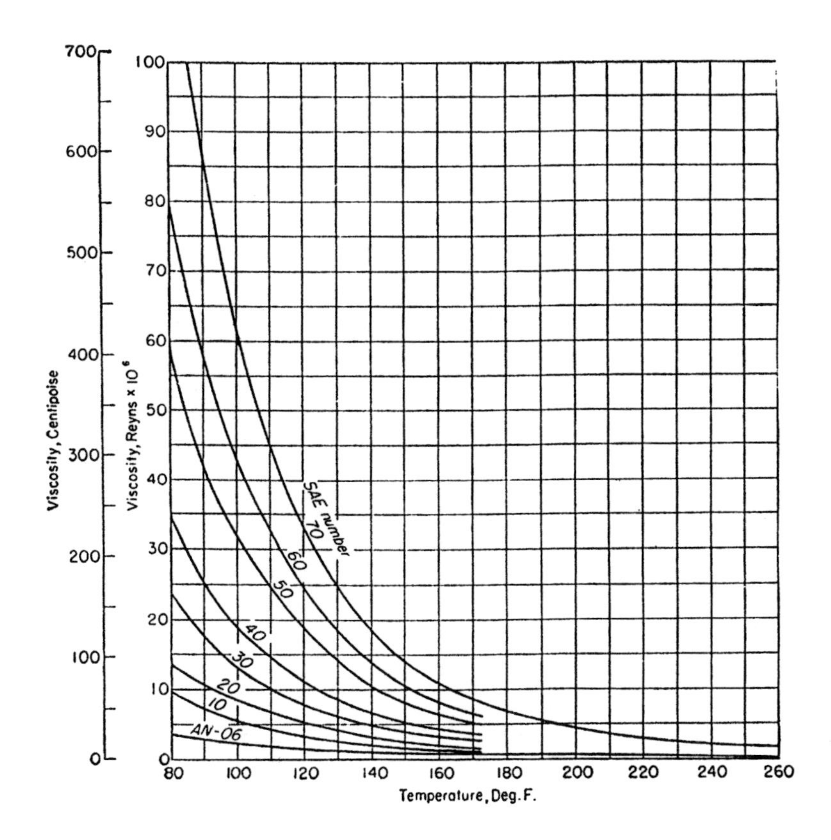

Figure 5.3. Viscosity-temperature curves for SAE oils.

3.0 FLUID STATICS

When a body is immersed in a fluid, it is subjected to an upward force equal to the weight of the displaced fluid that acts through the center of mass of displaced fluid. This is called the buoyant force. Figure 5.4 shows the free body diagram for a block of wood floating in water where W is the weight of the body, O_1 is its center of weight, B is the buoyant force, and O_2 is the center of buoyancy. Buoyant forces play an important role in the design and performance of surface vessels, submarines, and lighter-than-air craft (dirigibles).

Figure 5.5 illustrates another static stability problem. This figure shows a dam that will be stable only if the moment of the weight of the structure about O (= Wb) is greater than the moment of the force (P) due to water pressure about O (= Pa). The pressure of the water (p) increases with the depth below the surface ($p = \gamma y$, where γ is the specific weight of water and y is the depth below the surface). The total force (P) on the face of the dam per unit dam width due to water pressure will be $(h/2)(\gamma h) = h^2\gamma/2$.

If water should rise to the top of the dam, then the moment about O due to water pressure would increase to $P'a'$. It is important that this moment about O not exceed the opposing moment due to the weight of the dam. Otherwise, the dam would tend to rotate counterclockwise about O allowing water under high pressure to penetrate below the dam causing a catastrophic failure.

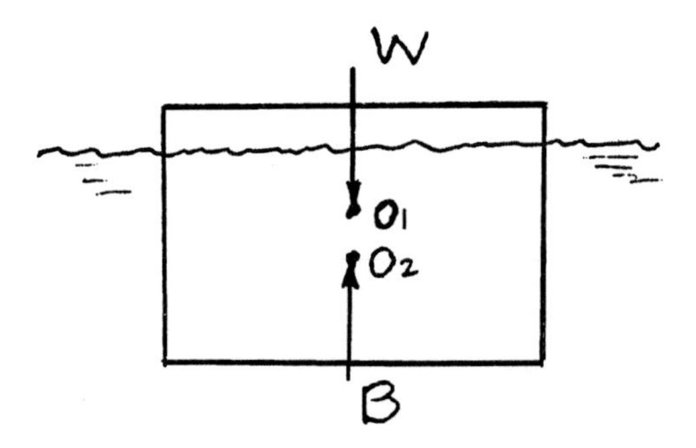

Figure 5.4. Free body diagram of wooden block floating in water; O_1 = center of weight of block and O_2 = center of buoyancy of displaced water.

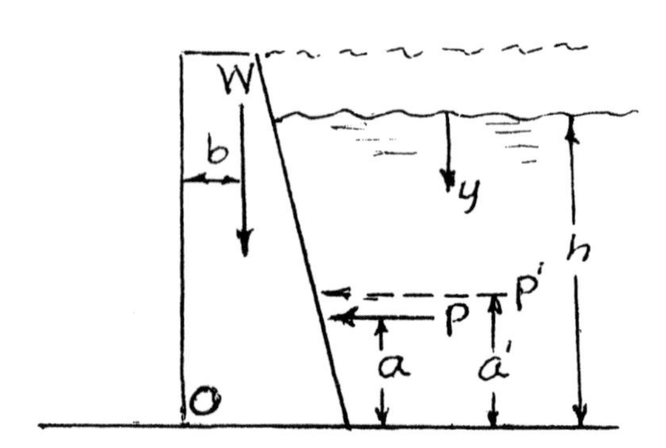

Figure 5.5. Dam with negative overturning moment about point O when $Wb > Pa$ for water level shown solid, and with positive overturning moment for dashed water level when $P'a' > Wb$.

4.0 SURFACE TENSION

The molecules or ions in the surface of a liquid or solid are in a different state of equilibrium than those lying in a parallel plane beneath the surface since they lack the influence of particles on one side of the surface. This causes surfaces to try to extend. Small droplets of liquid tend to become spherical by virtue of this effect. The liquid property that measures the tendency for a surface to extend is called surface tension (T_e) and the dimensions of T_e are $[FL^{-1}]$. The corresponding quantity for a solid is called surface energy and also has the dimensions $[FL^{-1}]$.

The surface tension of liquids gives rise to important forces only when thin films are involved. For example, this is the force which holds contact lenses in place on the surface of the eye, and which makes it possible to stack gage blocks in the workshop with negligible error.

Figure 5.6 (*a*) shows a cylinder of liquid with its axis perpendicular to the paper, and having a surface of small radius of curvature (r). The height of this cylinder perpendicular to the paper is (ℓ). Figure 5.6 (*b*) shows the forces acting on a free body having chordal length ($2r \sin \theta$). These forces consist of tensile forces in the surface equal to ($T_e \ell$), and a force on the chord due to internal pressure in the liquid which is assumed to be the same at all points. Equating force components in the vertical direction to zero for equilibrium:

Eq. (5.4) $$-2T_e \ell \sin \theta + p(2\ell r \sin \theta) = 0$$

or

Eq. (5.5) $$p = T_e / r$$

Thus, the pressure within the cylinder will be greater than in the air by an amount equal to the ratio of surface tension to the radius of the cylinder.

Figure 5.7 shows the edge of two gage blocks in contact having a mean spacing of $h = 2r$. A small oil film of surface tension $T_e = 30$ dynes/cm $= 1.72 \times 10^{-4}$ lb/in. will normally be present on such surfaces. If the radius of curvature of the meniscus (r) is about equal to the peak-to-valley surface roughness (about 10^{-6} in.), then:

$$p = -(1.72 \times 10^{-4})/10^{-6} = -172 \text{ psi } (-1.19 \text{ MPa})$$

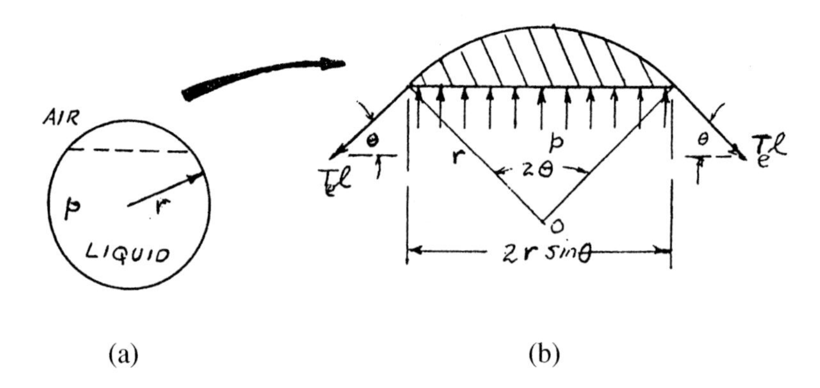

Figure 5.6. *(a)* Small cylindrical liquid particle, and *(b)* free body diagram of portion of particle at *(a)* showing surface tension forces and internal pressure (p) that will be greater than atmospheric pressure for the curvature shown.

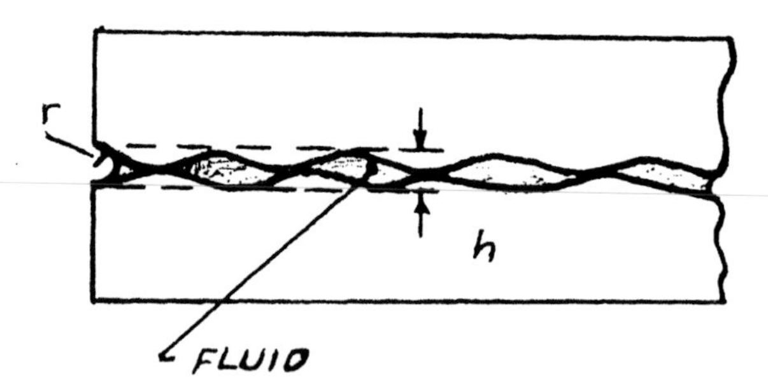

Figure 5.7. Gage blocks with oil film of thickness (h) between the surface finish peaks.

The pressure within the meniscus will be less than that of the atmosphere since curvature of the meniscus will be negative. Pressure in the oil film is thus seen to be negative and equal to several atmospheres. This large negative pressure will tend to force the surfaces of the blocks together until the peaks of asperities on the two surfaces are in contact.

Proof that it is pressure that holds gage blocks together and small contact lenses in place on the cornea of the eye is offered by the fact that the surfaces will float apart if submerged in the fluid responsible for the negative pressure generating meniscus. Galileo has pointed out that cause must precede effect and the force holding the surfaces together cannot be due to the tendency for a vacuum to develop on separation of the surfaces.

5.0 PIPE FLOW

The pressure drop per unit length of circular pipe ($\Delta p/\ell$) corresponding to the mean flow velocity (V) is a quantity of considerable practical importance, and was the first engineering application of dimensional analysis (Reynolds, 1883). A section of pipe sufficiently far from the inlet so that equilibrium has been established in the flow pattern is shown in Fig. 5.8. If inertia and viscous forces are included in the analysis but compressibility effects are ignored, then the important variables are those listed in Table 5.2.

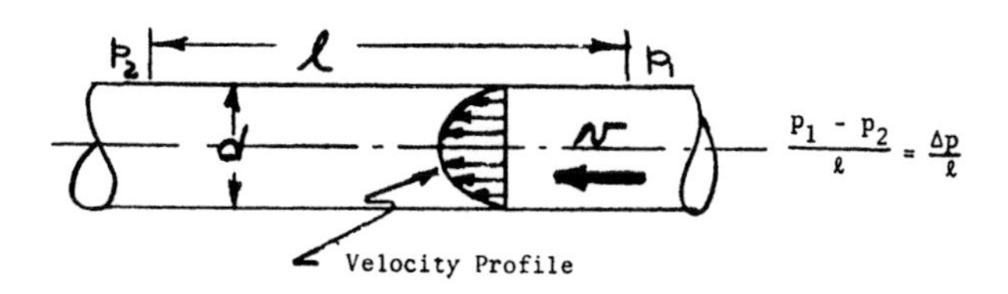

Figure 5.8. Velocity profile for pipe flow.

Table 5.2. Quantities Involved in Flow of Fluid in a Pipe

Quantity	Symbol	Dimensions
Pressure drop per unit length of pipe	$\Delta p/\ell$	$[FL^{-3}]$
Diameter	d	$[L]$
Mean fluid velocity	V	$[LT^{-1}]$
Fluid density	ρ	$[FL^{-4}T^2]$
Viscosity	μ	$[FTL^{-2}]$

Before dimensional analysis:

Eq. (5.6) $\Delta p/\ell = \psi_1(d, V, \rho, \mu)$

After dimensional analysis:

Eq. (5.7) $(\Delta p d)/(\ell \rho V^2) = \psi_2 [(\rho V d)/\mu]$

The nondimensional quantity $[(\rho V d)/\mu]$ is called the Reynolds Number (R).

The Reynolds Number may be interpreted as being proportional to the ratio of inertia and viscous forces acting on a fluid particle. When R is low, the viscous force is dominant and the fluid tends to move in straight lines, surface roughness plays a minor role, and the flow is called laminar. At high values of R, the inertia force is dominant and the motion of the fluid is random, roughness of the pipe is important, and the flow is called turbulent. The greater the roughness, the lower will be the Reynolds Number at which flow becomes fully turbulent.

The function (ψ_2) may be found by measurements on a single pipe size using water. When this is done, Fig. 5.9 results and two regimes are evident:

I For low values of R (called laminar flow)

II For high values of R (called turbulent flow)

The surface roughness of the pipe proves to be of importance in region II. The nondimensional quantity $[2(\Delta p d)/(\ell \rho V^2)]$ is called the friction factor (C_f), hence $C_f = 2\psi_2$.

Pipes constitute an important means of transport for gas, oil, and water. For example, the length of pipelines in the USA for transporting petroleum products is of the same order as the length of the railroads. It is important that an engineer be able to estimate the power required to pump a variety of fluids through pipes of different sizes and construction. A useful chart for estimating the friction factor is presented in Fig. 5.10. This chart differs from the simplified one in Fig. 5.9 in that it includes the effect of pipe-wall roughness. The nondimensional roughness parameter (ε/d) in Fig. 5.10 is the ratio of the peak-to-valley roughness to the bore diameter of the pipe. Typical values of (ε/d) are given in Table 5.3 for different commercial types of 1 in. diameter pipe. For a 10 in. diameter pipe, values of (ε/d) are approximately an order of magnitude lower (i.e., ε remains about constant as d increases).

In Fig. 5.10 there are four regimes of flow:

1. A laminar flow zone, (C_f) independent of roughness, R <2,000
2. A transition zone, $R = 2{,}000\text{–}4{,}000$
3. A low turbulence zone, C_f = function of R and ε/d
4. A high turbulence zone, C_f independent of R

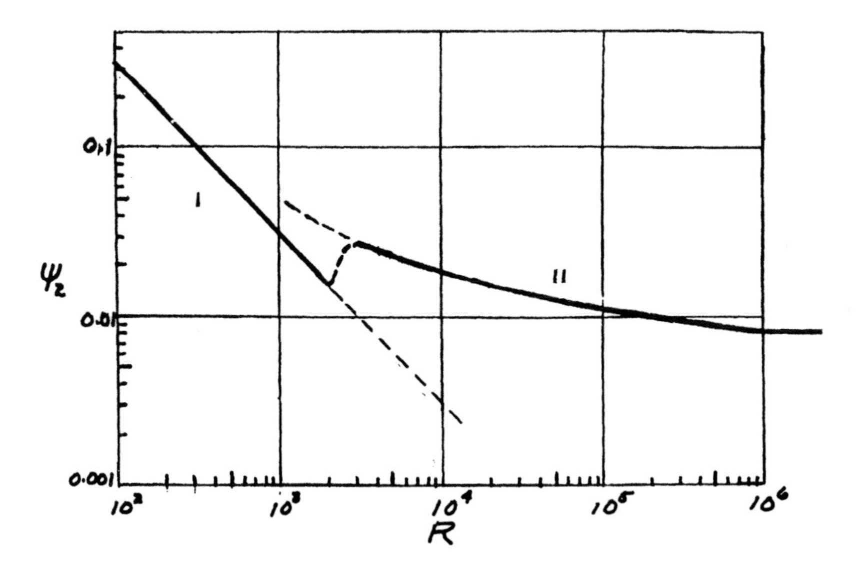

Figure 5.9. Variation of ψ_2 with Reynolds number for smooth pipe.

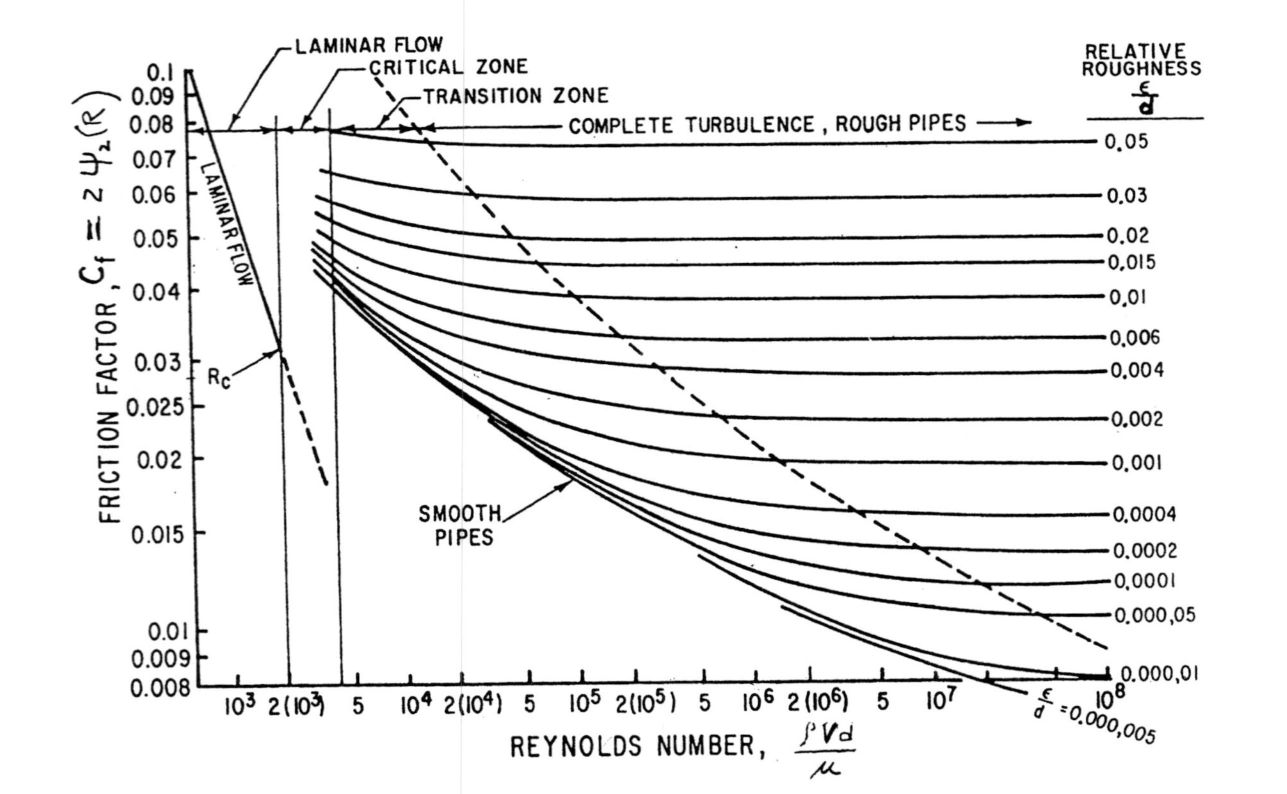

Figure 5.10. Variation of C_f with Reynolds Number and relative roughness for round pipe. (*After Moody, Trans ASME, 66, 671, 1944.*)

Table 5.3. Approximate Values of ε/d for 1 in. Diameter Pipe

Pipe Type	ε/d
Cast iron	0.015
Concrete	0.010
Galvanized iron	0.0060
Asphalted cast iron	0.0045
Commercial steel	0.0018
Wrought iron	0.0018
Drawn tubing	0.00006

Example—Pipe Flow. Consider the pressure drop and horsepower required to pump 3 ft^3 of water per second through a 12 in. (305 mm) diameter cast iron pipe for a distance of one mile (5,280 ft) (1,610 m).

The mean velocity $\bar{V} = Q/A = 3/(\pi d^2/4) = 3.282$ fps (1 m·s^{-1})

From Table 5.1 for water:

$$\rho = [(0.036)(12)3]/32.2 = 1.932 \text{ lb.s.}^2\text{ft}^{-4}$$

$$\mu = (1.5 \times 10^{-7})(144) = 21.6 \times 10^{-6} \text{ lb.s.ft}^{-2}$$

$$\text{R} = \rho\bar{V}d/\mu = [(1.932)(3.82)(1)]/(21.6 \times 10^{-6}) = 3.4 \times 10^5$$

From Table 5.3:

For 1 in. cast iron pipe, $\varepsilon/d = 0.015$

For 12 in. pipe, $\varepsilon/d = 0.0015$

From Fig. 5.10:

$$C_f = 0.022$$

From Eq. (5.7) and $\psi_2 = C_f/2$:

$$\begin{aligned}\Delta p &= (C_f/2)(\ell/d)(\rho \bar{V}^2)\\ &= (0.011)(5{,}280/1)(1.932)(2.82)^2\\ &= 1{,}637 \text{ lb/ft}^2\\ &= 11.37 \text{ psi } (0.078 \text{ MPa})\end{aligned}$$

$$\text{hp} = (\Delta pQ)/550 = [(1{,}637)(3)]/550 = 8.93 \text{ hp } (6.66 \text{ kW})$$

The most important engineering application of laminar flow at low Reynolds Number is hydrodynamic lubrication that is discussed next. Other applications, such as flow in capillaries of small diameter, are less important in engineering, and therefore, consideration of this type of flow is postponed to the end of this chapter.

6.0 HYDRODYNAMIC LUBRICATION

Bearing surfaces of one type or another restrain the motion of a part relative to that of its neighbor in practically every mechanism. The journal bearing (Fig. 5.11) is commonly used to support a rotating shaft against a radial force, and was first analyzed by Reynolds in 1886.

Fluid introduced at (A) is moved in a circumferential direction by the rotating journal, and eventually leaves the ends of the bearing. The difference between the bearing diameter and that of the journal (shaft) is very small, and is called diametral clearance (c). The journal rotating at (N) rpm acts as a pump to develop high pressure on the lower side of the clearance space. In a well-designed bearing, this pressure is sufficient to support the load on the journal (W) without solid contact. Viscosity of the fluid (μ) plays an important role in determining the magnitude of the pressure generated in such a hydrodynamic bearing. As the load is increased, the journal will approach the bearing surface, and the minimum film thickness (h) is a measure of the extent to which this has occurred.

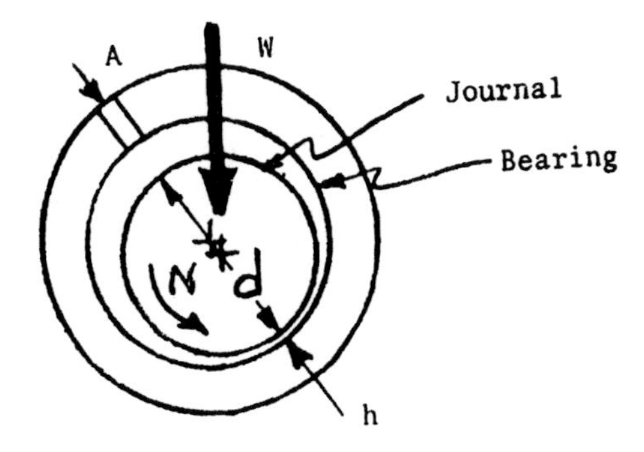

Figure 5.11. Hydrodynamic journal bearing.

In the design of a journal bearing, the minimum film thickness (h) is of major interest since the load capacity of the bearing is the value of W corresponding to the minimum allowable value of (h). Quantities to be considered in performing a dimensional analysis for (h) are listed in Table 5.4. Quantities to be considered in performing a dimensional analysis for (f) are listed in Table 5.5.

Table 5.4. Variables of Importance in Dimensional Analysis of Hydrodynamic Journal Bearing

Quantity	Symbol	Dimensions
Minimum film thickness	h	$[L]$
Journal diameter	d	$[L]$
Axial bearing length	ℓ	$[L]$
Diametral clearance	c	$[L]$
Specific load on bearing	P	$[FL^{-2}]$
Journal rpm	N	$[LT^{-1}]$
Viscosity of fluid	μ	$[T^{-1}]$
Density of fluid	ρ	$[FL^{-4}T^{2}]$

P is the load (W) per projected bearing area = $W/\ell d$

Table 5.5. Variables of Importance in Dimensional Analysis of Coefficient of Friction of a Hydrodynamic Journal Bearing

Quantity	Symbol	Dimensions
Coefficient of friction	f	$[0]$
Journal diameter	d	$[L]$
Bearing length	ℓ	$[L]$
Diametral clearance	c	$[L]$
Unit load	P	$[FL^{-2}]$
Journal speed	N	$[T^{-1}]$
Viscosity of fluid	μ	$[FTL^{-2}]$

After dimensional analysis:

Eq. (5.8) $$h/c = \psi_1\,[d/c, \ell/c, \mu N/P, \rho(Nc)^2/P]$$

The last nondimensional quantity $[\rho(Nc)^2/P]$ is a Reynolds Number. When R is evaluated for practical bearings, it is found to be small compared with unity. This suggests that inertia effects will be negligible compared with viscous effects, and that ρ need not have been included in the dimensional analysis. This is verified by experiment. Thus, Eq. (5.8) may be written:

Eq. (5.9) $$h/c = \psi_2[d/c, \ell/c, (\mu N)/P]$$

It is further found that ℓ has a small influence on h as long as ℓ/d is greater than one, which is usually the case. Thus ℓ/c may be omitted from Eq. (5.9). Also, d/c and $\mu N/P$ may be combined into one nondimensional group as follows:

Eq. (5.10) $$h/c = \psi_3[(d/c)^2\mu N/P]$$

This was found to be the case by Sommerfeld in 1904 and $[(d/c)^2\mu N/P]$ is called the Sommerfeld Number (S). Figure 5.12 is a plot of h/c vs the

Sommerfeld Number that is useful for estimating the minimum film thickness in journal bearing design.

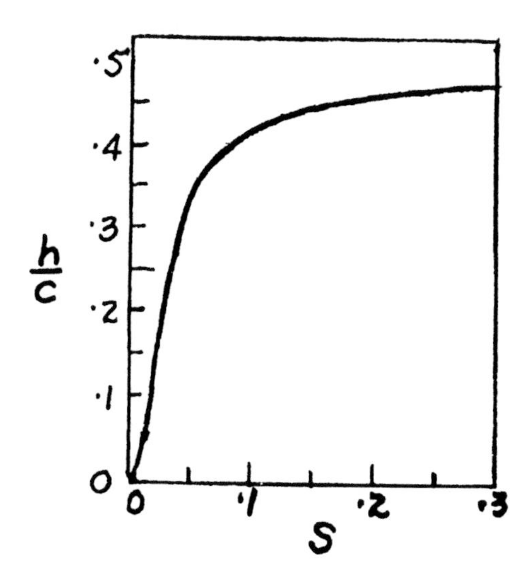

Figure 5.12. Variation of h/c with Sommerfeld Number (S) for a journal bearing of essentially infinite length (i.e., $d/\ell \cong 0$).

The coefficient of friction (f) for a journal bearing is:

Eq. (5.11) $\quad f = F/W$

where F is the tangential shearing force the lubricant exerts on the journal surface, and W is the load on the journal.

A lightly loaded bearing will operate with negligible end flow since negligible pressure is developed within the film when the journal and bearing are concentric. For this special case, the shear stress at the journal surface will be:

Eq. (5.12) $\quad \tau = \mu(du/dy) = \mu[V/(c/2)]$

The friction force on the journal will be:

Eq. (5.13) $$F = 2(\mu)(V/c)(\pi \ell d)$$

This is known as the Petroff (1883) equation. The value of h/c for a lightly loaded bearing will be 0.5.

The horsepower dissipated in a lightly loaded journal bearing will be:

Eq. (5.14) $$\text{hp} = (FV)/(12)(550)$$

Consider the following example:

$\mu = 10^{-5}$ Reyn (SAE 30 at 110°F, Fig. 5.3)

$d = 1.000$ in. (25.4 mm)

$\ell = 2.000$ in. (50.8 mm)

$V = 941$ ips. (1,800 rpm)

$c = 0.001$ in. (25 μm)

From Eq. (5.14), hp = 0.17 (127 W)

More detailed analysis of journal bearing friction reveals that the Petroff equation gives satisfactory results only above a Sommerfeld Number (S) of 0.15. For lower values of S the coefficient of friction will be higher than the lightly loaded approximation due to Petroff. The curve labeled Reynolds (based on a complete solution due to Reynolds) in Fig. 5.13 enables the coefficient of friction (f) to be estimated when the eccentricity of the journal in the bearing plays a significant role (that is, when $S < 0.15$).

7.0 BERNOULLI EQUATION

A useful energy approach to the solution of fluid mechanics problems employs the Bernoulli equation. This equates the total energy per unit mass (m) between two points on the same stream line when energy loss due to friction is negligible. A stream line gives the path of a fluid particle, and is in the direction of the resultant velocity vector at all points. The Bernoulli equation between points (1) and (2) is as follows:

Eq. (5.15) $$p_1/\rho + V_1^2/2 + gh_1 = p_2/\rho + V_2^2/2 + gh_2$$

where:
p/ρ = pressure energy per unit mass
$V^2/2$ = kinetic energy per unit mass, (KE)/m
gh = elevation energy per unit mass

As an example, consider the water clock used by Galileo to measure time (Fig. 5.14). This shows the path of a fluid particle from a point on the upper surface (1) that is maintained at a constant elevation (h_1) to a point (2) just beyond the orifice in the side of the vessel. Substituting into Eq. (5.15), and noting that $p_1 = p_2$, $\rho_1 = \rho_2$, and $V_1 = 0$:

Eq. (5.16) $$V_2 = [2g(h_1 - h_2)]^{0.5}$$ (Toriceli, 1640)

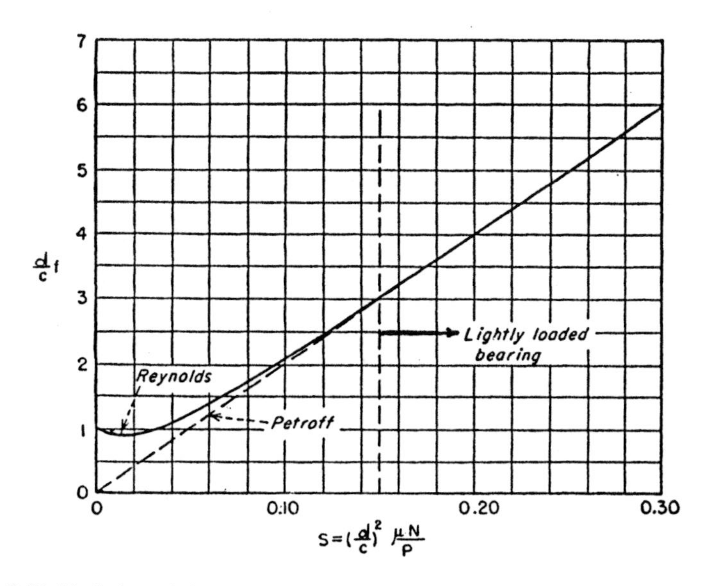

Figure 5.13. Variation of d/c vs Sommerfeld Number (S) for a journal bearing operating under a wide range of loads. The Petroff solution based on concentricity of the journal is seen to hold only for values of S >0.15. For lower values of S, a solution is required that takes journal eccentricity into account (labeled Reynolds).

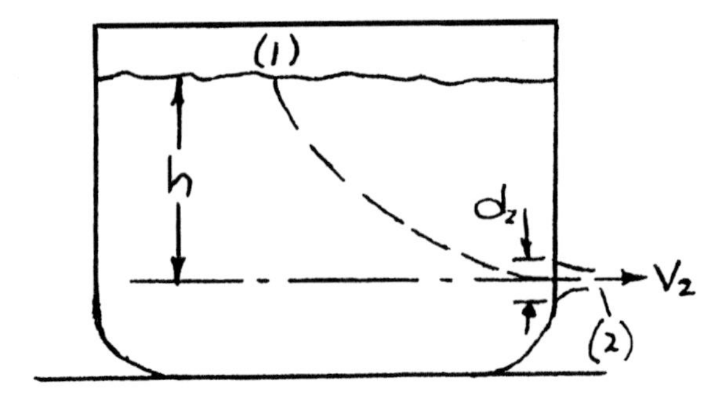

Figure 5.14. Flow of water from large reservoir with constant level through orifice of diameter, d_2 (similar to Galileo's water clock, where fluid is collected and weighed to measure time).

Thus, for an *h* of 20 in. (508 mm), from Eq. (5.15), *V* would be 124 in./s (315 m·s^{-1}) and for an orifice 1.00 in. (25.4 mm) in diameter, the rate of discharge would be 97.4 in.^{3}/s (1,596 cc.s^{-1}). Actually, the rate of discharge would be only about 60% of this value for a sharp-edged orifice due to friction losses.

8.0 GALILEO

It is suggested that at this point passages 107–126 of the Galileo text be read, where the nature of drag, surface tension, buoyancy, and the weight of air are discussed. Galileo points out that it is easy to find two bodies that fall with different speeds in water but with the same speed in air. This suggests that at least two effects are involved—buoyancy and drag.

He next discusses the equilibrium of fish and how the entrainment of more or less air may compensate for changes in the density of water. This anticipates the submarine. The basis for the hydrometer is also contained in the discussion of neutral equilibrium of bodies. The extreme sensitivity of a ball of wax impregnated with sand used by physicians to establish neutral equilibrium in the measurement of density is also described.

In searching for possible forces on falling bodies, Galileo considers the force that enables a droplet of water to stand high on a cabbage leaf. He reasons that this force is not of buoyant origin, since the drops should then stand even higher when surrounded by a denser material than air, such as wine, but this is experimentally not the case. He reasons that the effect is external to the droplet and assigns the name *antipathy* to this action which today we know to be surface tension. However, he warns that having named such an action does not mean that we understand its nature any better.

Galileo indicates that all bodies (lead and feathers, for example) should fall with the same velocity in vacuum. In order to study the speed of fall in a perfect vacuum, one should approach this situation as closely as possible. Hence, motions in air should be studied. The resistance the medium offers to being pushed aside (drag) will then be small.

In the absence of drag, Galileo reasons that only two forces will be present—the true weight of the body in vacuum (W) and that due to buoyancy (B). The effective weight of a body in air (W_e) will be the difference between its true weight in vacuum (W) and the buoyancy force (B) in air:

Eq. (5.17) $$W_e = W - B$$

Galileo identifies the concept of a weight in vacuum. This equation enables the weight in vacuum (W) to be obtained from the weight in air (W_e) and the buoyancy due to air (B).

Galileo assumes that the resistance to drag is proportional to the velocity of a body. Actually, this is true only for a low Reynolds Number (as discussed in Ch. 6). For turbulent flow (high Reynolds Number), the drag force varies as the square of the speed of the body (V).

In a typical digression, Galileo discusses the weight of air. Aristotle concluded that since a leather bottle inflated with air weighs more than the flattened bottle, air has weight. He concluded that air weighs one-tenth as much as water. Galileo presents two ingenious methods for determining the weight of air and concludes that water weighs 400 times as much as air. Actually, today we know that water is 800 times as heavy as air. The discrepancy in Galileo's determination of the weight of air was due to his failure to recognize that air is compressible.

In discussing the resistance (drag) of large and small bodies, Galileo concludes that the surface area and the surface roughness (rugosity) play important roles. It is reasoned that since the weight of a body ($\sim d^3$) is diminished in greater proportion than its area ($\sim d^2$) when size (d) is reduced, then a fine powder should be expected to fall with a lower velocity than a coarser one. The concept of a terminal or equilibrium velocity is also clearly recognized.

9.0 CAPILLARY FLOW

The flow of fluid in capillaries of very small diameter was first discussed by a German engineer named Hagen in 1839. Poiseuille, a Parisian physician, independently made similar experiments concerning the flow of blood through veins of the body and published his results from 1840–1846. The fundamental relationship for the rate of flow through a capillary is known as the Hagen-Poiseuille Law.

This problem may be approached by dimensional analysis. Before dimensional analysis:

Eq. (5.18) $$Q = \psi_1 (d, G, \mu)$$

where:
- Q = rate of fluid flow $[L^3T^{-1}]$
- d = diameter of capillary $[L]$
- G = pressure gradient along the capillary $[FL^{-1}]$
- μ = coefficient of viscosity $[FTL^{-2}]$

The pressure gradient is the pressure from one end of the capillary to the other divided by the capillary length ($G = \Delta p/\ell$)

After dimensional analysis, there is only one nondimensional group and:

Eq. (5.19) $$Q\mu/Gd^4 = \text{a nondimensional constant}$$

More detailed analysis reveals that the constant is $\pi/128$. Hence:

Eq. (5.20) $$Q = \pi/128(\Delta p/\ell)(d^4/\mu)$$

This is known as the Hagen-Poiseuille Law.

PROBLEMS

5.1 From Table 5.1, the specific weight of mercury is 0.462 lb in.$^{-3}$. What is the density (ρ) in lb in.$^{-4}$sec^2?

5.2 From Table 5.1, the viscosity of mercury is approximately 2.3×10^{-7} Reyn. If mercury is sheared between two flat plates as in Fig. 5.1 that are 0.005 in. apart and $V_S = 100$ fpm, estimate the resisting force on the upper plate per square inch.

5.3 Repeat Problem 5.2 if the fluid is SAE 70 lubricating oil at 100°F.

5.4 Repeat Problem 5.3 if the SAE 70 lubricant is at 195°F.

5.5 A cube of wood, measuring 4 in. (10.16 cm) on a side, floats in fresh water with 1/2 in. (12.7 mm) extending above the surface of the water.

a) What is the specific weight of the wood (lb/cu.in) (N/cc)?

b) For stability, should O_1 (center of gravity of block) be above or below O_2 (center of gravity of displaced water)?

5.6 A rubber toy balloon is filled with helium (specific weight = 6.44×10^{-6} lb/cu.in.) to a volume of one cubic foot.

a) What is the buoyant force?

b) If the weight of the balloon is 0.1 oz (0.278N), what will be the initial upward acceleration?

5.7 Repeat Problem 5.6 if the balloon is filled with H_2 (sp.wt = 3.24×10^{-6} lb/cu.in).

5.8 For the dam shown in Fig. 5.5, L = 50 ft. What is the force due to water pressure per unit foot of dam width?

5.9 If the weight of the dam of Problem 5.6 is 100,000 lbs/ft, what is the required distance (b) for neutral equilibrium of the dam?

5.10 Two very smooth, perfectly flat, gage blocks each have a surface measuring two inches on a side. If they are coated with a film of oil having a surface tension of 10^{-4} lb/in. and a thickness between the blocks after assembly of 10^{-5} in., estimate the normal force required to separate these blocks.

5.11 a) Estimate the pressure drop from one end to the other of a commercial steel pipe having an inside diameter of 1.5 in. and a length of 100 ft when the flow rate is 1 gallon of kerosene per second (1 gal = 231 cu.in.).

b) Estimate the horsepower required.

5.12 Repeat Problem 5.11 if the diameter of the pipe is 3 in. and all other conditions are the same.

5.13 Repeat Problem 5.11 if the length is 1,000 ft and all other conditions are the same.

5.14 A journal bearing has a diameter (d) of 2 in., length (ℓ) of 3 in., and diametral clearance (c) of 0.002. If it operates at 1,800 rpm and the viscosity of the oil is 2×10^{-6} Reyns and the load the bearing supports (W) is 1,000 lbs:

a) Estimate the minimum film thickness (h).

b) Estimate the coefficient of friction (f).

c) Estimate the horsepower dissipated in the bearing.

Use the Petroff equation in estimating b) and c).

5.15 Repeat Problem 5.14 if $W = 5{,}000$ lbs and all other quantities are the same.

5.16 Repeat Problem 5.14 if N is 1,200 rpm and all other quantities are the same.

5.17 Repeat Problem 5.14 if $W = 10{,}000$ lbs and all other quantities are the same.

5.18 A barometer consists of a glass tube about one meter long with one end permanently sealed. The tube is filled with Hg and all gas bubbles are removed. The open end of the tube is temporarily sealed and the tube is inverted. After submerging in a bath of Hg, the temporary seal is removed and the mercury falls to a level corresponding to atmospheric pressure

(p_a), establishing an essentially perfect vacuum in the permanently sealed end (see Fig. P5.18).

a) Using the Bernoulli equation, derive the equation that relates p_a and h.

b) If the atmospheric pressure is 14.97 psi, what is the value of h in. of mm of Hg using the appropriate values from Table 5.1?

c) If water were used instead of Hg, what is the value of h in feet?

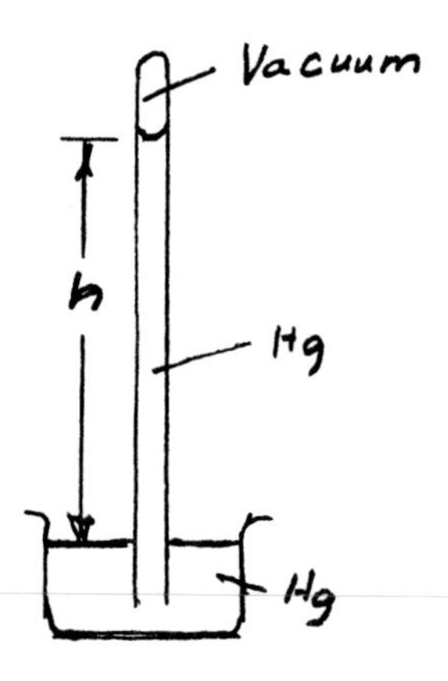

Figure P5.18.

5.19 Consider the two static bodies of fluid in Fig. P5.19 that are connected by a conduit containing the liquids indicated. If the pressure in vessel A is 10 psi, what is the pressure in vessel B? (Use data from Table 5.1 as required.)

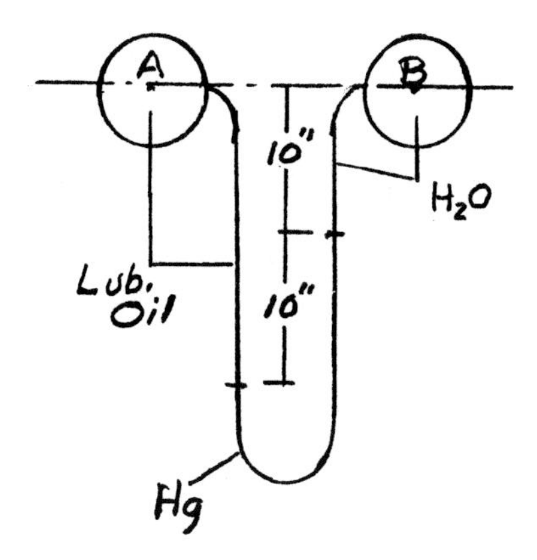

Figure P5.19.

5.20 A 10 lb projectile is fired straight up with an initial velocity of 1,000 fps. Neglecting air drag, calculate:

a) The kinetic energy at launch.

b) The kinetic and potential energies at maximum elevation.

c) The maximum elevation.

5.21 A small droplet of water stands on a waxed surface shown in Fig. P5.21. Its diameter is 10^{-4} inches.

a) Is the pressure inside the droplet larger or smaller than that in the atmosphere?

b) If the surface tension of the water is 73 dynes/cm (4.2×10^{-4} lb/in.) what will be the pressure in the droplet in psi?

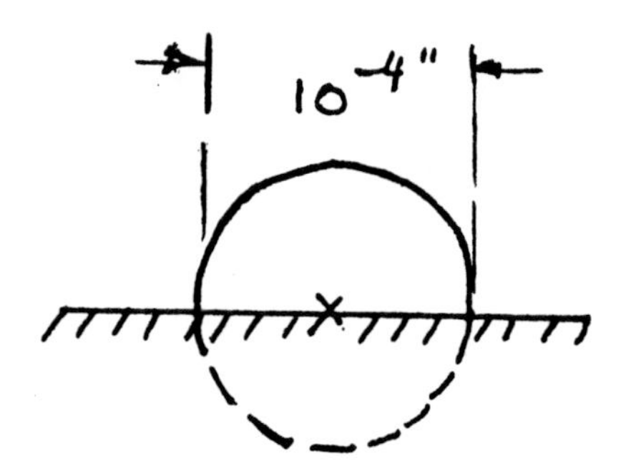

Figure P5.21.

5.22 An orifice plate and manometer (Fig. P5.22) provide a convenient means for measuring the volume rate of flow (Q). Assume the flow to be fully turbulent so that viscous forces may be ignored relative to inertia forces.

a) Perform a dimensional analysis for the volume rate of flow (Q) as a function of Δp ($\Delta p = \gamma_m h$ where γ_m = specific weight of manometer fluid), D, d, and ρ for the fluid in the pipe.

b) If Δp increases by a factor of 4, what is the corresponding increase in the rate of flow when D and d remain constant?

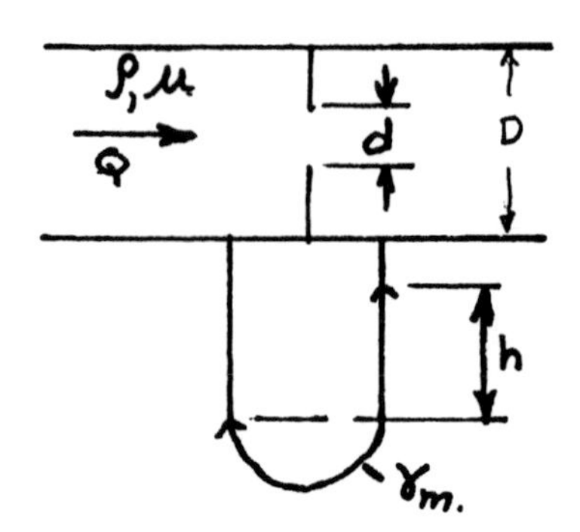

Figure P5.22.

5.23 A venturi is a constriction in a circular pipe used to measure the rate of flow of fluid in the pipe (Fig. P5.23). The difference in pressure Δp between points (A) and (B) is measured by a manometer which is a glass tube containing mercury and water as shown. The difference in pressure between (A) and (B) is the pressure of the column of mercury at (C). The flow rate (Q) will be a function of diameters (d) and (D), Δp, and the mass density of the water flowing in the pipe. That is:

$$Q = \psi(\rho, D, d, \Delta p)$$

a) Perform a dimensional analysis.

b) If, for some reason, you know that Q varies as d^2, use this fact to simplify the result found in a) if D/d remains constant.

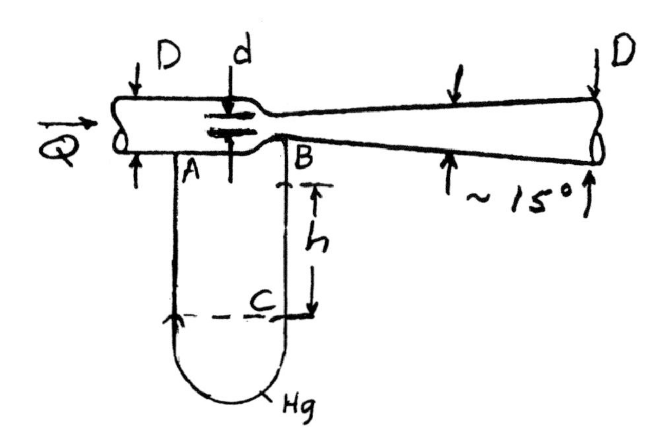

Figure P5. 23.

5.24 A siphon is a means for transporting fluid from a high reservoir (A) to a lower reservoir (B) without a pump as shown in Fig. P5.24. If the pipe is completely filled before flow begins and the pressure at all points remains above absolute zero, the difference in elevation (h_2) provides the pumping energy. It is even possible to move the fluid over a hump (h_1) before moving downward to reservoir (B).

a) Neglecting pipe friction, what is the maximum possible value of h_1 for water?

b) Neglecting pipe friction, what is the exit velocity from the pipe at B if $h_2 = 20$ ft?

c) If the diameter of the pipe is 4 in., what is the flow rate under the conditions of b)?

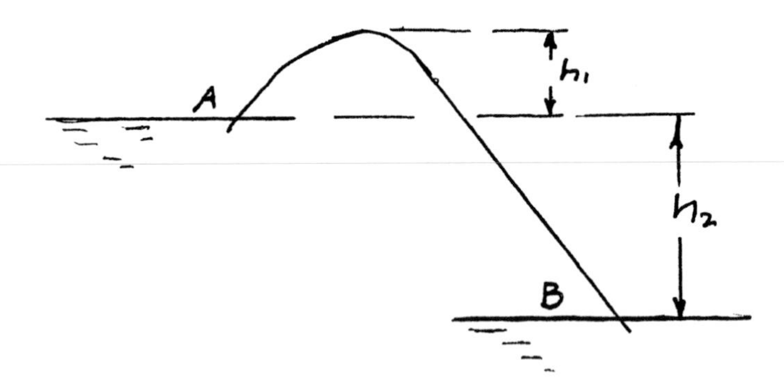

Figure P5.24.

5.25 A weir (Fig. P5.25) is a notched plate that is used to measure the rate of flow (Q) of fluid in a stream or open channel by measuring height (h). Friction between the fluid and the plate can be ignored in this problem.

a) Identify the variables of importance and perform a dimensional analysis. Use of a triangular opening enables more accurate values over a wide range of flow rates.

b) If the value of h doubles for a weir having a given α, what is the percent increase in Q?

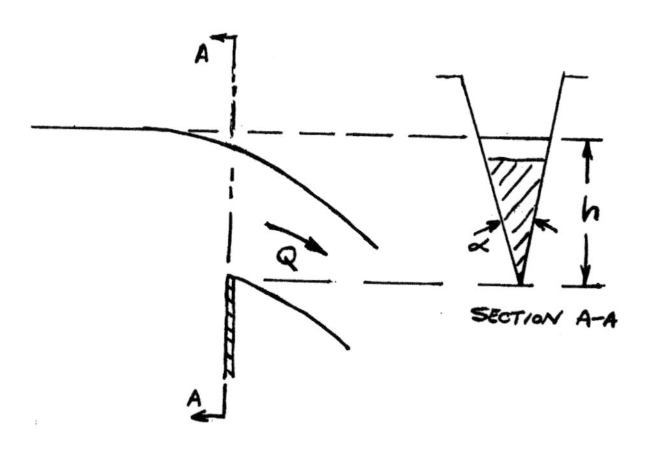

Figure P5.25.

5.26 The Navajo Indian Irrigation Project in New Mexico employs three miles of the world's largest diameter (17½ ft inside) prestressed concrete pipe. Each 20 ft section of pipe weighs 140 tons and contains 17½ miles of prestressing wire. The outside diameter of the pipe is 20.25 ft. The design flow rate is 1,800 ft^3/sec. Find:

a) The pressure drop in psi required over the three mile distance.

b) The required horsepower to overcome the resistance to flow in the three miles of pipe.

Note: You are apt to be surprised at the answers obtained.

5.27 An Ostwald viscometer is constructed as shown in Fig. 5.1. Ten cc of liquid is caused to flow through the capillary as the upper surface level goes from (*A*) to (*B*), and the lower level goes from (*C*) to (*D*). If the length of the capillary and the mean distance between the upper and lower liquid surfaces are each 10 cm, estimate the diameter the capillary should have for an outflow time of 100 sec when water having the following properties is used:

$$\gamma = 0.036 \text{ lb/in.}^3$$

$$\mu = 1.5 \times 10^{-7} \text{ lb sec in.}^{-2}$$

If an oil has an outflow of 1,000 sec in this viscometer, what is its kinematic viscosity (μ/ρ) in square inches/sec?

6

Aerodynamics: The Boundary Layer and Flow Separation

1.0 INTRODUCTION

In the preceding chapter, it was emphasized that for low values of Reynolds Number, the inertia forces on a fluid particle may be considered zero, while, for high values of Reynolds Number, viscous forces may be ignored. From this it might be reasoned that the viscous drag on a stationary body in a fluid moving at high speed (i.e., having a high Reynolds Number) should be negligible. However, this is not so. The fluid in contact with any body has exactly the velocity of that body. There is no slip between fluid and body. This means that the high-speed fluid must fall to zero in the vicinity of a stationary body, and the region in which this occurs is called the boundary layer. Figure 6.1 shows the velocity gradient in the vicinity of a solid wall and the extent of the boundary layer. Usually, the boundary layer is considered to extend outward from the body to the point where the velocity is within 1% of the free stream velocity (V).

In this chapter, some of the complexities of lift and drag of airfoils and other bodies are considered, mainly in qualitative terms.

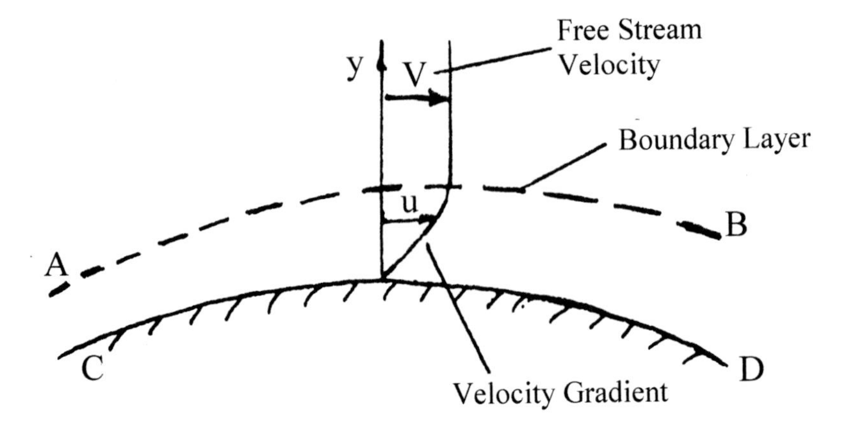

Figure 6.1. The boundary layer.

2.0 STAGNATION POINT

If there is no change in elevation, then the Bernoulli equation says that when the fluid slows down, the pressure must rise and vice versa. Figure 6.2 shows paths taken by fluid particles (stream lines) past a sphere when viscosity is zero. The central stream line comes to rest at the leading point of the sphere (*A*) and this point is called a stagnation point. The pressure at *A* will be a maximum according to the Bernoulli equation. There will be a second stagnation point at the trailing point (*A´*). The velocity along stream line 1-1 will be a maximum at *B* and the pressure will be a minimum at this point. Due to the symmetry pertaining about horizontal and vertical centerlines, the pressures at corresponding points (*C*) will be the same. There will be no net lift or drag forces on a stationary sphere in an ideal fluid flow stream (one involving fluid of zero viscosity).

However, in practice, it is found that while there is no lift force on such a sphere in a viscous air stream there is an appreciable drag force which varies in a complex way with fluid velocity or Reynolds Number (as will be discussed presently in connection with Fig. 6.5). It was not until the concept of the boundary layer was introduced by Prandtl in the early 1900s that this discrepancy between theory and experiment was explained.

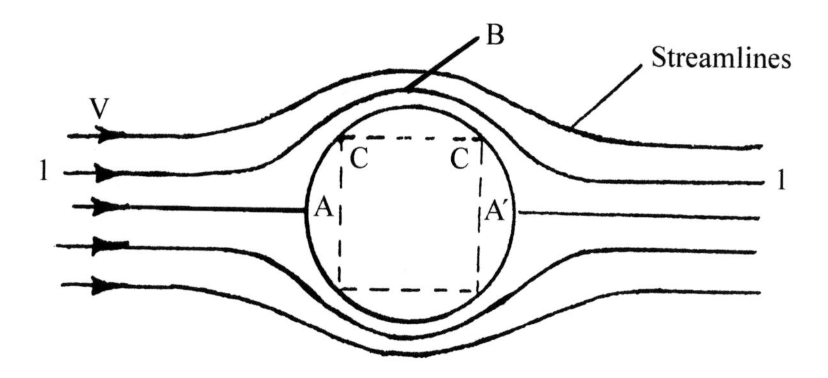

Figure 6.2. Flow of ideal (frictionless) fluid past a sphere.

3.0 VISCOUS DRAG COMPONENT

Within the boundary layer, viscous forces cannot be ignored even in the case of air. At the surface of a body, the velocity goes to zero and will be subjected to a tangential stress that is proportional to the velocity gradient there in accordance with Newton's law of viscous shear [Eq. (5.3)]. The horizontal components of the tangential shear stresses acting at all points of the sphere will give rise to a force known as the viscous drag force.

4.0 FLOW SEPARATION AND PRESSURE DRAG

However, the viscous drag force is not the only component of drag. Since the boundary layer involves friction losses, the Bernoulli equation [Eq. (5.15)] no longer holds. Proceeding along the surface of the sphere, the total energy will decrease as flow energy is irreversibly converted into thermal energy. Consequently, velocities at points on the rear half of the sphere will be lower than corresponding points on the front half of the sphere. The second stagnation point (A') will occur before the trailing edge is reached (Fig. 6.3) and flow separation will take place at this point. The pressure in the separated region will be low. In general, the pressure at corresponding points (such as C in Fig. 6.2) will be lower for the rear half of the sphere than for the front half. This will give rise to a component of drag known as pressure drag. The total drag force on the sphere is the sum of the viscous and pressure drag components.

The pressure at corresponding points on either side of the horizontal center line will be the same with or without a boundary layer, and there will be no net lift force (force perpendicular to the free stream velocity direction) in either case.

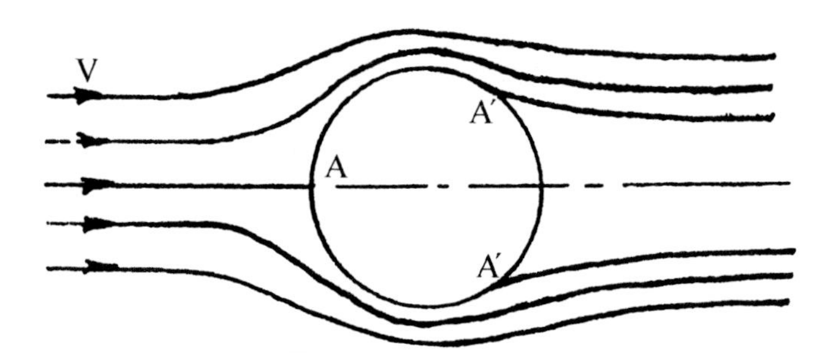

Figure 6.3. Fluid flow past a stationary sphere with flow separation.

5.0 LAMINAR-TURBULENT TRANSITION IN THE BOUNDARY LAYER

The boundary layer can be either laminar or turbulent. At the forward stagnation point, it will always be laminar. However, as energy is removed in the form of heat, the boundary layer thickens, which, in turn, gives rise to an increase in the mean Reynolds Number for the boundary layer. Upon reaching a critical value of Reynolds Number, the boundary layer becomes turbulent.

If the boundary layer is laminar, fluid particles move parallel to the dashed line in Fig. 6.1. There is no opportunity to replace energy converted to heat within the boundary layer by energy transport from the free stream. However, when the boundary layer is turbulent, there are transverse velocity components along the dashed line of Fig. 6.1. This enables the energy going into heat to be replenished from the main stream which, in turn, delays flow separation. The pressure drag will, thus, be less for a turbulent boundary layer than for a laminar one.

While the viscous drag is greater with a turbulent boundary layer, the pressure drag will be less. At certain values of Reynolds Number, the decrease in pressure drag due to boundary layer turbulence far exceeds the increase in friction drag. In such cases, the total drag decreases with an increase in turbulence in the boundary layer.

6.0 STREAMLINING

A streamlined body is one that is so shaped as to decrease the area subjected to flow separation which, in turn, gives rise to a decrease in pressure drag. Figure 6.4 is a streamlined body of revolution having the same frontal area as the sphere of Fig. 6.3. The streamlined body gives rise to a lower deceleration of the fluid to the rear of the point of maximum velocity (*B*) which, in turn, postpones flow separation to a point of smaller diameter. Streamlined bodies will give lower drag than nonstreamlined bodies under conditions of high speed flow (high Reynolds Number). However, for low speed (laminar) flow, there is no flow separation, and no change in pressure drag. The larger wetted area for the streamlined body gives rise to a larger viscous drag component. Thus, streamlined bodies give lower drag only under conditions of flow that postpone separation. In the laminar region of flow, they give larger drag than nonstreamlined bodies. In the laminar regime of flow, both viscous and pressure drag components are of importance.

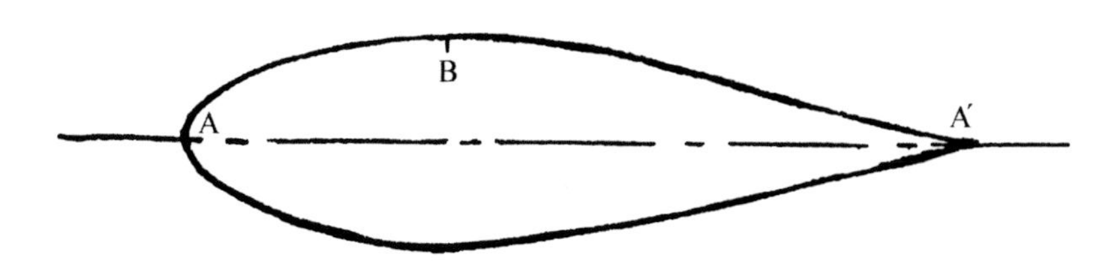

Figure 6.4. Streamlined body.

7.0 DRAG ON A SPHERE

The force (D) that a flowing fluid exerts upon a totally immersed body of fixed geometry, such as a sphere, will be considered next in terms of dimensional analysis. If compressibility of the fluid is ignored, the quantities of Table 6.1 suffice to define the system.

After performing a dimensional analysis with ρ, d, and V as the dimensionally independent set:

Eq. (6.1) $$D/(\rho d^2 V^2) = \psi[(\rho V d)/\mu] = \psi \mathrm{R}$$

where R is the Reynolds Number and the nondimensional quantity on the left side of the equation is the Euler number (E).

If the Reynolds Number is very large, viscous forces will be negligible compared with inertia forces and μ may be dropped from the dimensional analysis. In such a case, $\psi(\mathrm{R})$ reduces to a constant (C_1):

Eq. (6.2) $$D/(\rho d^2 V^2) = C_1$$

Table 6.1. Quantities Involved in Drag on a Sphere

Quantity	Symbol	Dimensions
Drag force	D	$[F]$
Diameter of sphere	d	$[L]$
Velocity of fluid	V	$[LT^{-1}]$
Density of fluid	ρ	$[FL^{-4}T^2]$
Viscosity of fluid	μ	$[FTL^{-2}]$

A fluid for which μ may be considered zero is called an ideal fluid. Air flowing at high speed approximates an ideal fluid and Eq. (6.2) may be used to find the drag on a body in such a case. The constant is approximately 0.1 for a sphere, but will have other values for bodies having other shapes. At very low values of R, inertia forces will be small compared with viscous forces and ρ need not be considered in the dimensional analysis.

Then after dimensional analysis:

Eq. (6.3) $D/\mu Vd = C_2$ (a constant)

Stokes (1856) found C_2 to be 3π for a sphere falling slowly (Reynolds Number <1) in a large volume of viscous liquid.

In engineering practice, it is customary to write Eq. (6.2) as follows:

Eq. (6.4) $D/(\rho AV^2/2) = \psi R$

where A, which is used in place of d^2, is the projected area of the body normal to the direction of fluid flow. The function ψR is called the drag coefficient (C_D) and,

Eq. (6.5) $D = C_D(\rho AV^2/2)$

where C_D is a function of Reynolds Number as shown in Fig. 6.5, for a smooth sphere. The deviation from a straight line at R = 1 is due to a change from laminar to turbulent flow in the fluid near the surface of the sphere.

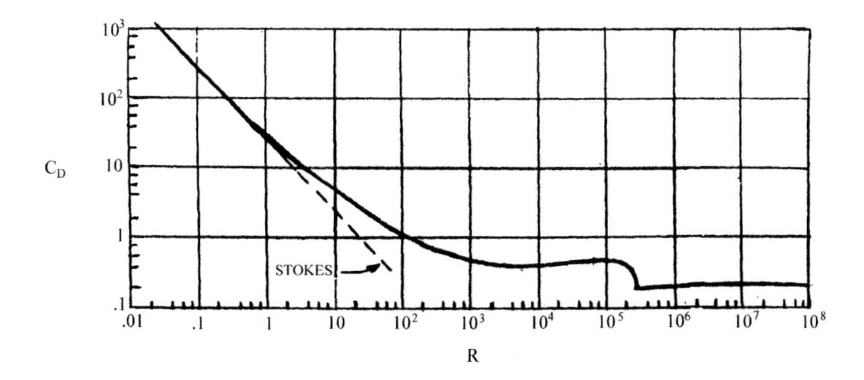

Figure 6.5. Variation of drag coefficient (C_D) with Reynolds Number (R) for a smooth sphere.

8.0 PARADOXES

The boundary layer concept is needed to explain the following paradoxes associated with fluid flow:

1. An increase in fluid velocity sometimes causes a decrease in the drag force on a body.
2. In a certain range of fluid velocity, a smooth sphere may experience a higher drag force than a rough sphere and vice versa.
3. Streamlining the shape of a body will sometimes cause a reduction in drag and sometimes an increase in the drag force on a body.

An example of the first paradox is evident in Fig. 6.5 at a Reynolds Number of about 2×10^5. At this particular Reynolds Number, the boundary layer becomes turbulent which, in turn, postpones flow separation, and greatly reduces the pressure drag with little change in viscous drag.

When a rough sphere is substituted for the smooth one of Fig. 6.5, turbulence occurs at a lower value of Reynolds Number. Hence, the drop in drag coefficient, which occurs at a Reynolds Number of 2×10^5 for the smooth sphere, may be made to occur at 0.5×10^5 for a rough sphere.

9.0 AIRFOILS

When considering the force on a body such as an airplane wing (Fig. 6.6), it is necessary to include an additional parameter in the analysis—the angle of attack (α). When this is done, Eq. (6.1) becomes:

Eq. (6.6) $$D/(\rho d^2 V^2) = \psi(\mathrm{R}, \alpha)$$

It is customary to resolve the resultant force the fluid exerts on the airfoil into two components—one parallel to velocity vector (V) which is called the drag force (D) and the other normal to V which is called the lift force (L). These forces may be computed from the following equations, where A is the wing area (chord × span).

Eq. (6.7) $D = C_D(\rho AV^2)/2$

Eq. (6.8) $L = C_L(\rho AV^2)/2$

where C_D and C_L, called the drag and lift coefficients, are strong functions of the angle of attack (α) but weak functions of R for normal aeronautical conditions. Values of C_D and C_L for different values of α are determined experimentally on models in a wind tunnel in a manner to be discussed in Ch. 7, Sec. 3.0. The airfoil section is an example of a streamlined body, and, as such, will have a C_D order of magnitude less than that for a sphere.

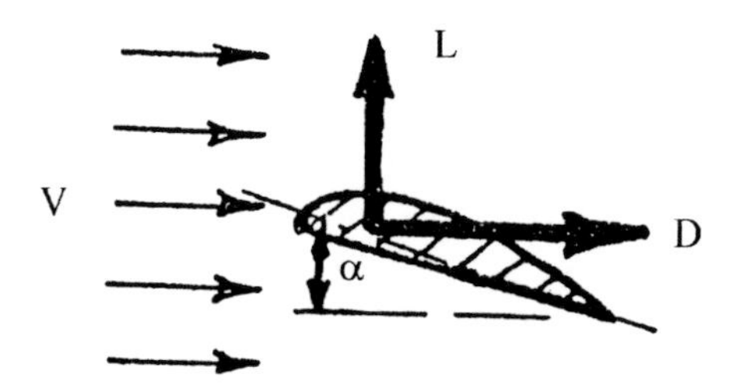

Figure 6.6. Lift (L) and drag (D) forces on an airfoil.

Airfoil sections are special streamlined bodies designed to provide lift to an aircraft. Different shapes provide different lift and drag characteristics. A great many different geometries have been studied, empirically, in wind tunnels by the National Advisory Committee for Aeronautics (NACA) which is now NASA (National Aeronautics and Space Administration). Lift and drag coefficient curves for a typical airfoil (NACA 4412) are given in Fig. 6.7. It should be noted that both lift and drag coefficients increase with angle of attack (α).

10.0 STALL

As the angle of attack (α) is increased, the flow velocity on the upper side of the airfoil is increased, and that on the lower side is decreased. Since the Bernoulli equation [Eq. (5.15)] applies outside the boundary layer, this gives rise to a low pressure above the wing where the velocity is high, and a high pressure on the underside of the wing. The result is a lift force that increases with angle of attack (α), as shown in Fig. 6.7. However, at a critical angle of attack, known as the stall angle, flow separation moves rapidly from the trailing edge of the airfoil to near the leading edge. This causes a sudden increase of pressure above the wing and a consequent loss of lift (referred to as stall). When an airfoil stalls, the plane will dive. This represents a dangerous situation. The position of stall corresponds to the maximum value of C_L which is at about 15°, in Fig. 6.7.

Since the variation in pressure across the boundary layer is slight, the pressure along the outer surface of the boundary layer (*AB* in Fig. 6.1) closely approximates that on the airfoil (*CD* in Fig. 6.1). Hence, potential flow theory, that assumes zero viscosity, gives a good approximation for lift.

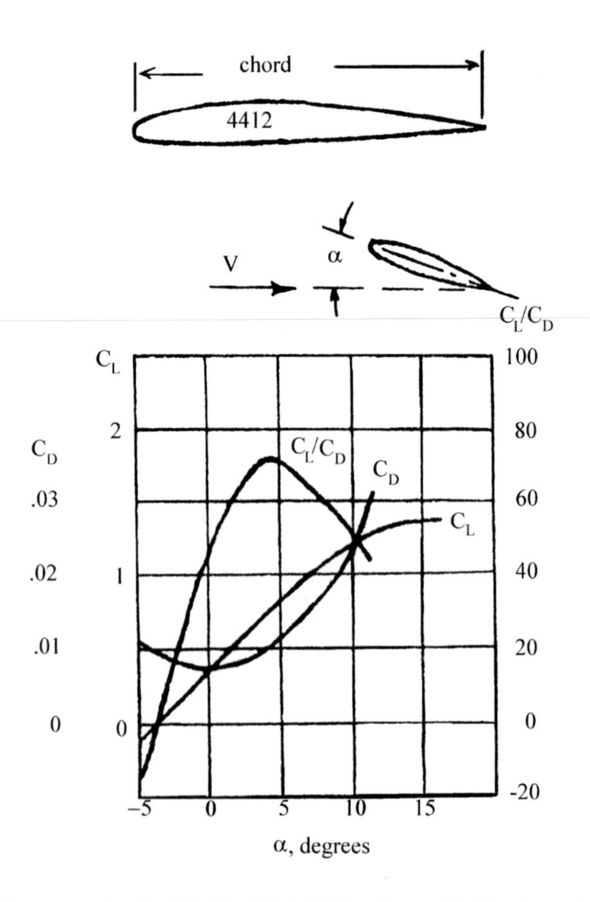

Figure 6.7. Characteristics for NACA airfoil 4412 at Reynolds Number = 10^6.

11.0 STEADY AIRPLANE PERFORMANCE

For an airplane in steady flight, its weight (W) is just balanced by the lift (L), or from Eq. (6.8):

Eq. (6.9) $$W = C_L(\rho/2)V^2A$$

At the same time, the thrust (T) is just balanced by the drag (D) or from Eq. (6.7):

Eq. (6.10) $$T = C_D(\rho/2)V^2A$$

The power (P) required to propel the aircraft at constant velocity (V) will be:

Eq. (6.11) $$P = TV$$

12.0 MAGNUS EFFECT

While the lift on a stationary cylinder in an air stream is zero [Fig. 6.8 (*a*)], that for a rotating cylinder [Fig. 6.8 (*b*)] is not zero. Air is dragged along with the rotating cylinder. This circulation, when combined with the translational flow, causes the velocity on the top side of the cylinder to be higher than that on the bottom side. As a consequence of the Bernoulli equation [Eq. (5.15)], the pressure on the bottom side of the cylinder will be higher than that on the top side, giving rise to an upward lift (L).

For a cylinder of diameter (d) and axial length (ℓ) (projected area perpendicular to $V = A_P = \ell d$), rotating at N rps ($V_C = \pi dN$) in air of density (ρ), the lift force (L) will be as follows before dimensional analysis:

Eq. (6.12) $$\begin{array}{lllll} L = \psi_1(\ \ell, & V_C, & V, & A_P\ ) \\ [L] & [FT^2L^{-4}] & [LT^{-1}] & [LT^{-1}] & [L^2] \end{array}$$

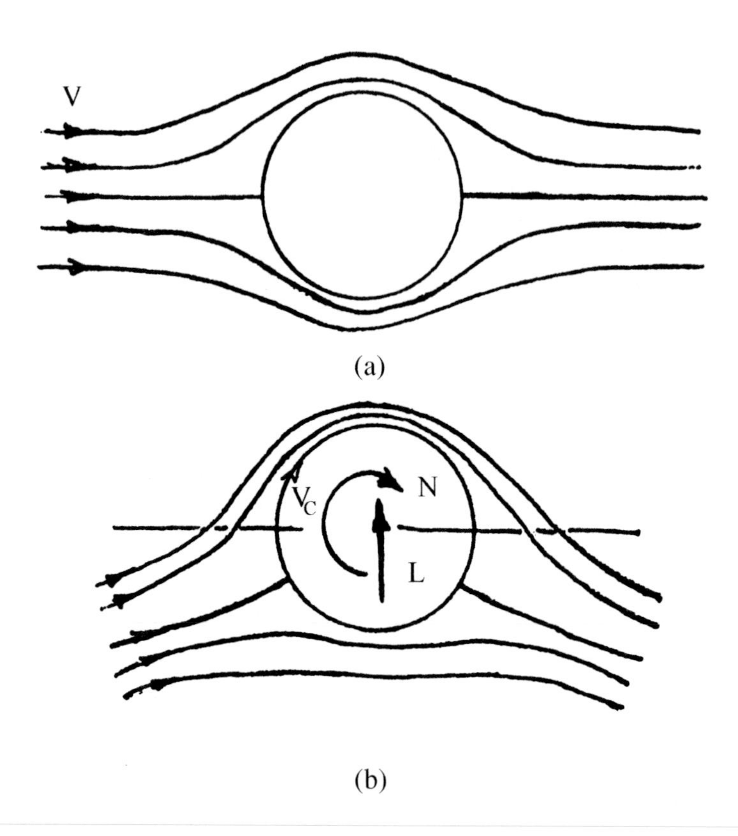

Figure 6.8. Stream line flow in free stream beyond boundary layer *(a)* stationary cylinder and *(b)* rotating cylinder.

After dimensional analysis, taking ρ, V, and A_P as the dimensionally independent set:

Eq. (6.13) $\quad L/(\rho V^2 A_P) = \psi_2(V_C/V)$

It is found experimentally that with other variables held constant, L is approximately proportional to V_C/V and:

Eq. (6.14) $\quad L/(\rho V^2 A_P) \sim V_C/V$

Substituting $A_P = \ell d$, and $V_C = \pi dN$ into Eq. (6.14):

Eq. (6.15) $$L \sim \rho V \ell d^2 N$$

Similarly, for a sphere:

Eq. (6.16) $$L \sim \rho V d^3 N$$

The lift associated with a rotating cylinder in a translating air stream is called the Magnus effect and it is this effect that causes a slice or hook in golf, a "cut" ball to move off to the side in tennis or ping pong, or a curve ball in baseball.

13.0 PERIODIC VORTICES

At a particular range of Reynolds Number (<2,500), vortices are shed alternately from opposite sides of a body at a definite frequency as shown in Fig. 6.9. The vortices in one row are staggered with respect to those in the other. Vortex shedding may also occur at high values of Reynolds Number.

Figure 6.9. Vortices being shed alternately from opposite sides of a body at a frequency $f = v/s$.

Such vortices were apparently first observed by Leonardo da Vinci about 1480 in a river behind a bridge pier. They were found to be spaced a distance s about four times the "diameter" of the bridge pier (actually $s/d = 4\frac{1}{4}$). In the middle of the 19th century, these vortices were studied by Strouhal who showed, empirically, that their frequency (f) on one side was related to the free stream velocity (V) and the width of the body (d) as follows:

Eq. (6.17) $(fd)/V = \text{a constant} = S_t$

Strouhal found the Strouhal Number (S_t) to be about ¼. Actually the constant varies between 0.2 and 0.5 depending on the shape of the body, but has a constant value for a given shape. In 1911, von Karman showed analytically that the only stable vortex configuration was that given by the Strouhal Number. These vortices are, therefore, sometimes called the Karman Vortex Street.

When a vortex leaves, the body is subjected to a reactive force. If the frequency of vortex formation corresponds to the natural frequency of the body (the frequency with which it will vibrate if bumped) large amplitudes of vibration may result.

In 1939, a large bridge at Tacoma, Washington was shaken to destruction and fell into the water as a result of exciting forces associated with a Karman Vortex Street. Subsequent investigation revealed that side plates on the bridge caused eddies to form at a frequency corresponding to the natural frequency of the bridge when a 42 mph wind blew across it. When the bridge was rebuilt, the side plates that had been responsible were eliminated, and the structure stiffened. This changed the natural frequency, and no subsequent difficulty has been encountered. Today all suspension bridges are tested by means of models in a low speed wind tunnel to be sure they are dynamically stable at all wind velocities.

Structures other than bridges are often subjects for potential Strouhal-Karman difficulties. These include high smokestacks and tall buildings. The St. Louis Arch (the Gateway to the West) was questioned from this point of view. Model tests and calculations performed by den Hartog at MIT revealed that the arch would be in trouble at wind velocities of 60 ±10 mph when the wind blows from due north or south ±5°. As a matter of safety, the arch has been instrumented to monitor its vibration and people are not allowed to go up into the arch if the wind velocity exceeds 40 mph.

The largest reported Karman Vortex Street was one observed by den Hartog. It was formed by clouds passing over a volcanic crater in the Hawaiian Islands that was five miles in diameter. This was photographed and the spacing of vortices found to be 24 miles on each side.

14.0 CONCLUDING REMARKS

In this chapter, it has been stressed how the engineer uses the simplest picture possible in the solution of problems. In general, inertia may be ignored in problems for low Reynolds Number (R) and viscous effects may be ignored when R is high. While this approximation is excellent for problems in the laminar regime, it is only partially permissible for high-speed flows. In the latter case, viscosity may be ignored only in calculations for lift and then only when flow separation is negligible (i.e., for a perfectly streamlined body). The boundary layer concept is needed to explain the appreciable viscous drag on bodies operating at high values of Reynolds Number. But, even this is not enough to explain drag in some high-speed flows. The additional concepts of flow separation and the role of turbulence in making it possible for energy transport into the boundary layer to postpone separation is needed to complete the picture.

It took over a hundred years to develop this body of theory. The first part to appear was that for very low Reynolds Number flows and that for an ideal fluid. This was, subsequently, followed by an introduction of the boundary layer and flow separation concepts.

More recently, shock waves and other details associated with flows at speeds greater than that of sound in the fluid (supersonic flows) have been introduced. Still more recently, the interaction of magnetic fields with the flow of charged particles has been extensively studied and formulated into a body of theory known as magnetohydrodynamics. In connection with the space effort and the performance of vacuum systems, it became necessary to study the flow of fluids at such low pressures that the mean free path of molecular motion plays an important role. At the other extreme, the flow of fluids must be studied under such high pressures that densification sufficient to cause solid-like behavior results. This latter area is still to be explored.

There has also been interest in the flow characteristics of powdered coal and other solid particles suspended in water as a slurry. The pumping of such slurries, which consist of about 50% by weight of solid particles, is of interest since it provides an inexpensive means of transporting powdered materials over large distances.

PROBLEMS

6.1 An NACA 4412 airfoil having an area (A) of 400 ft^2 operates at a velocity of 400 mph and an angle of attack of 10°. Estimate:

a) The lift force.

b) The drag force.

6.2 Repeat Problem 6.1 if $\alpha = 0°$ and other variables are the same.

6.3 Repeat Problem 6.1 if $V = 200$ mph and all other variables are the same.

6.4 For steady flight conditions of Problem 6.2, determine the horsepower required to propel the plane at constant V.

6.5 For steady flight conditions of Problem 6.2, determine the total weight that may be carried.

6.6 If the standard baseball were to be made 10% larger in diameter, what effect would this have on the lift force tending to produce a breaking curve if all other conditions (mass density of air, speed of pitch, and rotational spin) were maintained the same?

6.7 If both the linear and angular speed of a pitched ball could be increased 10%, what effect would this have on the lift force tending to produce a breaking curve?

6.8 A flexible rod 1/8 inch in diameter has its axis perpendicular to a stream of air having a velocity of 40 mph.

a) Estimate the frequency at which vortices will leave alternate sides of the rod.

b) What should the natural frequency (Hz) of the wire be in order that it be caused to vibrate with a large frequency?

6.9 Find the Strouhal Number (S_t) corresponding to the 24 mile spacing of vortices on one side of the five mile diameter crater in the example given in the text.

6.10 A pitot static tube (Fig. P6.10) is used to measure the volume rate of flow in a pipe (Q). For the fully turbulent condition viscous forces may be ignored. Perform a dimensional analysis for the volume rate of flow (Q) as a function of Δp (between A and B which is equal to $\gamma_m h$ where γ_m = specific weight of manometer fluid), D, and r for the fluid in the pipe. If Δp increases by a factor of 2, what is the corresponding increase in the rate of flow?

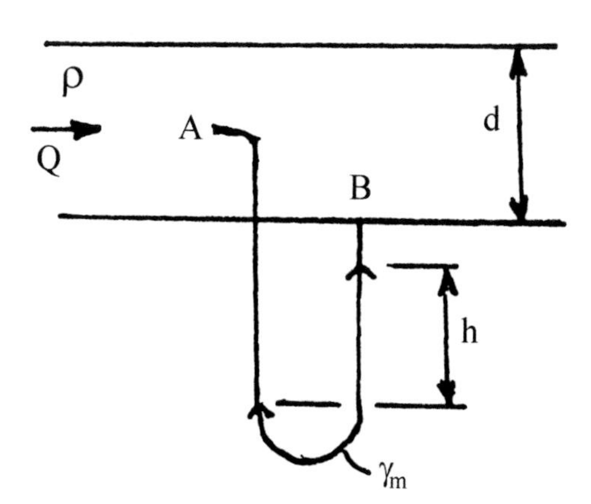

Figure P6.10.

6.11 A one inch diameter sphere of wood is lighter than water, and when released at considerable depth, it rises toward the surface until it reaches a constant velocity of 10 in./sec. Estimate the specific weight of the wood. (The volume of a sphere is $\pi/6d^3$.)

6.12 By what factor is the wind resistance on an automobile changed when its speed is increased from 30 to 60 mph?

6.13 Will a streamlined body of revolution operating at a low Reynolds Number encounter more drag than an equivalent sphere? (A sphere with diameter equal to maximum chordal length of streamlined body.)

6.14 A pitched ball spins as shown in Fig. P6.14 through still air with velocity (V) in the direction shown. Will the combined motion cause the ball to move along path (a) or path (b)?

Figure P6.14.

6.15 Levigation is a process of sizing very fine powders (so fine that they cannot be classified into different size ranges by screens or sieves). In the levigation process, particles of different size are put into a liquid whose density is less than that of the solid. The particles fall through the liquid with speeds that vary with their size. If a tall settling column is used, liquid drawn from a given level after a long time will contain particles of the same size. Abrasives are classified as to size in this way. Consider aluminum oxide (Al_2O_3) particles in water. Aluminum oxide has a specific weight three times that of water.

a) What three forces will act upon a solid particle that is falling through liquid with uniform velocity, i.e., with no acceleration?

b) If the particles are one micron (40×10^{-6} in.) in diameter when falling in water of viscosity 1.5×10^{-7} Reyn, estimate their terminal (equilibrium) velocity in inch/sec. The size of the vessel is very large compared with the solid particles. You may wish to recall that the volume of a sphere is $\pi/6$ (diameter)^3, and the specific weight of water is 0.036 pounds per cubic inch.

c) What is the time to fall 1 inch?

d) What is the Reynolds Number in terms of the diameter (d) of the spherical particle?

The acceleration due to gravity is 386 in./sec^{-2}, and a viscosity of 1 Reyn is 1 lb.sec.in^{-2}.

6.16 a) Find the terminal velocity of a two inch diameter hailstone.

b) Estimate the equivalent free fall height (h) such that the energy content on impact is the same as in a). The specific gravity of hail is about 0.9.

6.17 An airplane has an NACA 4412 airfoil with a 20 ft span and chord of a 3 ft (area = 60 ft^2).

a) Determine the weight (W) this airfoil is capable of lifting when the angle of attack (α) corresponds to the maximum lift-to-drag ratio and the air speed is 100 mph.

b) Find the horsepower required to overcome drag under these conditions. The density of the air is 0.00251 $lb.sec^2ft^{-4}$.

6.18 The Sportavia-Putzer Co. in the Eifel Mountains of West Germany manufactures a motorized sailplane. The built-in engine (39 hp) is capable of launching the 88 lb plane, and a cruising speed of 110 mph is claimed. With the aid of favorable thermal currents, the plane may climb to 12,000 ft and with the engine switched off is capable of gliding at least 60 miles. The wing span is 35 ft and the wing area is 105 ft^2. Assume that the airfoil characteristics of the NACA 4412 given in the text pertain to this craft and that the density of the air (ρ) is 0.00242 $lb.sec^2ft^{-4}$.

a) Calculate the hp required to make the 4412 airfoil cruise (min C_D) at the advertised speed of 110 mph (161 fps).

b) Is the advertised cruising speed of 110 mph reasonable assuming a propeller efficiency of 80%?

6.19 A weight of 100 lbs hangs from the slender cantilever beam shown in Fig. P6.19. This beam is 10 in. long and its width and height are 1 in. and 2 in. respectively. The beam material is steel ($E = 30 \times 10^6$ psi). The weight of the beam is considered to be negligible.

a) Find the static deflection of the beam (δ).

b) Find the frequency with which this beam will oscillate in cycles per second if it is pulled down and released (i.e., find its natural frequency).

c) If this beam is in a stream of fluid flowing at high velocity (V) in the direction shown, estimate the velocity (V) in fps for which the amplitude of vibration excited by the action of the fluid will be a maximum. (The Strouhal Number for vortices peeling from a rectangle such as that shown in Fig. P6.19 will be about 0.16).

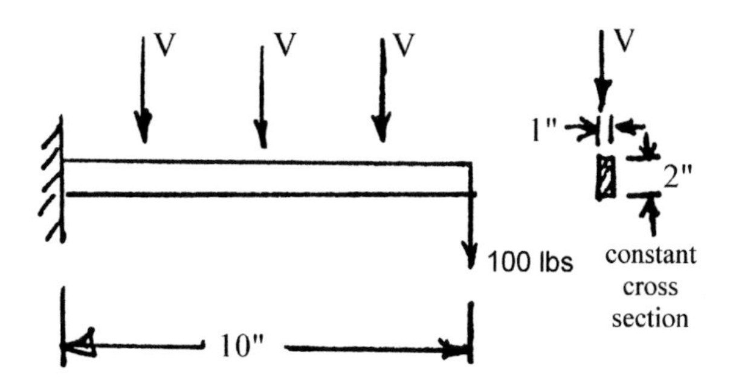

Figure P6.19.

7

Similitude

1.0 INTRODUCTION

Many engineering problems are so complex that a complete analytical solution is impractical, either due to lack of time or because the general problem cannot be resolved into components for which known solutions exist. Even if a problem can be divided into subproblems that can be solved, the composite solution will sometimes be inadequate due to important interactions among the subproblems. In such cases, it often proves expedient to study the performance of a physical model of the full-scale prototype. One of the most important uses of dimensional analysis lies in the design of engineering models and the interpretation of their performance.

A *true model* of an engineering situation is one for which all Pi quantities associated with the problem are equal for both model and prototype. When this is the case, a condition of dimensional similitude is said to exist. In designing a model, certain variables may be assigned arbitrary values, but the number of these must not exceed the number of dimensionally independent quantities. The remaining variables must then be modeled so that all Pi quantities are the same for model and prototype in order that a true model pertains.

Usually, scale models are used for which all dimensions stand in a fixed ratio and all angles are identical in model and prototype. Such a model is said to be geometrically similar to the prototype. When geometrical similitude is established, one of the arbitrary scales is fixed.

2.0 EXAMPLE: HYDRODYNAMIC BEARING

The similitude concept may be illustrated by considering an example. The problem of journal bearing load capacity is convenient for this purpose. In this problem, the minimum film thickness (h) is related to a number of other variables, already defined in Ch. 5, Sec. 6.0 as follows:

Eq. (7.1) $$h = \psi_1(d, c, \ell, P, N, \mu)$$

After dimensional analysis:

Eq. (7.2) $$h/c = \psi_2(d/c, \ell/c, (\mu N)/P)$$

Consider a prototype bearing having the following characteristics:

Journal diameter (d)	8 in. (203 mm)
Diametral clearance (c)	0.010 in. (0.254 mm)
Bearing length (ℓ)	10 in. (254 mm)
Specific load (P)	1,000 psi (6.895 MPa)
Speed (N)	500 rpm
Viscosity (μ)	10×10^{-6} lb sec/in.^{2} (0.69 poise)

It is desired to test a 1 to 10 scale model in order to predict the minimum film thickness (h) from measurements on a model. In this problem, there are three dimensionally independent quantities and three arbitrary scale factors. One of these has already been fixed:

$$\text{Size factor: } K_1 = (d_p/d_m) = 10$$

where subscripts p and m refer to prototype and model respectively. Assume it is convenient to operate the model at 1,800 rpm and to use an oil having a viscosity of 2×10^{-6} lb sec/in.^{2}. Thus, the remaining two scale factors are:

$$\text{Speed factor: } K_2 = N_p/N_m = 500/1{,}800 = 0.28$$

$$\text{Viscosity factor: } K_3 = \mu_p/\mu_m = 10 \times 10^{-6}/2 \times 10^{-6} = 5$$

The remaining specifications for the model may be found as follows:

$$(d/c)_p = (d/c)_m; \qquad c_p/c_m = d_p/d_m = 10$$

$$(\ell/c)_p = (\ell/c)_m; \qquad \ell_p/\ell_m = c_p/c_m = 10$$

$$(\mu N/P)_p = (\mu N/P)_m; \qquad P_p/P_m = (\mu N)_p/(\mu N)_m = (5)(0.28) = 1.4$$

$$(h/c)_p = (h/c)_m; \qquad h_p/h_m = c_p/c_m = d_p/d_m = 10$$

The corresponding dimensions for prototype and model are summarized in Table 7.1.

Table 7.1. Summary of Dimensions of Model and Prototype for Journal Bearing Load Capacity

Quantity	Prototype	Model
d, in.	8	0.8
c, in.	0.010	0.001
ℓ, in.	10	1
P, psi	1,000	715
W, lb	80,000	572
N, rpm	500	1,800
μ, lb sec/in^2	10×10^{-6}	2×10^{-6}

If the minimum film thickness (h) is found to be 0.0001 in. when the model is tested, then without actually knowing ψ_2 in Eq. (7.2), the prototype bearing should operate with a minimum film thickness of (10)(0.0001) = 0.001 inch.

One of the important characteristics of a model test based on similitude is that the influence of changes in one quantity may be studied by changing another more easily adjusted quantity. For example, in the bearing problem of Eq. (7.2), the influence of changes in speed (N) may be observed by changing only the load (P). Function ψ_2 and (h/c) is influenced by changes in the dimensionless group (μN)/P regardless of whether these changes in ($\mu N/P$) result from a change in μ, N or P or a combination. Similarly, h/c will remain unchanged if both N and P are changed, provided these changes are such that ($\mu N/P$) remains constant. This observation can often be used to advantage in planning experimental work.

While the word *model* is frequently used for a device that is the same as the prototype in all regards but size, such a restricted definition is not to be inferred here. An engineering model may differ from a simple scale model only in that air is substituted for water as the fluid. On the other hand, a model may bear no outward resemblance to its prototype. In this latter situation, the model is usually called an *analog*. An example of an analog model is an electrical network that is used to study the flow of fluid in a system by utilizing the fact that the flow of current and fluid are governed by similar equations.

3.0 WIND TUNNEL

It is not always possible to employ a perfect model for which all Pi quantities are the same as those in the prototype. In such a case, the least important Pi quantities must be identified and these are the ones that are not modeled. Model testing in a wind tunnel illustrates this difficulty.

The drag force on a prototype airfoil operating at 325 mph is to be determined from wind tunnel measurements on a 40 to 1 scale model. The dimensionless equation for this case is:

Eq. (7.3) $$D/\rho V^2 d^2 = \psi(\rho V d/\mu)$$

There are three dimensionally independent quantities and three scale factors may be fixed:

Size factor: $K_1 = d_p/d_m = 40$

Viscosity factor: $K_2 = \mu_p/\mu_m = 1$

Density factor: $K_3 = \rho_p/\rho_m = 1$ (K_2 and K_3 are for std. air)

To have a perfect model, the model speed must be such as to make the Reynolds Number the same for prototype and model:

$$(\rho V d/\mu)_p = (\rho V d/\mu)_m; \; V_p/V_m = d_m\mu_p\rho_m/d_p\mu_m\rho_p = 1/40$$

The required air speed in the wind tunnel will thus be:

$$V_m = 40V_p = 40(325) = 13{,}000 \text{ mph}$$

This is, obviously, not attainable.

An alternative solution is to compress the air in the wind tunnel. Since ρ for air varies directly with pressure, while μ is essentially independent of pressure, a pressure of 40 atm would enable $R_p = R_m$ when V_m is 325 mph. However, a pressure of 40 atmospheres is too difficult to employ.

The solution usually adopted is to use a distorted model. It has already been demonstrated that the Reynolds Number has a relatively slight influence on drag when R is large. This will be the case for a wide range of speeds for both model and prototype in the present problem. The wind tunnel test could be performed with standard air and the drag force on the model measured at a convenient speed such as 100 mph. The corresponding drag on the prototype would be obtained as follows:

$$(D/\rho V^2 d^2)_p = (D/\rho V^2 d^2)_m; \; D_p = (40)^2(325/100)^2 D_m = 16{,}900 \; D_m$$

Wind tunnel determinations of lift coefficient (C_L) will generally be more accurate than those for drag coefficient (C_D) when the model tests have to be run at a Reynolds Number different from that for the prototype.

4.0 TOWING TANK

The measurement of drag on a surface vessel in a towing tank is another problem that is difficult to model exactly. For surface vessels which operate at a water-air interface, the acceleration due to gravity (g) must be included in a dimensional analysis for drag, since water is lifted vertically from the level surface against gravitational attraction during the formation of a bow wave. The variables of importance in this case are:

$$D = \psi_1(\rho, V, \mu, \ell, g)$$

where ℓ is the length of the vessel.

The end result of a dimensional analysis for a surface vessel of fixed shape will be:

Eq. (7.4) $$D/(\rho V^2 \ell^2) = \psi_2(\rho V \ell/\mu, V^2/g\ell)$$

The dimensionless quantity $(V^2/g\ell)$ is called the Froude Number (F). In this instance, three scale factors may be arbitrarily fixed. These will naturally be associated with size, density, and viscosity, since water is the only practical fluid to use in the towing tank.

If the scale factor is 50 to 1, then for the Froude Number equality:

Eq. (7.5) $$(V^2/g\ell)_p = (V^2/g\ell)_m;\ V_m = (\ell_m/\ell_p)^{0.5} V_p = V_p/7.05$$

while for Reynolds Number equality:

Eq. (7.6) $$(\rho V \ell/\mu)_p = (\rho V \ell/\mu)_m;\ V_m = (\ell_p/\ell_m) V_p = 50\ V_p$$

Both of these requirements on the model speed (V_m) cannot be satisfied simultaneously. A way out of this difficulty was found by Froude (1879), who found, to a good approximation, the wave-making drag is a function of the Froude Number (F) only while the skin-friction drag is a function of the Reynolds Number only. This enables wave-making drag and skin-friction drag to be studied separately and to a good approximation:

Eq. (7.7) $$D = D_W + D_f = \psi_3(F) + \psi_4(\mathrm{R})$$

where D, D_W, and D_f are total drag, wave-making drag, and skin-friction drag respectively.

Two models are used at a model speed (V_m) given by Eq. (7.5):

- A scale model to obtain the total drag on the model (D_m)
- A special model to obtain the skin-friction drag $(D_f)_m$

The second model is a thin rectangular plank having the same length and wetted area as the scale model. The drag in this case is the skin-friction drag $(D_f)_m$, since there is no bow wave. The wave-making drag for the model is obtained by difference:

Eq. (7.8) $$(D_W)_m = D_m - (D_f)_m$$

The $(D_f)_m$ value is adjusted for Reynolds Number equivalency using Eq. (7.6), and both $(D_W)_m$ and $(D_f)_m$ appropriately scaled up to prototype values $(D_W)_p$ and $(D_f)_p$. The total drag on the prototype (D_p) is the sum of these two values:

Eq. (7.9) $$D_p = (D_W)_p + (D_f)_p$$

This procedure is outlined merely to indicate the complexity involved when it is not possible to proceed with conditions of complete similitude.

5.0 SOIL BIN

Just as wind tunnels are used to design airplanes, trains, automobiles, and model basins are used to design ship's hulls, soil bins are used to test off-road vehicles, lunar roving vehicles, caterpillar tractors, and military tanks.

Consider the caterpillar vehicle shown in Fig. 7.1. The draw bar force (D) required to drive a unit of fixed track design (length, width, and tread arrangement) at uniform speed (V) over sand of a given composition, particle size and moisture content will be a function of the following variables:

d = wheel diameter, $[L]$

W = vehicle weight, $[F]$

ρ = sand density, $[FL^{-4}T^{2}]$

s = sand particle size, $[L]$

V = vehicle speed, $[LT^{-1}]$

g = acceleration due to gravity, $[LT^{-2}]$

That is,

Eq. (7.10) $D = \psi_1(d, W, \rho, s, V, g)$

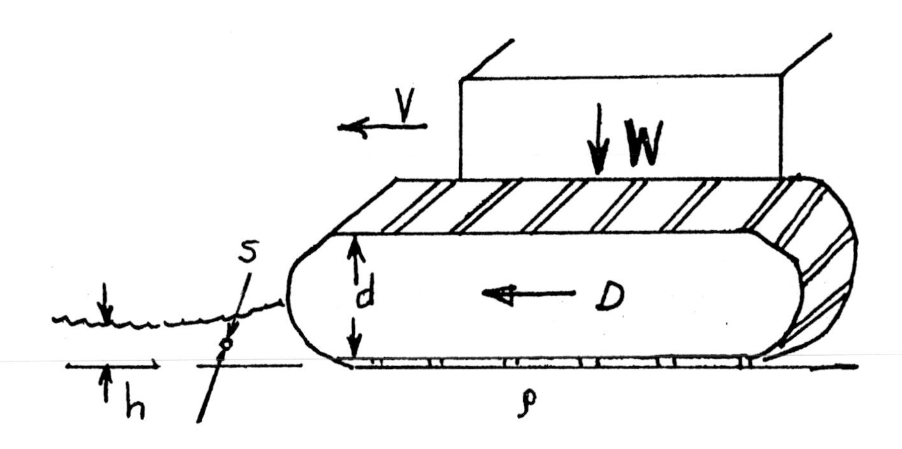

Figure 7.1. Model of tracked vehicle operating in a soil bin with soil particles of mean diameter (s).

After dimensional analysis, taking d, ρ, and V as the dimensionally independent quantities:

Eq. (7.11) $D/(\rho V^{2}d^{2}) = \psi_2[W/(\rho V^{2}d^{2}), s/d, V^{2}/gd]$

If a one tenth scale model is to be tested using sand of the same density as that on which the prototype will operate, and under the same acceleration due to gravity (g), the following relations pertain:

$$s_p/s_m = d_p/d_m = 10$$

$$(V_p/V_m)^2 = (g_p/g_m)(d_p/d_m) = 10$$

$$W_p/W_m = (\rho_p/\rho_m)(V_p/V_m)^2(d_p/d_m) = 100$$

$$D_p/D_m = (\rho_p/\rho_m)(V_p/V_m)^2(d_p/d_m) = 100$$

where subscripts p and m refer to prototype and model respectively.

If the model is to be tested on earth but the prototype is to be used on the moon, then $g_p/g_m = 1/6$. The values of the four quantities listed above would be:

$$s_p/s_m = d_p/d_m = 10$$

$$(V_p/V_m)^2 = (g_p/g_m)(d_p/d_m) = 10/6$$

$$W_p/W_m = (\rho_p/\rho_m)(V_p/V_m)^2(d_p/d_m) = 100/6$$

$$D_p/D_m = (\rho_p/\rho_m)(V_p/V_m)^2(d_p/d_m) = 100/6$$

6.0 HYDRAULIC MACHINERY

Most hydraulic machinery such as hydraulic turbines, pumps, fans, compressors, propellers, torque converters, fluid couplings, etc., operate at such high values of Reynolds Number that viscous forces are small compared with inertia forces. It is, therefore, customary and satisfactory to ignore Reynolds Number effects in the treatment of such problems. Hydraulic turbines are representative of this class of equipment and will be briefly discussed.

Hydraulic turbines are used to convert the gravity potential energy of a body of water into shaft work. Such machines are of three general types

(Fig. 7.2) depending on the vertical distance the water falls (i.e., the head available, h) and the available flow rate (Q). For heads greater than about 800 ft, a Pelton (1880) impulse turbine would be used; for intermediate heads (15–800 ft), a Francis (1849) radial flow turbine would normally be used; while, for low values of head (<15 feet) a Kaplan axial flow propeller type turbine would be most economical. The flow rate will usually vary inversely with the head employed. The efficiency (ratio of output to input energy) will normally be between 85 and 90% for a Pelton wheel and between 90 and 95% for the other two types. The type of turbine to be used at a given site will depend upon the head, flow rate, and investment capital available as well as the efficiency of the available machines. The unit power cost including investment charges is normally the criterion on which the final choice is based. The efficiency of the unit plays a very important role in such a study. An improvement of one percent in the efficiency of a 50,000 hp (37,500 Kw) machine is worth $320,000 per year in power saved, if the unit value of the power generated is 1 cent per kilowatt hour.

The power developed (P) by a hydraulic turbine is an important dependent variable. A dimensional analysis for this quantity leads to the following equation, where all variables involved are defined in Table 7.2. Since three fundamental dimensions (F, L, T) are involved, there will be three dimensionally independent quantities which have been taken to be ρ, N, and d.

Eq. (7.12) $$P/(\rho N^3 d^5) = \psi_1[Q/Nd^3, gh/(N^2d^2), (\rho Nd^2)/\mu]$$

It should be noted that both h and g enter this problem as a product because that is the way they influence the potential energy per unit mass available to do useful work. Treatment of (gh) as a single variable yields one less Pi quantity in the final result.

As already mentioned, the Reynolds Number ($\rho Nd^2/\mu$) plays a minor role and may be omitted from the analysis. Therefore:

Eq. (7.13) $$P/(\rho N^3 d^5) = \psi_2[Q/Nd^3, gh/(N^2d^2)]$$

A similar analysis for the efficiency (e) of the turbine would yield

Eq. (7.14) $$e = \psi_3[Q/Nd^3, gh/(N^2d^2)]$$

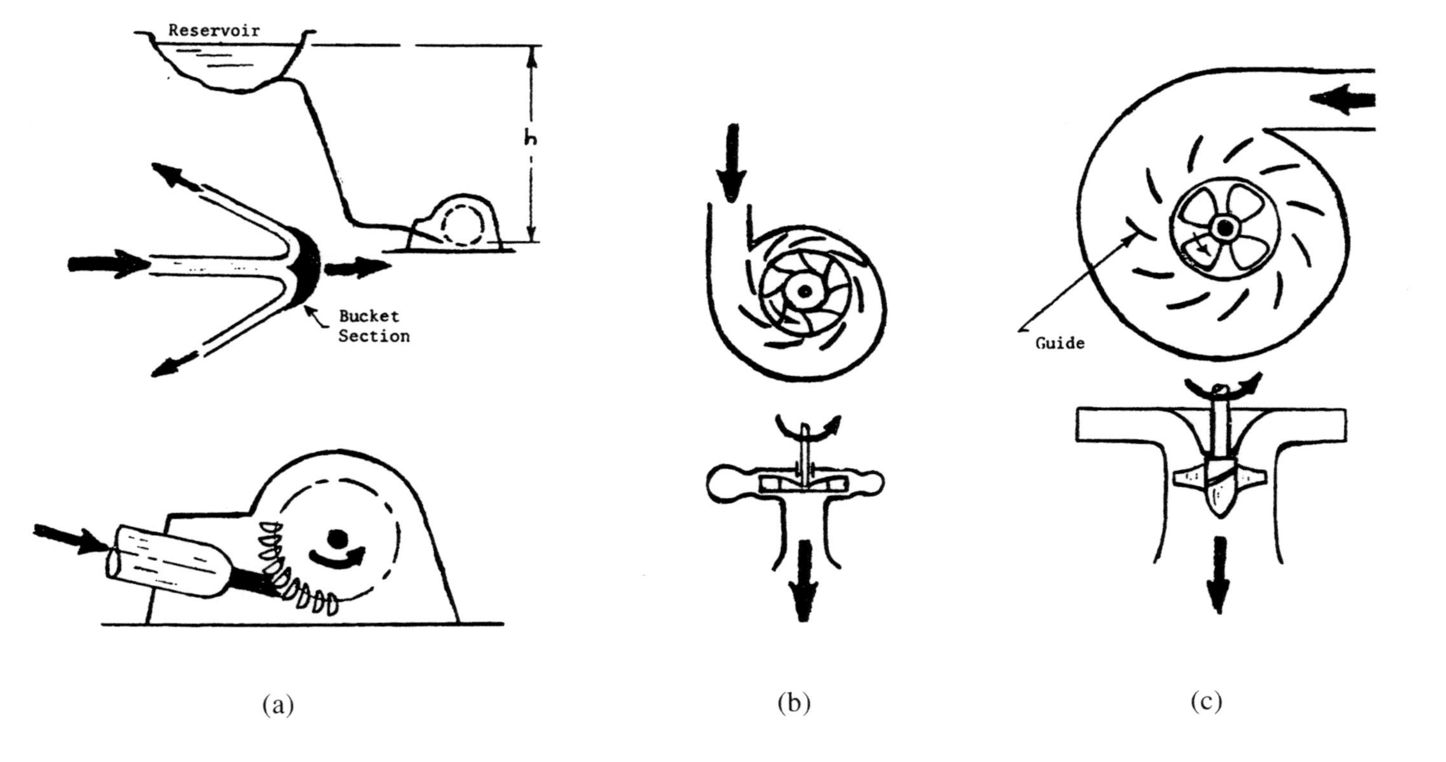

Figure 7.2. Types of water power turbines *(a)* Pelton water wheel, *(b)* Francis turbine, and *(c)* Kaplan turbine.

Table 7.2. Variables Involved in Hydraulic Turbine Analysis

Quantity	Symbol	Dimensions
Power	p	LFT^{-1}
Impeller diameter	d	L
Impeller speed	N	T^{-1}
Flow rate	Q	L^3T^{-1}
Head	gh	L^2T^{-2}
Fluid density	ρ	$FL^{-4}T^2$
Fluid viscosity	μ	FTL^{-2}

7.0 STRUCTURAL MODELS

Models are extensively used for civil engineering structures since most of these are built only once, represent an enormous capital investment, and are not easily corrected should the prototype fail to meet design requirements.

The towers of the George Washington Bridge were modeled in celluloid at a scale of 140 to 1 in order to study the distribution of strain. Even though these were distorted models, it was subsequently found that predicted values were within 10% of those measured later on the actual structure. The Hoover Dam was modeled at a scale of 240 to 1 before construction, in order to identify and study points of stress concentration. The canyon walls were modeled in concrete while the dam was made of plaster ($E = 90{,}000$ psi) which is a convenient modeling material for brittle materials. The water pressure on the dam was modeled, using mercury in plastic bags.

The Empire State Building was modeled at a scale of 144 to 1 in order to study wind forces and the stiffening effect of the masonry surrounding the steel. In this instance, the data were collected for future design use. The steel work was modeled using brass strips. The natural frequency of the model was compared to that of the prototype and the observed difference attributed to the stiffening effect of the masonry. The wind force was next estimated by comparing the deflection of the model (corrected for lack of masonry) with that of the actual building, for future design use.

Elevated water towers that are located in areas where earthquakes are apt to occur must be designed to withstand the shaking motion of the earth. In the earthquake of March 10, 1933, centered in the Los Angeles area, several water tanks failed. It is important that the natural frequency of the tower and tank in any degree of fullness not correspond to the main frequency of the earth's vibration. The dynamic response of a 46.5 to 1 model water tank filled with mercury and mounted on a shaking table was determined. The results of this study indicated that conditions could be significantly improved by making the structure more flexible.

In the previous chapter, a brief account of the Tacoma Narrows bridge disaster was considered. Tests on scale bridge models are being used today to be sure a proposed structure is not dynamically unstable to excitation due to a Karman Vortex Street that may develop in a high wind.

8.0 SIMULATION

Simulation involves use of a model to predict performance. There are two types—analytical and physical. Analytical simulation involves use of a mathematical model to predict performance often with use of a computer. Results are only as good as the model adopted. Physical simulation involves doing experiments on one system in order to understand another better.

9.0 GALILEO REGARDING SIMULATION

It is suggested that passages 139–150 of the Galileo text be read at this point. A good example of physical simulation is Galileo's observations concerning use of a combination of pendulums of different length and different oscillating frequencies to simulate combinations of musical notes.

Combinations of musical notes where maximum amplitudes coincide most of the time are pleasant while those that are nearly always out of phase and random are discordant to the human ear. Two notes that are an octave apart (the diapason = one having twice the frequency of the other) is an example of such a pleasing combination. The diapente (the fifth) is another example of a pleasant combination of notes where frequencies stand in the ratio of 3:2. In this case, there are also relatively few points in time (three) between points of reinforcement and these are uniformly spaced. Combinations that are less rhythmic are less pleasant.

Galileo indicates that a pendulum, like a musical string has a natural frequency that is independent of the weight of the bob, but varies inversely as the square root of the length of cord supporting the bob. Also, a pendulum may be excited to vibration if blown upon at a frequency equal to its natural frequency, just as a musical string can be caused to resonate at its natural frequency if another string vibrating at the natural frequency of the first string is brought close to the first one.

Galileo suggests that an analogy may also be drawn between the appearance of pendulums of different frequency (different lengths) swinging from the same axis, and combinations of musical notes. Pendulums that line up frequently are pleasing to observe while those that line up infrequently are less pleasing. Two pendulums having lengths in the ratio of four to one are pleasing to observe since their frequencies will stand in the ratio of 2 to 1, and they will be in phase every other cycle.

Similarly, three pendulums having lengths of 16, 9, and 4 units will be pleasing since the frequencies of the two extreme ones will differ by two (one octave) while the frequency of the middle one will be halfway between the other two (corresponding to the diapente). The three frequencies will stand in the ratio 4:3:2. All strings will line up for every other swing of the one of lowest frequency, and the combination will be pleasing to observe. When the pendulum lengths do not correspond to such a simple relationship, they will be out of phase for long periods of time. This is equivalent to the simultaneous sounding of discordant notes that do not reinforce at frequent intervals.

10.0 GALILEO REGARDING MUSICAL STRINGS

Galileo observed that the frequency of a musical string may be changed to that of a higher note by:

- Shortening the string
- Increasing tension in the string
- Decreasing the size of the string

By observing waves on the surface of water excited by a submerged glass harmonica (a tumbler excited to vibration by rubbing its top edge), it was found that the frequency changed by a factor of two when the sound emitted changed by one octave. The frequency (f) of a string was found to change by one octave when its length (ℓ) was halved, when the force on the string (W) was increased by a factor of four, or when its "size" was increased by four. Thus:

Eq. (7.15) $$f \sim (W)^{0.5}/[\ell(\text{size})^{0.5}]$$

The meaning of size is not clear, although it may be assumed to be diameter and may be checked by dimensional analysis.

At the outset, the frequency of a plucked string might be assumed to vary with the variables listed in Table 7.3. That is:

Eq. (7.16) $$f = \psi_1(\ell, d, W, \rho, E)$$

where E is Young's modulus.

Table 7.3. Variables Assumed to Influence Frequency of Musical String

Variable	Symbol	Dimensions
Frequency	f	$[T^{-1}]$
Length	ℓ	$[L]$
Diameter	d	$[L]$
Tension	W	$[F]$
Density	ρ	$[FL^{-4}T^2]$
Young's modulus	E	$[FL^{-2}]$

After performing a dimensional analysis:

Eq. (7.17) $$f(\rho/E)^{0.5}d = \psi_2[\ell/d, W/(d^2E)]$$

To satisfy Eq. (7.15) relative to W and ℓ, Eq. (7.17) must be written:

Eq. (7.18) $f(\rho/E)^{0.5}d = K(d/\ell)[W/(d^2E)]^{0.5}$

where K is a constant. Hence:

Eq. (7.19) $f = (K/\ell)[W/(\rho d^2)]^{0.5}$

This is in agreement with Eq. (7.15), provided size is interpreted as ρd^2 (the mass of the string per unit length) instead of diameter (d). The modulus E is seen to cancel. This is because Galileo probably used a single material in obtaining the results expressed in Eq. (7.15). If different materials were to be used, the same frequency would be obtained, provided W or d or both were varied, with E such that W/d^2E remained unchanged.

Theoretically, the constant K may be shown to be 0.5 for ideal end conditions. Results given in Table 7.4 were obtained for experiments in which weights were hung on wires of different diameters stretched over pins of variable spacing (ℓ) (Fig. 7.3) until the frequency of the string was either middle C (256 Hz) or twice this frequency. In all cases, the wires were of steel with $\rho = 72.5 \times 10^{-6}$ lb in.$^{-4}$sec^2. Tuning forks and a person with a musically sensitive ear were used in obtaining the data of Table 7.4. The mean value of K from these experiments was found to be 0.32. This differs from the value 0.5 due primarily to friction between the wires and the pins.

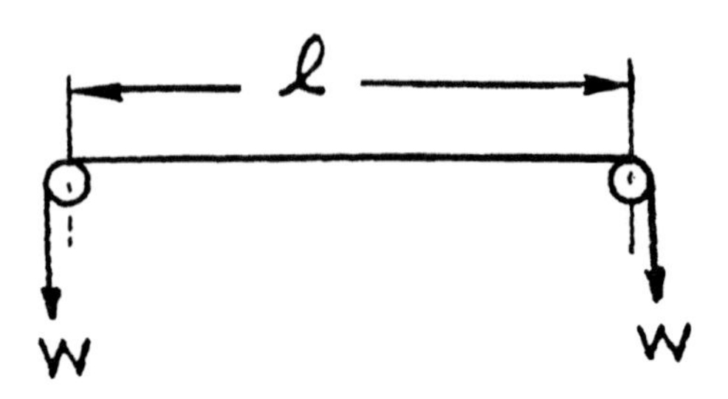

Figure 7.3. Wire stretched over pegs and loaded with weights (W).

Galileo discusses adoption of the factor two in defining an octave, and concludes that it is so because two notes differing in frequency by a factor of two are usually pleasing to hear.

Table 7.5 shows the eight notes constituting the musical scale, beginning with middle C. The fifth note (G) has a frequency approximately 50% greater than the fundamental (C) while the eighth note C (one octave above middle C) has a frequency exactly twice that of middle C. Each note in the scale has a frequency 1.104 times that of the preceding one. To obtain the note D, the length of the string for middle C should be divided by 1,104, while to obtain the fifth (G) the length of the string should be divided by 1.5 (i.e., it should be two-thirds the length for middle C).

Table 7.4. Experiments to Determine Contributions of ℓ, d, and W to Give Frequency of 256 Hz (As Determined by Use of Tuning Forks)

$f(\mathrm{sec}^{-1})$	ℓ, in.	d, in.	W, lb	K
512	16.75	.010	4	.322
256	33.00	.010	4	.317
512	15.38	.016	8	.336
256	30.75	.016	8	.336
512	13.75	.024	17	.309
256	27.00	.024	17	.304
256	21.25	.016	4	.330
256	27.25	.024	17	.311
				Av = 0.32

Table 7.5. Frequencies of Notes on Musical Scale

Note	Frequency, sec^{-1}	
Middle C		256
D	$(1.104)256 =$	282
E	$(1.104)^2 256 = 1.209(256) =$	312
F	$(1.104)^3 256 = 1.345(256) =$	344
G	$(1.104)^4 256 = 1.485(256) =$	380
A	$(1.104)^5 256 = 1.64(256) =$	420
B	$(1.104)^6 256 = 1.81(256) =$	464
C	$(1.104)^7 256 = 2(256) =$	512

PROBLEMS

7.1 Complete the following table to satisfy dimensional similitude [Eq. (7.2)].

Quantity	Prototype	Model
d, in.	8	0.8
c, in.	0.010	0.001
ℓ, in.	10	1
P, psi	1,000	500
W, lbs	80,000	400
N, rpm	500	?
μ, lb.s.in.$^{-2}$	10^{-7}	10^{-7}

7.2 Complete the following table to satisfy dimensional similitude.

Quantity	Prototype	Model
d, in.	10	2
c, in.	0.010	(a)
ℓ, in.	15	(b)
P, psi	1,000	(c)
W, lbs	(e)	(d)
N, rpm	500	1,800
μ, Reyn	50×10^{-6}	5×10^{-6}

7.3 Repeat the hydrodynamic bearing example with the coefficient of friction (f) as the main dependent variable. Before dimensional analysis:

$$f = \psi_1(d, c, \ell, P, N, \mu)$$

a) Derive the expression for (f) after dimensional analysis.

b) For a model with size factor $K_1 = d_p/d_m = 10$, speed factor $K_2 = N_p/N_m = 0.28$, and viscosity factor $K_3 = 5$, find the ratio:

$$\frac{\text{coefficient of friction for prototype}}{\text{coefficient of friction for model}} = \frac{f_p}{f_m}$$

7.4 A 1:50 scale model of an airfoil is tested in a wind tunnel with standard air flowing at a velocity (V) of 300 mph. If the lift force on the model (L_m) is 20 lbs, estimate the lift force on the prototype operating at 400 mph, assuming that a difference in Reynolds Number between model and prototype is not important.

7.5 A 1:30 scale model of an airfoil is tested in a wind tunnel with standard air flowing at $V = 300$ mph. If the drag force (D_m) on the model is 1 lb, estimate the drag force on a prototype operating at 400 mph, assuming the difference in Reynolds Number between model and prototype is unimportant.

7.6 a) A 1:25 scale model airplane wing is tested in a wind tunnel at an angle of attack of 5° and found to have a drag coefficient of 0.015. If the air in the wind tunnel has the same properties as that for the prototype, estimate the horsepower required to make the prototype fly at a speed of 200 mph and an angle of attack of 5°, if the wing area is 80 ft^2. The specific weight of the air is 44.3×10^{-6} pounds per cubic inch.

b) If the speed of the plane were to be doubled and all other items remained the same, by what factor would the horsepower increase?

7.7 It is desired to estimate the power developed in a large Kaplan turbine that will have a rotor diameter of 20 ft by building a 1:20 scale model and operating it under conditions of dimensional similitude, such that the measured power for the model may be used to predict the horsepower of the turbine. The prototype, of course, will operate with water, and the model will also be tested with water.

a) If the flow rate for the prototype will be 10,000 ft^3/sec when operating at 1,000 rpm, what flow rate should be used for the model if it is to be operated at 1,800 rpm under conditions of dynamic similitude?

b) What head (h) should be used in the model test if the prototype is to operate with a head of 10 ft of water?

c) If the model turbine produces 0.015 hp, what would the predicted horsepower of the prototype turbine be?

7.8 A prototype Francis turbine is to have the following characteristics:

Impeller speed (N) = 500 rpm

Impeller diameter (d) = 20 ft

Head (h) = 440 ft of fresh water

Flow rate (Q) = 240 ft^3/sec

a) Determine the head (h) and flow rate (Q) for a geometrically similar model turbine that is to have an impeller diameter of 20 in. and a speed (N) of 3,600 rpm, and which satisfies the principle of dimensional similitude.

b) If the measured horsepower for the model is 2, predict the power developed by the prototype turbine.

7.9 Before building a prototype Francis type hydroturbine, it is decided to build a 1:10 scale model and test it using water falling through a distance of 10 ft. The water will fall through a distance of 100 ft in the case of the prototype. The prototype turbine is to operate under the following conditions:

Flow rate: $Q_P = 3{,}000$ ft^3/sec

Speed: $N_P = 500$ rpm

a) Determine the speed (N_m) (rpm) at which the model turbine should be run and the flow rate Q_m (ft^3/sec) that should be used in the model test for dynamic similitude.

b) If the horsepower of the model turbine is found to be 9.70 hp, estimate the horsepower that would be developed by the prototype turbine.

c) If the efficiency (e) of the model turbine is found to be 90%, estimate the efficiency for the prototype turbine.

Notes: a) Friction is to be ignored in the pipe leading from the reservoir to the turbine.

b) In this problem, it is permissible and desirable to treat the product (gh) as a single variable, where g is the acceleration due to gravity and h is the head in feet.

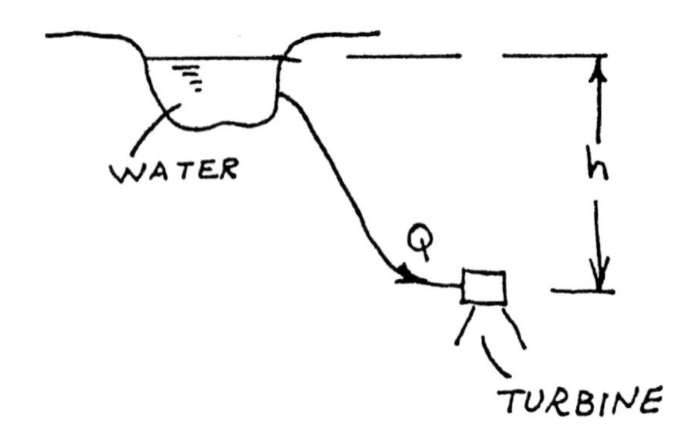

Figure P7.9.

7.10 A centrifugal pump is shown diagrammatically in Fig. P7.10 (*a*). Fluid enters (1) at low pressure and is discharged at (4) at a much higher pressure. Flow is induced by the centrifugal force associated with the rotary motion caused by the vaned impeller of diameter (D).

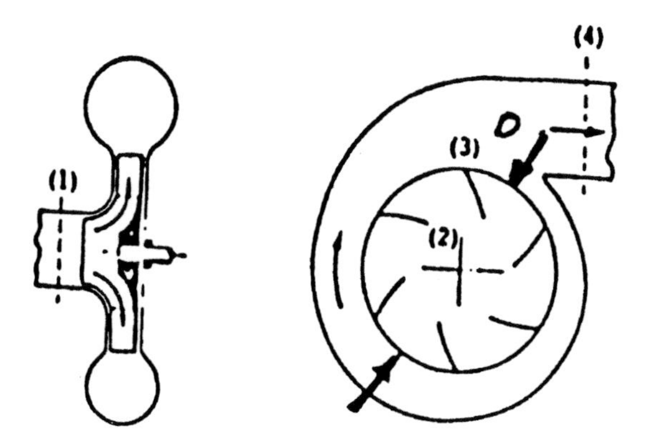

Figure P 7.10 (*a*).

For such a machine, there are three important dependent variables: required horsepower (P), specific output energy (gh), and efficiency (e). The quantity (gh) is the energy output per unit mass of fluid outflow. Each of these is a function of the following independent variables:

ρ = fluid density

N = impeller speed

D = impeller diameter

Q = volume rate of flow

Viscosity (μ) has a negligible influence.

The curves shown in Fig. P7.10 (*b*) apply to an existing pump when pumping water at a speed of 2,300 rpm. It is proposed that a geometrically similar pump be built that is 50% larger than that corresponding to Fig. P7.10 (*b*). This pump is to be driven by an electric motor having a rating of 1,000 hp at 1,800 rpm and a mechanical efficiency of 90%. The fluid to be used in the new pump is oil having a specific gravity of 0.9.

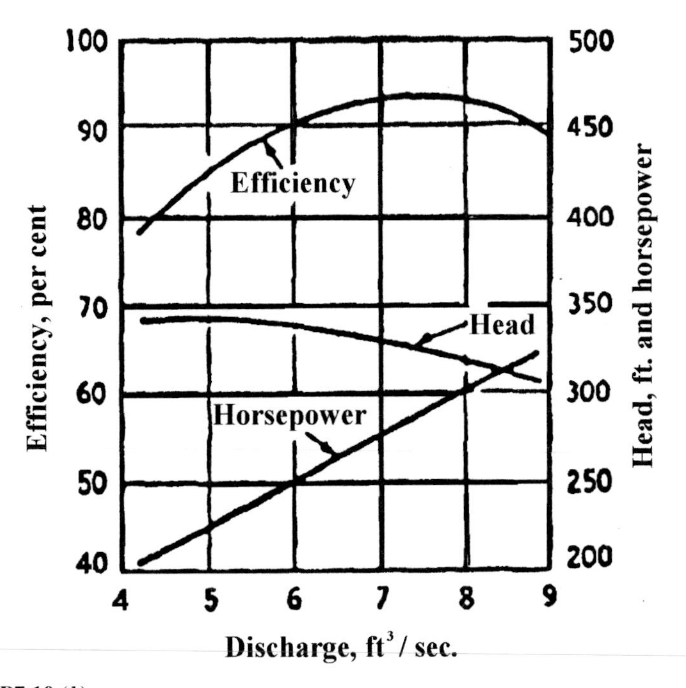

Figure P7.10 (*b*).

Determine the head (h, ft) the new pump is capable of developing when the motor is operating at its rated capacity, and estimate the efficiency (e) and flow rate (Q, ft^3/sec) of the new pump.

7.11 A model centrifugal pump with a constant speed motor has the following characteristics:

Impeller diameter = 10 in.

Speed = 1,800 rpm

When pumping water at the rate of 7 ft^3/sec, against a head of 10 ft, 250 hp is required. It is desired to construct a larger, geometrically similar pump to pump water against a head of 160 feet that has:

Impeller diameter = 20 in.

Speed = 3,600 rpm

a) Find the flow rate at which this new unit should be operated if the efficiency is to be the same as that for the model pump.

b) What is the horsepower required when the new pump is operating under these conditions?

7.12 A propeller for an outboard motorboat has three blades and a fixed pitch-to-diameter ratio (the pitch of a propeller is the axial distance it will advance through the water in one revolution, if there is no slip). The power (P) required to operate the propeller at a speed (N rpm) will depend upon the mass density of the water (ρ), the outside diameter of the propeller (d), and the axial velocity of the propeller through the water (V). Viscosity may be neglected (high R).

a) Perform a dimensional analysis for power (P).

b) Write an expression for Reynolds Number that will pertain in this problem if the viscosity were to be included in the list of significant variables.

7.13 When a 10 in. diameter boat propeller of a given design is tested, the horsepower (P) is found to be 40, when N is 6,000 rpm and V is 25 mph. A larger, geometrically similar propeller (d = 12 in. diameter) is to be used at N = 4,000 rpm and a speed of 20 mph. Estimate the power required in this second case if both propellers are operated in fresh water.

7.14 a) List the important variables that influence the elastic deflection (δ) of a thin ring (Fig. P7.14).

b) Perform a dimensional analysis with δ as the main dependent variable.

c) A model is to be used to predict the deflection of a large ring of steel that is 10 times as large as the scale model which is made of a plastic having a Young's modulus of 200,000 psi. If the load on the prototype is to be 10,000 pounds, what load should be placed on the model for dynamic similitude?

d) If the deflection for the model under the load found above is 0.01 in., find the corresponding deflection of the prototype.

e) If you know that:

$$\delta \sim 1/b$$

$$\delta \sim t^{-3}$$

$$\delta \sim W$$

Use this information to simplify the result found in b).

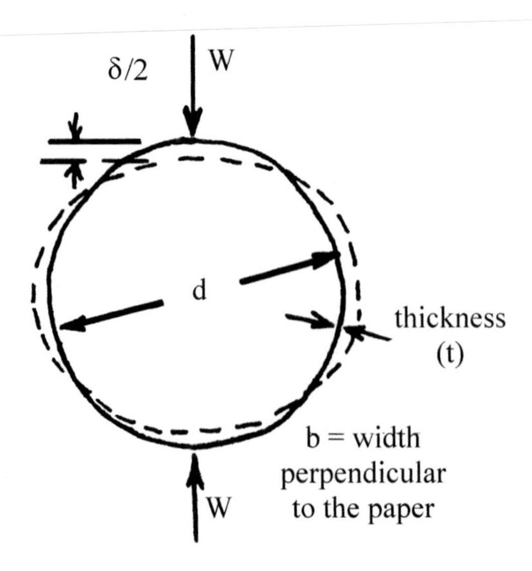

Figure P7.14.

7.15 When a plastic foam coffee cup is loaded as shown, the diameter of the cup (d) changes to (d'), giving rise to a deflection $\delta = d' - d$. The deflection (δ) in inches is found experimentally to vary with load (W) in pounds as follows: $\delta = CW$, where $C = 1$ for the plastic foam cup. In this problem, it is found that deflection (δ) is a function of the following variables:

E = Young's modulus

ℓ = height of cup

t = wall thickness

d = largest diameter

α = half cone angle

W = load

It is desired to make a steel part that is geometrically similar to the plastic cup, but with all dimensions twice as large. The Young's modulus of the steel is 5,000 times that of the plastic. Find the value of C in the above equation for the steel unit.

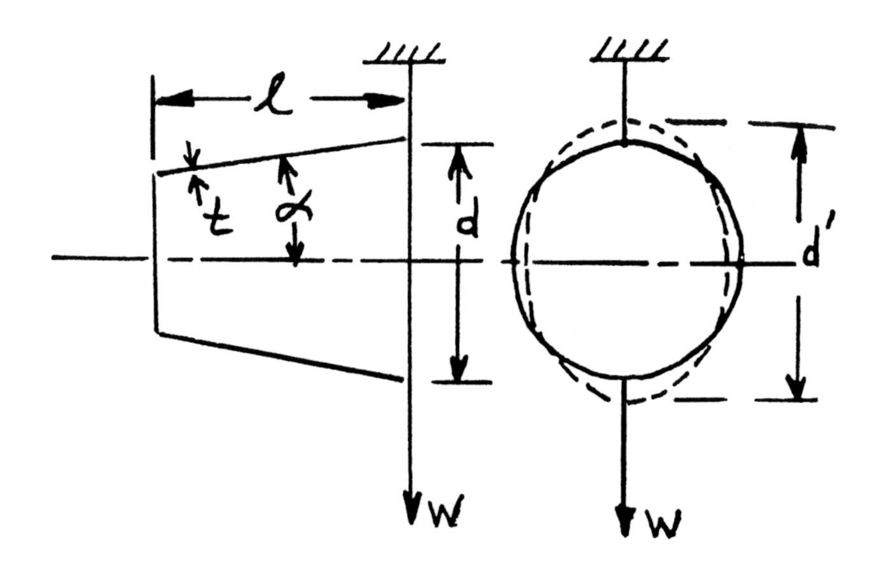

Figure P7.15.

7.16 A large prototype beam of steel ($E = 30 \times 10^6$ psi) has built-in ends and an intermediate support (A) as shown in Fig. P7.16. When subjected to a load (W), it deflects as shown by the dotted curve.

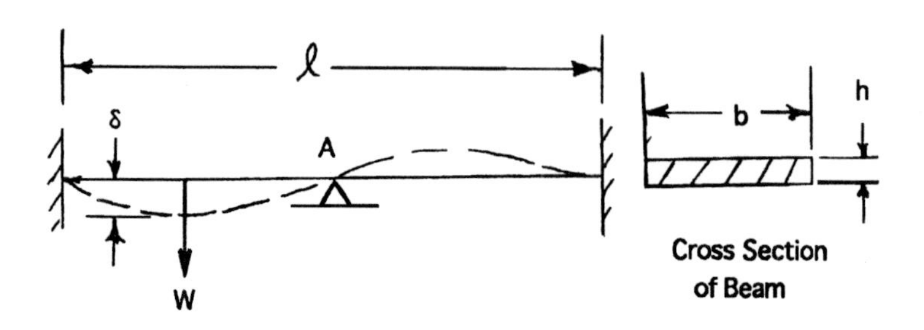

Figure P7.16.

a) Perform a dimensional analysis for the deflection at the load (δ). (This beam is long compared with other dimensions so that shear deflection may be ignored, and the shear modulus (G) plays a negligible role).

b) Using the result of this dimensional analysis, a 1:10 scale model is to be constructed from aluminum ($E = 10 \times 10^6$ psi). If W on the large beam is 1,000 lbs, how large should W on the model beam be in order that the measured deflection at the load on the model may be used to predict the deflection for the prototype by use of the Principle of Similitude? (Use results from Ch. 4, Sec. 6.)

7.17 A weight (W) is dropped from a height (y) upon the cantilever beam in Fig. P7.17 causing a permanent deformation (δ), where δ is a function of h, b, ℓ, and σ_f in addition to W and y (where σ_f is the plastic flow stress of the beam material). Thus before dimensional analysis:

$$\delta = \psi_3(W, y, h, b, \ell, \sigma_f)$$

and after dimensional analysis:

$$\delta/\ell = \psi_4\,[W/(\sigma_f\,\ell^2),\ y/\ell,\ h/\ell,\ b/\ell]$$

However, further reflection reveals that in place of W and y, the energy of impact ($U = Wy$) may be used, and instead of h and b, the moment of inertia about the neutral axis (I_N) = 1/12 bh^3 [in.^{4}] may be employed. Starting over:

$$\delta = \psi_3\,(U, I_N, \ell, \sigma_f)$$

a) Perform a dimensional analysis.

b) When a 1:10 scale model test is performed under geometrically similar conditions where $\sigma_p/\sigma_m = 2$, it is found that when $U_m = 10$ in. lbs, $\delta_m = 0.1$ inch. Find the corresponding values of U_p and δ_p.

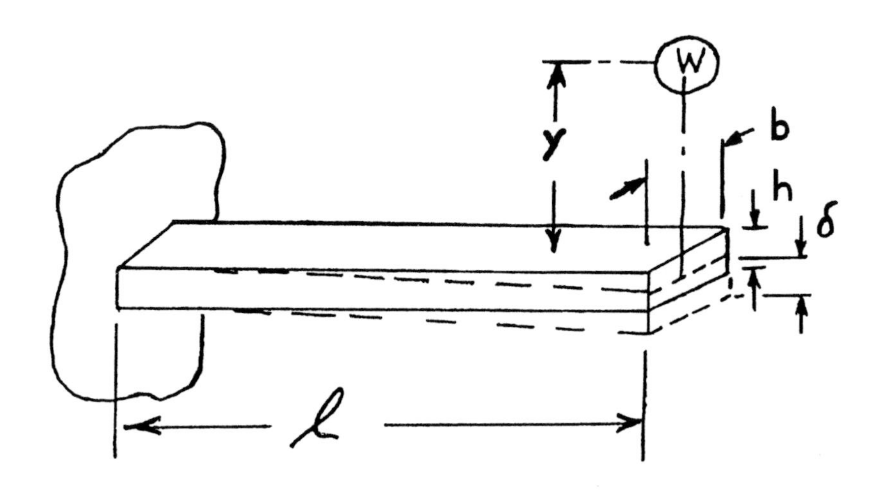

Figure P7.17.

7.18 A 0.010 in. diameter steel string ($\rho = 72.5 \times 10^{-6}$ lb in.$^{-4}$sec^{2}) is hung vertically as shown in Fig. P7.18, and a weight (W) is suspended from its end. Find W, such that the fundamental frequency of this string is middle C (256 cps) on the musical scale when the string is plucked. Friction may be ignored at the support and at the weight. (Solution of this problem involves Galileo's discussion of strings.)

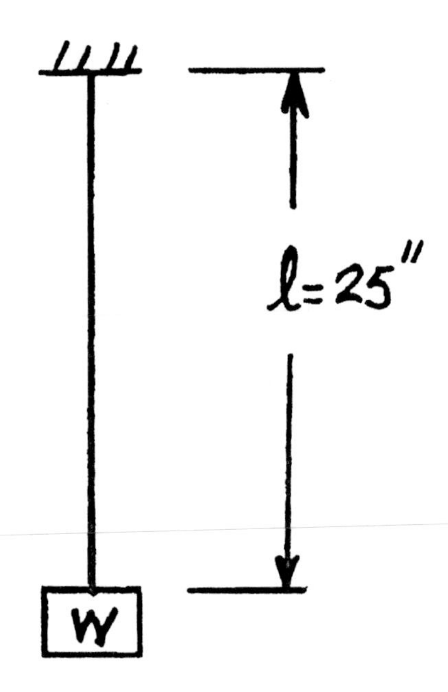

Figure P7.18.

8

Materials Science

1.0 INTRODUCTION

Materials science is concerned with the structure of materials, the binding forces that hold materials together, and the relation between structure, properties, and behavior. Two types of structure may be distinguished—atomic structure and microscopic structure.

2.0 ATOMIC STRUCTURE

All materials are composed of *atoms*, which in turn consist of a nucleus and a number of shells of planetary electrons. *Electrons* are negatively charged particles that are extremely small and essentially weightless. An atom consists mostly of empty space. This aspect of atomic structure was anticipated by Galileo. An atom is electrically neutral and for each planetary electron there is a positively charged particle (*proton*) in the nucleus. The protons have about 2,000 times the mass of an electron. The weight of the atom is due primarily to the protons and neutrons in the nucleus. A *neutron* is an uncharged proton (or proton + electron).

The simplest of all atoms is hydrogen which consists of one proton and one electron. Helium has two planetary electrons and two protons plus two neutrons in the nucleus. The atomic weight of helium is 4 (it has 4 times the mass of a hydrogen atom). Helium is chemically inert, as are all of the noble gases. It is used in dirigibles and balloons since it has very low density, but, unlike hydrogen, it is nonflammable and will not explode. Oxygen has an atomic number of eight. This atom has eight protons plus eight neutrons in the nucleus, hence its atomic weight is 16. It also has eight planetary electrons, two in the outer shell (valence electrons) and six in an inner shell.

The simplest representations of atoms are known as *Bohr models*. Such models are somewhat oversimplified and are incapable of explaining all atomic characteristics. However, the simplicity of the Bohr approach and the fact that it explains the main characteristics of the atom of interest to engineers makes it a very useful tool.

A periodic table of the elements summarizes a number of important characteristics of neutral atoms. The iron atom (Fig. 8.1) has 26 planetary electrons in 4 shells and 26 protons plus 26 neutrons in the nucleus. This electrically neutral particle is about 0.01 microinches (μ in.) in diameter. The outermost planetary electrons (valence electrons) are special. They determine the chemical properties of the element. Metals have few valence electrons but the nonmetals have several. When a metal loses its valence electrons, it becomes an ion. A positively charged ion is called a *cation*. When a nonmetal takes on valence electrons to bring the total number to eight, it too becomes an ion (in this case a negative particle called an *anion*). Cations (positively charged particles) are considerably smaller than neutral atoms, while anions are relatively larger than neutral atoms.

Radicals are combinations of atoms that are particularly stable and which are involved in chemical reactions without being decomposed into their constituent elements. They carry a negative charge as anions do. Relative sizes of a few anions, cations, and radicals are shown in Fig. 8.2. The diameters of ions in a metal structure are all about the same size (≅ 1–4 Angstrom units [Å], 1 Å = 10^{-10}m = 0.1 nm = 0.004 μ in.)

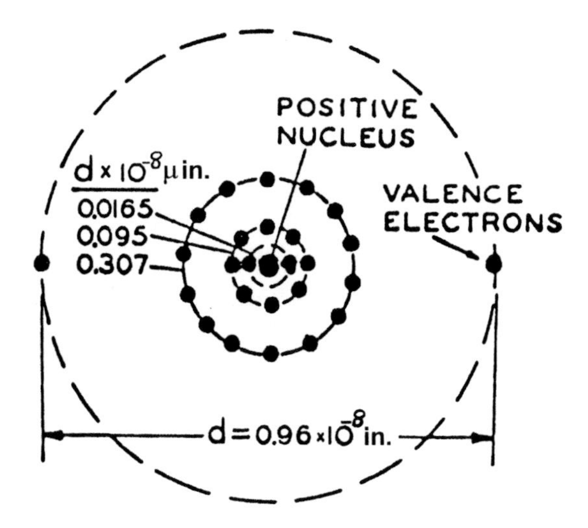

Figure 8.1. Bohr representation of iron atom which has 26 planetary electrons two of which are valence electrons.

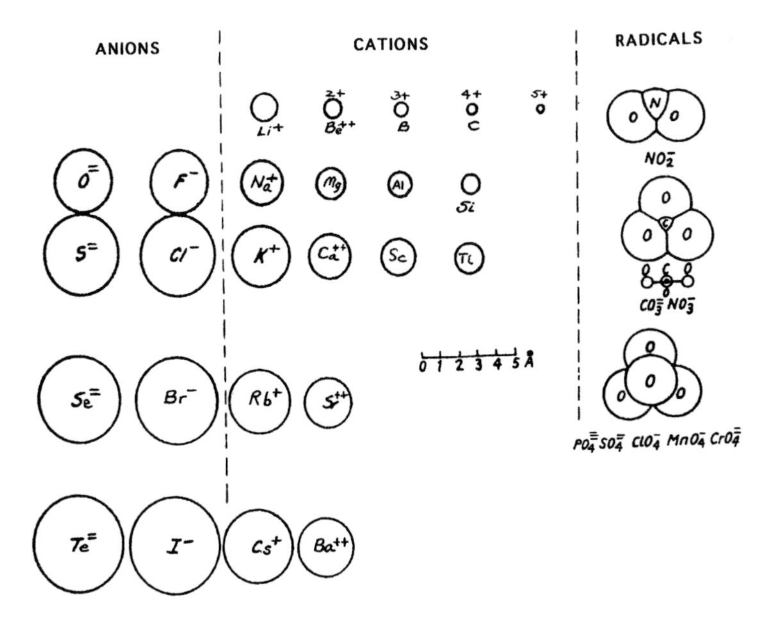

Figure 8.2. Relative sizes of a few anions, cations, and radicals.

3.0 BONDING FORCES

The valence electrons are responsible for the bonding forces holding materials together, and there are four principal types of bonds:

1. Brittle materials such as inorganic salts are held together by electrostatic forces. In this case, the valence electrons from the metal transfer to the nonmetal, and the two resulting positively and negatively charged particles are held together electrostatically (ex., table salt, NaCl, Fig. 8.3).

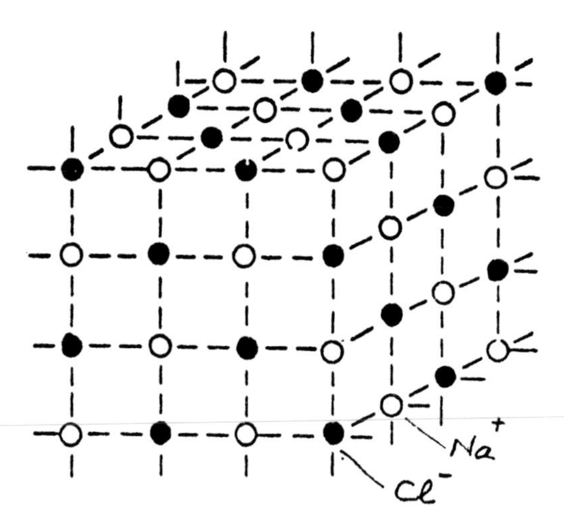

Figure 8.3. Ionic structure of sodium chloride that derives its strength from electrostatic bonds associated with transfer of electrons.

2. Organic materials (carbon and silicon compounds) are held together by sharing of electrons. They are weak and have low melting points (ex., wax and CCl_4).
3. Metal atoms in the solid state are completely ionized. The valence electrons are not associated with any particular ion, but are uniformly distributed as a sea of electrons. The positively charged ions repel each other while the action of the negative sea of electrons on the collection of positive ions tends to condense the structure. The result

is a very close packing of ions which behave as though they were relatively rigid spheres (ex., iron at room temperature).

4. *Molecules* (collections of atoms) are electrically neutral, but the electrical centers for plus and minus charges do not generally coincide. These particles have dipoles which provide weak electrostatic bonds between molecules when regions of opposite signs come close together (ex., adsorbed film of water on a glass surface).

Iron at room temperature has cations arranged as shown in Fig. 8.4 (*a*). This is the, so-called, body centered cubic (bcc) arrangement and constitutes nearly the closest packing of spheres possible. The solubility of carbon (a very small atom) is very low in this type of iron (~ 0.05 wt %), since the center of this unit cell is occupied by an iron atom. Iron at temperatures above about 1,450°F has a different ionic arrangement [Fig. 8.4 (*b*)]. This is the face centered cubic (fcc) structure that corresponds to the closest packing of spheres possible. The solubility of carbon in this lattice arrangement is high since the center of this unit cell is not occupied. The valence (or free) electrons will be uniformly distributed for both the structural arrangements shown in Fig. 8.4.

The bonding forces discussed above are the forces Galileo visualized in passage 67 of his text. Such bonds are broken at the melting point and reestablished when the temperature returns to the freezing point.

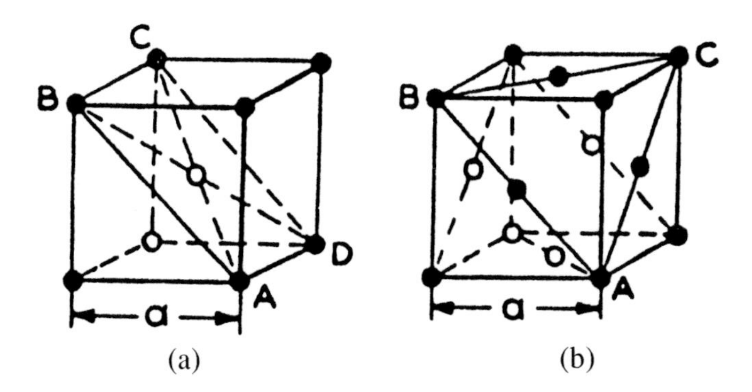

Figure 8.4. Arrangements of ions in a metal lattice. *(a)* Body centered cubic arrangement assumed by iron at room temperature and *(b)* face centered cubic arrangement assumed by iron at temperatures above about 1,450°F.

4.0 MICROSCOPIC STRUCTURE

In addition to an atomic picture of metals, a microscopic one is also of value in explaining the behavior of materials. Metals are not normally single crystals, but consist of a collection of crystallites (grains) ranging in size from 10^{-3} to 10^{-2} in. (25–250 μm). In a stress-free material, a collection of grains will tend toward spheres (Fig. 8.5), but when a material is deformed, grains are elongated and considerable internal strain energy is present in the metal. If the metal is heated to a certain temperature (recrystallization temperature), the grain boundaries shift and again become spherical.

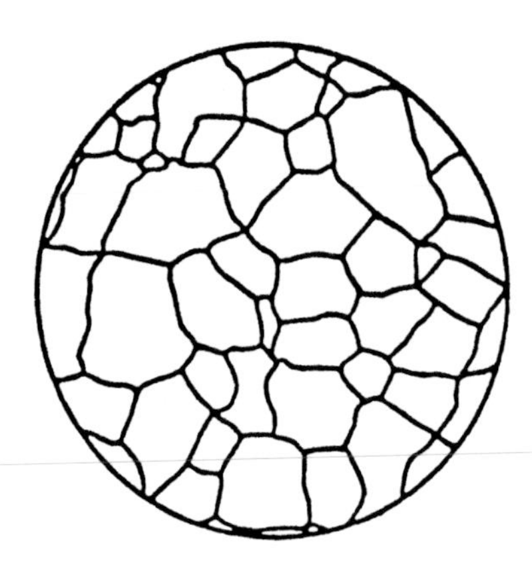

Figure 8.5. Appearance of pure iron at a magnification of 1,000 × showing grain boundaries. The atoms in any one grain are uniformly aligned whereas the alignment varies from grain to grain. The grain boundaries consist of atoms with alignment that changes from that of one neighbor to the other.

The structure of metals is studied microscopically by polishing the surface of a small specimen of metal until it is flat and free of scratches. This surface is then lightly etched with dilute acid (for example, one wt% HNO_3 in alcohol) and examined with a reflecting microscope at powers ranging from 25–1,000 times. Crystals oriented in different directions as well as different phases are etched at different rates, and reflect different amounts of light (Fig. 8.6). Thus, it is possible to obtain a good deal of information concerning the structure of a metal or alloy by photomicroscopy.

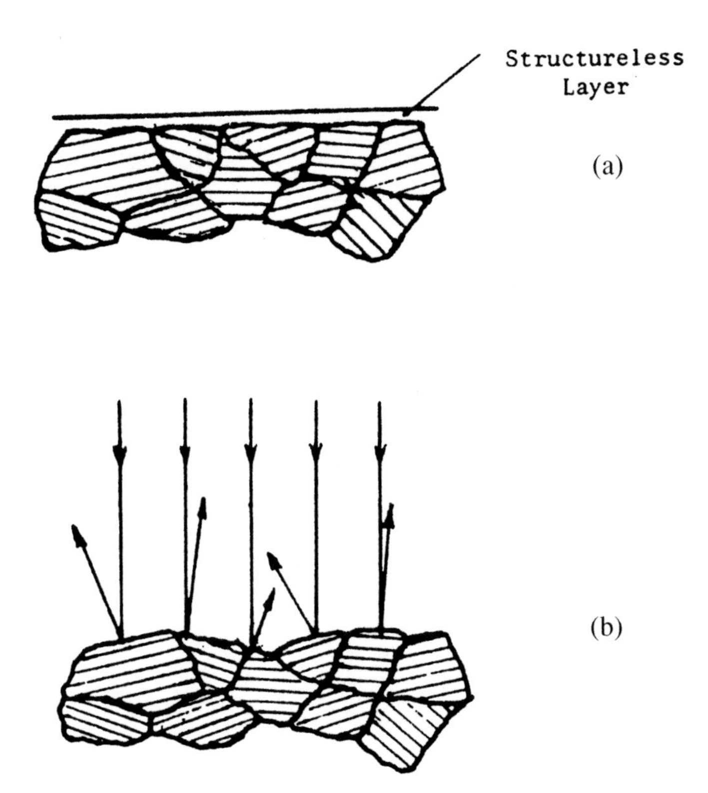

Figure 8.6. Light (high reflection) and dark (low reflection) regions on the surface of etched specimen when viewed with vertical illumination. *(a)* Polished surface showing a structureless, highly reflecting layer that forms upon polishing (the so-called Beilby layer) and *(b)* surface etched by dilute acid showing a variable inclination of surface crystals due to different rates of etching for different lattice inclinations in adjacent crystals.

5.0 THEORETICAL STRENGTH OF METALS

Metals are generally ductile and may be plastically deformed. The fact that the volume of a metal does not change when it is heavily deformed, as correctly demonstrated by Galileo in passages 96–100 of his text, suggests that the mode of plastic deformation is shear rather than tension (slip rather than stretching).

Figure 8.7 (*a*) shows two rows of a perfect array of ions in a metal that are being subjected to shear. The upper layer will resist being displaced from its equilibrium position relative to the lower layer. In Fig.

8.7 (*b*), the resisting stress is assumed to vary linearly with strain. The stress, when an upper ion is midway between ions below, should be zero, since attractive and repulsive forces will tend to be the same and cancel each other. By symmetry, the stress should be zero at points 1, 3, and 5 where the distance 1 to 5 is the horizontal atom spacing (a_1). The maximum stress this model is capable of sustaining should be (τ_0) which will occur at point 2, corresponding to a displacement of $a_1/4$.

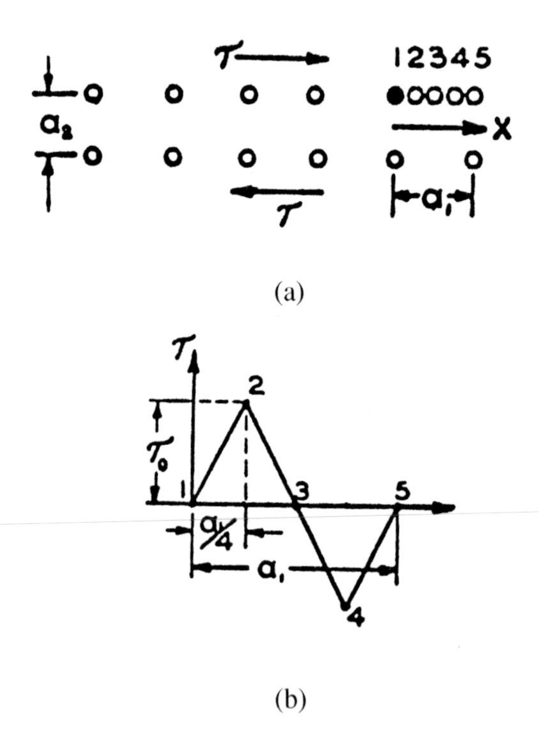

Figure 8.7. Variation of shear stress with deformation when one row of ions in a perfect lattice is moved relative to another in shear; *(a)* adjacent rows of ions and *(b)* assumed variation of stress with displacement.

From Fig. 8.7(*b*):

Eq. (8.1) $\tau = \tau_0[x/(a_1/4)]$

But, by Hooke's Law:

Eq. (8.2) $\tau = G\gamma$

where G is the shear modulus and γ is the shear strain which will be:

Eq. (8.3) $$\gamma = x/a_2$$

where a_2 is the vertical spacing of ions. Therefore,

Eq. (8.4) $$\tau = G(x/a_2) = \tau_0[x/(a_1/4)]$$

or,

Eq. (8.5) $$\tau_0 = (G/4)(a_1/a_2)$$

Since a_1 is about equal to a_2:

Eq. (8.6) $$\tau_0 = G/4$$

When this value for steel ($12 \times 10^6/4 = 3 \times 10^6$ psi = 20.7 GPa) is compared with actual values of maximum strength for ordinary samples (10^5 psi), a discrepancy of 30 is observed. However, very small metal whiskers have been specially grown which do have a strength approaching 3×10^6 psi = 20.7 GPa.

If shear stress (τ) is assumed to vary sinusoidally with strain (γ), instead of linearly, then the theoretical strength would be:

Eq. (8.7) $$\tau_0 = G/2\pi$$

This clearly indicates that the discrepancy of 30 is not due to the linear stress-strain relation that was assumed.

6.0 THE DISLOCATION

The discrepancy between the theoretical and actual strength of metals was not understood until 1934. Three scientists (Orowan, Polanyi, and Taylor) independently suggested that real metals do not have a perfect lattice structure as assumed in Fig. 8.7, but contain a special form of defect called a *dislocation* (Fig. 8.8).

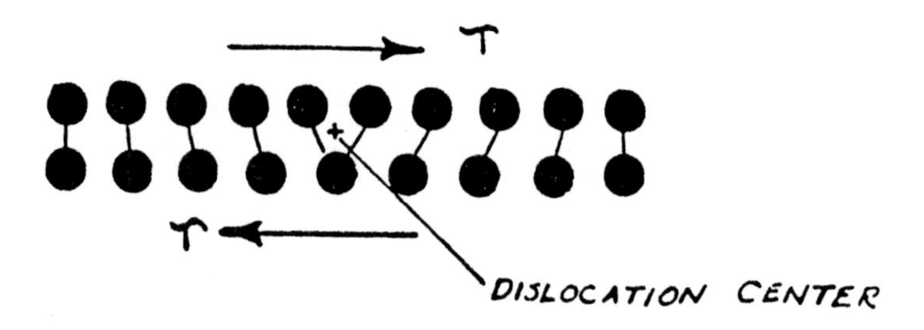

Figure 8.8. A dislocation.

The dislocation is a crystal imperfection in which there is one more atom in the upper row than in the lower row. When shear stresses are applied, there are as many atoms resisting displacement on one side of the dislocation center as there are tending to promote it on the other. Hence, it takes much less energy to cause a dislocation to move across a crystal, than to move one layer of a perfect array of atoms over another. Dislocations are present as defects occurring during solidification, or are generated at cracks or other points of stress concentration when a metal is stressed.

A useful analogy to the role a dislocation plays in reducing the force required for shear deformation in a crystal is shown in Fig. 8.9. Producing a hump (ruck) in the rug and moving it to the other side requires much less force than when the flat rug is moved across the floor all at the same time.

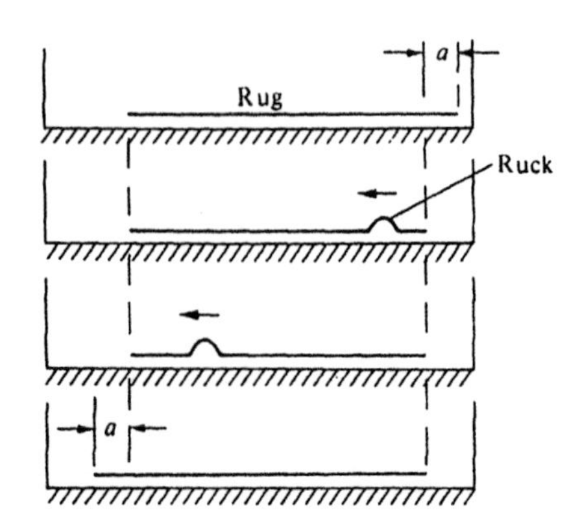

Figure 8.9. Movement of a rug a distance (a) by moving a ruck in the rug across the rug rather than sliding the flat rug across the floor all at once.

When a dislocation is caused to move across a crystal, one layer will have been displaced relative to its neighbor by one atom spacing. Very large shear displacements are possible when a defect (grain boundary, crack, impurity, etc.) is present. This defect is capable of generating many dislocations one after the other when shear stress is applied.

Dislocations may be demonstrated by means of a soap bubble analogy. This consists of a raft of small bubbles, all the same size, generated on the surface of a soap solution. The bubbles represent atoms and are subjected to two forces just as the atoms in a metal are. Surface tension causes the bubbles to attract each other and form a dense array, while pressure within the bubbles prevents them from approaching each other closer than a characteristic distance. When a raft of bubbles is formed on a fluid surface, grain boundaries are evident as well as dislocations (Fig. 8.10). When such a raft of bubbles is sheared, deformation is seen to occur as dislocations move across the "crystal." While the forces at work are not identical to those associated with atoms, the soap bubble model is a useful analogy. A film has been produced by Sir Lawrence Bragg who devised the soap bubble analogy, and this film lends considerable credibility to the relatively sophisticated dislocation concept.

Figure 8.10. Bubble model showing grain boundary running diagonally across center of field and dislocation just above center of field.

7.0 BEHAVIOR OF REAL MATERIALS

All metals in practical use exhibit a yield point at which elastic behavior ceases and irreversible plastic flow begins. If deformation occurs at a relatively low temperature, dislocations will accumulate and each successive dislocation will require a higher stress to make it move across a crystal. This is called strain hardening. The temperature beyond which strain hardening ceases is best defined in terms of a nondimensional homologous temperature (T_H).

The atoms in a crystal are in constant vibrational motion, the amplitude of which increases as temperature rises above an absolute zero temperature (-273°C) where all motion ceases. At the melting point of a material, the vibrational motion is sufficient that atoms no longer retain a fixed relationship to their neighbors. The *homologous temperature* is the ratio of the absolute temperature of a material to absolute temperature at the melting point. The homologous temperature at absolute zero is zero while that at the melting point is one.

A homologous temperature of 0.5 is a special value. At values above 0.5, dislocations do not accumulate and cause strain hardening, but disappear as fast as they are generated. These dislocations move into grain boundaries causing an increase in grain boundary area, and a decrease in grain size. Deformation that occurs above $T_H = 0.5$ is called hot working. When deformation occurs below $T_H = 0.5$ this is called cold working. Dislocations then accumulate with attendant strain hardening. If a metal is cold worked and then held at a temperature above $T_H = 0.5$, the accumulated dislocations will move into grain boundaries and the strain hardening produced during cold working will be removed. This heat treatment is called recrystallization.

Metals are made harder and stronger by structural defects that interfere with the movement of dislocations across a crystal. These defects may consist of grain boundaries, impurities, or precipitates that are insoluble in the base metal. The greater the number of interfering defects, the greater will be hardening or strengthening effects.

A material that contains no dislocations, or defects capable of generating dislocations under stress, will behave in a perfectly brittle manner. It will remain elastic all the way to the point of fracture even though the strain at fracture may be two orders of magnitude or more greater than the point where irreversible flow normally occurs.

Materials fracture in either a brittle or a ductile fashion. In brittle fracture, the design criterion is the maximum normal strain and fracture will occur on a plane normal to that of maximum normal strain. Ductile fracture involves shear, the design criterion is then maximum shear stress, and fracture occurs on a plane of maximum shear strain.

8.0 GALILEO

In passages 96–100 of his text, Galileo discusses the fact that gold is capable of being reduced enormously in thickness which suggests it consists of particles that are extremely small. He mentions that an extremely thin layer of gold, when placed on a silver cylinder and the combination is drawn, repeatedly, to a diameter equal to that of a human hair (0.003 in. = 0.076 mm), the gold will be reduced to a phenomenally small thickness. As an example, he considers a silver cylinder 3 in. (76 mm) in diameter by 9 in. (229 mm) long (3 thumb widths by 1/2 cubit) drawn to a diameter of 0.003 in. (0.076 mm). Equating the volume of silver before and after drawing, the final length of the composite will be 9×10^6 in. (229 km). Then, equating the volume of gold before and after drawing, the ratio of initial thickness to final thickness of gold would be 10^3. Thus, if the initial ten layers of gold leaf measured 10^{-3} in. thickness, the final thickness of the gold would be 10^{-6}in. (about 4,000 atom diameters).

9.0 WEAR

Wear, along with corrosion, is responsible for the loss of huge amounts of money each year. There are many types of wear depending upon the conditions of sliding contact, the materials, and the environment. The most common types of wear are abrasive wear and adhesive wear. In the former type, hard particles in one of the sliding elements (or in the form of hard third body particles) cut grooves in the softer material in a manner similar to that involved in fine grinding. In the case of adhesive wear, the high points (asperities) on mating surfaces come close enough to establish strong bonds. If a weak point in the bulk material happens to lie in the vicinity of a contacting asperity, a wear particle will be generated.

It has been found, empirically, that for both abrasive and adhesive wear, the wear volume (B, in.^{3}) is a function of the applied load (W, lbs), the sliding distance (L, in.), and the hardness of the softer of the two sliding elements (H, psi). *Hardness* is the resistance a material offers to indentation, and its dimensions are $[FL^{-2}]$. Thus:

Eq. (8.8) $$B = \psi_1(W, L, H)$$

After performing a dimensional analysis:

Eq. (8.9) $$B/L^3 = \psi_2[(HL^2)/W]$$

It is further found, empirically, that:

Eq. (8.10) $$B \sim L$$

from which it follows that:

Eq. (8.11) $$B/(LW) = K/H$$

where K is a constant depending on the materials in sliding contact and the sliding conditions (speed, load, etc.). The quantity on the left (B/LW) is the wear parameter. This equation indicates that the wear parameter varies inversely with material hardness (H).

The nondimensional quantity (K) is called the wear coefficient which depends upon the materials in sliding contact, their hardness, and the lubricant present. Dimensionless (K) values are given in Table 8.1 for several representative adhesive wear situations for dry surfaces sliding in air. The presence of a lubricant greatly reduces the wear rate. For example, the first case (52,100 against 52,100) with engine oil as a lubricant gives a value of K of about 10^{-10}. Values of K for dry abrasive wear in air will be considerably higher than those given in Table 8.1.

Table 8.1. Values of Wear Coefficient for Several Dry Sliding Pairs of Metals

Materials		*K* for Wear Surface
Wear Surface	**Mating Surface**	
BB Steel	BB Steel	10^{-10}
Mild Steel	Mild Steel	2×10^{-3}
Tool Steel	Mild Steel	10^{-5}
Cast alloy (Co) Steel	Tool Steel	2×10^{-5}
Polytetrafluoroethylene	Tool Steel	4×10^{-8}
Copper	Mild Steel	30×10^{-4}
Stainless Steel	Mild Steel	2×10^{-5}

BB = Hard Ball Bearing Steel (52,100)

10.0 SOLIDS AND LIQUIDS

In passages 85 and 86 of his text, Galileo differentiated between solids and liquids depending on whether a finely divided solid would stand up in a pile, like sand, or flow into a puddle, like pitch. This is a perfectly good point of view, but depends on the length of observation time. Different conclusions will be obtained if the time involved is a few minutes, on the one hand, or a few years on the other.

In 1867, Maxwell suggested that there is a characteristic time required for a force to be reduced to a small fraction of its initial value due to viscous flow. For example, when glass is annealed, residual stresses are removed by holding the object at an elevated temperature long enough so that sufficient viscous flow can occur to enable the residual stresses initially present to be reduced to about the 10% level. The higher the annealing temperature, the shorter the time required to lower the internal stresses.

This relaxation time (T_R) should be expected to depend on the viscosity (μ) and the magnitude of the stress present, which in turn will depend on the shear modulus (G). Thus, we should expect that:

Eq. (8.12) $$T_R = \psi(G, \mu)$$

After dimensional analysis:

Eq. (8.13) $$(T_R G)/\mu = \text{a constant}$$

Maxwell found the constant to be approximately unity.

It has been suggested that materials might be classified as solids or liquids depending on their relaxation times relative to the average span of human life (10^9 sec $\cong$ 30 years). If all materials are assumed to have the same shear modulus (10^{10} dynes/cm^2), then Eq. (8.13) becomes:

Eq. (8.14) $$T_R = \mu 10^{-10}$$

Table 8.2 gives values of relaxation time for different materials. It is evident that water at room temperature is very much a liquid (since $T_R = 10^{-12}$ sec is very small compared with 10^9 sec). On the other hand, glass at room temperature is very much a solid (since $T_R = 10^{10}$ is significantly longer than the mean span of human life = 10^9 sec). Glacier ice is a liquid as is glass at the annealing temperature. The time required to anneal glass, in practice, is about 1,000 sec which corresponds with the relaxation time in Table 8.2. It is interesting to note that the annealing temperature for glass is defined as the temperature at which the viscosity is 10^{13} poise. This form of definition is logical from the point of view of a relaxation time.

Whether a material should be considered to be a solid or a liquid depends on the duration of the event involved. If an action taking place involves molasses and takes less than 10^{-7} sec, then the equations of elasticity should be applied instead of the viscous equations of fluid mechanics. In this case, the entire event would be over in less than the relaxation time (10^{-7} sec).

Table 8.2. Estimated Values of Relaxation Time for Several Materials

Material	Viscosity, poise	T_R, sec
Water	10^{-2}	10^{-12}
Molasses	10^{3}	10^{-7}
Glacier Ice	10^{13}	10^{3}
Glass (Room Temperature)	10^{20}	10^{10}
Glass (Annealing Temperature)	10^{13}	10^{3}

Classification of materials into solid or liquid may also be made in terms of structure. According to this point of view, a solid has long-range order in its structure, while a liquid contains many holes or imperfections and has only short-range order. The liquid moves due to migration of holes which continues as long as the slightest force field is present on the liquid. A true solid is capable of remaining at rest indefinitely as long as the applied load is below some value generally referred to as the flow stress. The flow stress is usually not a fixed value, but increases with strain as a consequence of the phenomenon known as strain hardening.

PROBLEMS

8.1 What is the diameter of the atom shown in Fig. 8.1?

a) In microinches (μ in.).

b) In microns (μ m).

c) In Angstrom units (Å).

8.2 Name and distinguish three types of atomic bonding forces.

8.3 Illustrate the molecular bonding forces that bind the atoms constituting water (H_2O) together.

8.4 When the temperature of an iron-carbon alloy is increased above a certain temperature (the transformation temperature), the atomic structure changes from (bbc) to (fcc). Which of these has the greatest solubility for carbon (ability to absorb carbon)?

8.5 Copper has a melting point of 1082°C and lead has a melting point of 327°C. What is the nondimensional homologous temperature for each of these materials at 100°C?

8.6 a) Will copper plastically deformed at 100°C be cold or hot worked?

b) Will lead plastically deformed at 100°C be cold or hot worked?

8.7 What is the approximate recrystallization temperature

a) For copper?

b) For lead?

8.8 Show that Eq. (8.9) follows from Eq. (8.8) when a dimensional analysis is performed.

8.9 Show that Eq. (8.11) follows from Eqs. (8.9) and (8.10).

8.10 If a metal is cold worked and is then held at a homologous temperature of 0.7 for several hours, would you expect to find:

a) A finer grain size?

b) A reduction in hardness from that pertaining immediately after cold work?

8.11 If hard ball bearing steel rubs against identical hard ball bearing steel in the absence of a lubricant, would you expect the adhesive wear rate to increase or decrease with:

a) Reduction of hardness of one of the items?

b) Decrease in the applied load?

c) Introduction of a lubricant?

8.12 Estimate the adhesive wear in mm^3 per Km of sliding distance for hard ball bearing steel operating against hard ball bearing steel in the absence of a lubricant for a load of 100 Kg and a hardness of both members of 600 Kg/mm^2.

8.13 If an event involving glacial ice takes 10 seconds, should equations of elasticity or plasticity be employed?

8.14 If an event involving room temperature glass takes 10 years, should equations of elasticity or plasticity be employed?

9

Engineering Materials

1.0 INTRODUCTION

There is a wide variety of materials an engineer may specify when designing a product and these will be discussed briefly in this chapter. They include the following:

- Metals
- Polymers
- Glasses and Ceramics
- Rock and Concrete
- Composites

2.0 METALS

Metals are particularly important since they are relatively strong, good conductors of heat and electricity, and may be made relatively ductile (deformable without fracture) by controlling their structure. Ductility makes it possible for them to be given a desired shape by plastic forming as well as by casting and stock removal by cutting and grinding.

2.1 Carbon Steels

Carbon steel is an alloy of iron and carbon. Iron at room temperature has a very low carbon solubility (0.05 wt %). Therefore, a steel containing 0.40 wt % carbon will contain precipitated carbon in the form of a hard compound (Fe_3C) called cementite. Iron above a critical temperature (about 1,400°F) changes its atomic structure from bcc (called ferrite or α iron) to fcc (called austenite or γ iron) and the carbon solubility is greatly increased. The atomic arrangements for α iron (bcc room temperature form) and γ iron (fcc high temperature form) have been shown in Fig. 8.4. By heating and cooling steel at different rates, carbon can be put into solution and precipitated. The size and distribution of the carbides depends on the rate of cooling. The carbides may be in the form of:

1. Large spheres (a structure called spheroidite)
2. Alternate plates of Fe_3C and α iron (pearlite)
3. Small Fe_3C particles (tempered martensite)

The spacing of carbides in a steel gives rise to a mean ferrite path (mean distance between adjacent carbides), which in turn gives rise to a variety of properties. The strength and hardness of steel increases as the mean ferrite path decreases due to increased interference to the motion of dislocations (Fig. 9.1). In general, as the structure of a metal is changed to provide greater hardness and strength, ductility decreases. This makes the choice of structure that is best for a given application, a challenging problem for the design engineer.

2.2 Alloy Steels

Elements other than carbon are alloyed with iron for one or more of the following reasons (the elements used for each of these purposes are given in parenthesis):

1. To delay the rate at which fcc iron transforms to bcc iron upon quenching allowing sufficient time for thick sections to be hardened throughout without the use of a quench that is so drastic as to induce damaging cracks in the steel. (Mn, Cr, Mo, W, Ni, Si)
2. To provide unusually hard complex carbide abrasive particles to improve wear resistance. (V, Mo, Cr, W)

3. To increase hardness at high temperatures (greater hot hardness). (Mo, Cr, V, W)
4. To provide greater strength at elevated temperatures. (Mn, Ni, Si, Co, Cr, Mo)
5. To inhibit grain growth during heat treatment. (V, Al)
6. To provide corrosion resistance. (Cr, Ni)
7. To combine with air entrained in liquid metal to prevent blowholes. (Si, Al, Ti)
8. To combine with sulfur or phosphorus which otherwise cause brittleness. (Mn)
9. To improve machining properties. (S, P, Pb)
10. To improve magnetic properties (Si)

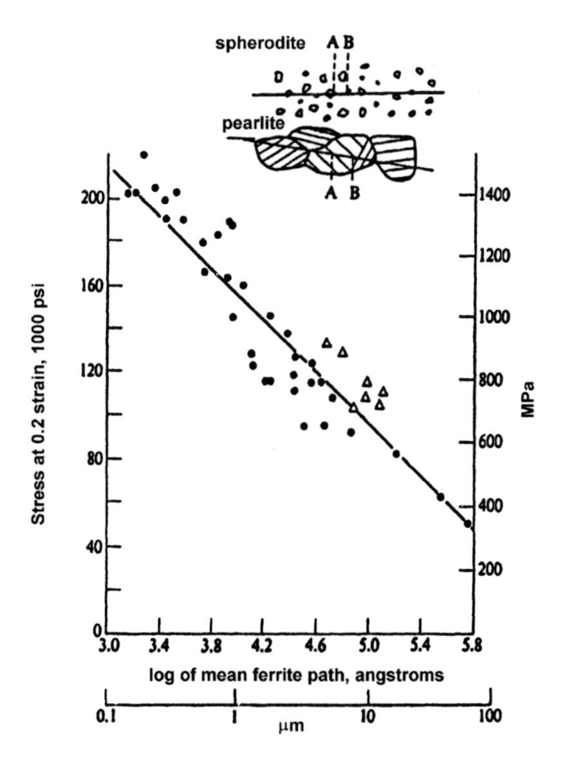

Figure 9.1. Variation in tensile stress at strain ($\in$) of 0.2 (~ ultimate stress) with log of mean ferrite path (mean of AB in insert) for steels of pearlitic (●) and spherodial (Δ) structure. [*After Gensamer, Pearsall, Pellini, and Low,* Trans ASM,*30:1003(1942)*]

These alloying elements perform their intended function in one or more of three basic ways:

- By forming carbides that are insoluble in fcc iron (W, Mo, V, Cr)
- By going into solid solution in the fcc iron by replacing iron atoms (Si, Mn, Ni, Co, Cr)
- By forming inclusions which are insoluble in fcc iron (Si, Al, Ti, Mn, S, P, Pb)

All of the materials going into solid solution in fcc iron increase the tendency for strain hardening. Of the materials which form inclusions, silicon, aluminum, and titanium will be in the form of oxides; while S or P will be present as (MnS) or (MnP). Lead will be present as small globules in elementary form.

The elements manganese, nickel, and cobalt have a pronounced tendency to decrease the rate of transformation of γ iron to α iron on cooling, and if present in sufficient quantity may enable fcc iron to be stable at room temperature. Examples of alloy steels that have an austenitic (fcc) matrix at room temperature are Hadfield's steel which contains 12 wt % manganese and 18-8 stainless steel which contains 18 wt % chromium and 8 wt % nickel. Other austenitic materials are the high temperature alloys that are used in aircraft gas turbines.

Stainless steels are steels that contain up to about 20 wt % chromium. The chromium concentrates at the surface where a protective chrome oxide layer forms. The higher the Cr content, the more corrosion resistant will be the material. It is important that the carbon content of a stainless steel be low; otherwise chromcarbides form and the Cr is not available to migrate to the surface. Also, significant amounts of nickel are usually present to further prevent the Cr from being tied up as carbides. Nickel decreases the tendency for fcc iron to transform to bcc iron on cooling. Face centered cubic iron has a relatively high solubility for carbon, and carbon in solid solution has less of a tendency to react with Cr to form chromium depleting chromcarbides. A widely used stainless steel is 18-8 (18 wt % Cr and 8 wt % Ni).

Tool steels are high carbon steels (up to about 1 wt % C) that contain appreciable amounts of tungsten or molybdenum. These elements increase the ability of a tool steel to retain its hardness to relatively high temperature. This property is important since most metal cutting tools operate at high speed, hence high temperatures. Widely used tool steels are T-1 (18 wt % W, 4 wt % Cr, and 1 wt % V) and M-2 (5 wt % Mo, 6 wt % W, 4 wt % Cr, and 2 wt % V).

Cast irons are relatively low melting alloys of carbon and iron. While steels have a carbon content less than about 1.3 wt %, cast irons will have a C content of 3 or more wt %. These materials are usually cast to shape and the resulting structure depends upon the rate of cooling. Slow cooling yields graphite as the form in which C precipitates while rapid cooling yields iron carbide. Graphitic cast iron (gray cast iron) is relatively soft and easily machined while rapidly cooled material (white cast iron) is extremely hard and brittle. Silicon added to cast iron promotes graphite formation.

2.3 Nonferrous Alloys

There are a large number of engineering materials that do not have iron as their base. Aluminum alloys are relatively light and corrosion resistant. They are good conductors of heat and electricity, but less strong and hard than alloys of steel. They may be hardened by addition of an element (usually Cu) that changes its solubility with temperature. A widely used alloy is 95.4 wt % Al, 4.5 wt % Cu. If such an alloy is held at an elevated temperature, all of the Cu will go into solid solution and diffuse to give a uniform dispersion. Upon quenching, the material cools to a supersaturated product. Upon heating and holding at an elevated temperature (called aging), precipitation occurs, the size and spacing of the precipitate dependent on the time at temperature. This hardening mechanism depends on the difference in solubility with temperature and is called precipitation hardening. The end result is the same as in heat treating steel—to produce a material containing many hard particles that interfere with the motion of dislocations. A wide variety of combinations of hardness and brittleness may be produced depending on the time at temperature (called tempering in the case of steel). The difference in solubility, in the case of steel, depends upon the transformation from a bcc to an fcc structure. Hardening of steel by heat treatment is, therefore, called transformation hardening while hardening of aluminum is precipitation hardening. Aluminum has good corrosion resistance, due to the tendency to form a dense, impervious oxide film in air.

Copper has the highest electrical conductivity of all metals and is used in highly purified form when electrical conductivity is important. Metals that have high electrical conductivity also have high thermal conductivity since the same characteristic is responsible for both. Many alloys of copper and zinc are used. These are called brasses. Alloys of copper and tin are called bronzes. Like aluminum, copper is relatively corrosion resistant due to a protective oxide film that has an esthetic greenish color.

Some of the many other metal alloys in use include:

- Nickel and cobalt base alloys (gas turbines)
- Low melting alloys (Pb, Bi, Sn, Zn)
- Refractory metals (Mo, Nb, W, Ti)
- Precious metals (Au, Pt, Ag)

2.4 Hardening and Softening

Metals may be hardened by:

- Solid solution hardening
- Reducing grain size
- Strain hardening by cold work
- Transformation hardening (steels)
- Precipitation hardening (aluminum)
- Dispersion hardening
- Surface hardening

All but the last two of these have been previously considered. Dispersion hardening involves mixing a metal powder with hard particles insoluble in the metal powder, pressing in a die, and then heating to cause the mixture to unite into a solid by a process called sintering (described below). An example of dispersion hardening is sintering copper powder containing a small amount of Al_2O_3 powder that is insoluble in Cu.

Surface hardening involves selective hardening of just the surface of a part or changing the composition of the surface by diffusing a small atom into the surface (usually C or N), followed by heat treatment of the surface. An example is case hardening of steel by increasing the carbon content of the surface by heating the part for about one hour in contact with some source of carbon (such as methane gas) followed by heat treatment.

The advantage of surface hardening over through-hardening is that the bulk of the part may remain relatively soft for shock resistance while the surface is hardened for wear resistance.

Metals may be softened by:

- Recrystallization to remove strain hardening due to cold work
- Annealing by heating and holding above the transformation temperature followed by slow cooling
- Grain growth by heating to, and holding at, a high temperature
- Overaging, in the case of precipitation hardening

2.5 Titanium and its Alloys

Titanium is a metal that has become available for special engineering applications relatively recently. It is an element in the IVB column of the periodic table. Titanium is abundant in nature as the oxide, but this oxide is relatively difficult to purify. One of the most popular ways of refining titanium is by conversion to the chloride:

Eq. (9.1) $$TiO_2 + 4Cl + C \rightarrow TiCl_4 + CO_2 \uparrow$$

The titanium chloride is reduced by magnesium:

Eq. (9.2) $$TiCl_4 + 2Mg \rightarrow Ti + 2MgCl_2$$

Titanium is separated from the chloride by leaching with water, and then vacuum melting and alloying. One of the most widely used alloys is 6 wt % aluminum, 4 wt % vanadium, and the remainder titanium (Ti-6-4).

The chief characteristics of titanium are:

- Low density (midway between aluminum and steel)
- High strength (equivalent to alloy steel)
- Excellent corrosion resistance (due to a very impervious and coherent oxide film)
- Extreme affinity for oxygen, nitrogen, and chlorine

- Low specific heat and heat conductivity
- Moderate Young's modulus (16×10^6 psi, compared with 10×10^6 for aluminum and 30×10^6 for steel)
- Relatively high cost

Titanium alloys are used for mechanical parts in the aircraft and space industries due to their high strength-to-weight ratio. Titanium is difficult to machine due to its low thermal properties, and its tendency to form strong bonds with metal oxides.

3.0 POLYMERS

Polymers are relatively light, weak organic materials with low melting points that are held together by bonds that are covalent for the most part. There are three basic types of polymers:

- Thermoplastics
- Thermosets
- Elastomers

Thermoplastics consist of long chain molecules that are produced from small molecules (called *mers*) by a polymerization process. An example of the formation of a typical polymer is given in Fig. 9.2 where polyethylene is produced from the monomer ethylene by application of heat, pressure, and a catalyst. When the resulting polymer is cooled, it solidifies. When reheated, a thermoplastic such as this will become liquid again. Due to this reversible liquid-solid conversion, scrap thermoplastic material may be recycled. Other thermoplastics are polypropylene, polyvinylchloride (PVC), polystyrene, polytetrafluoro-ethylene (PTFE), polyesters, polycarbonates, nylons, cellulosics.

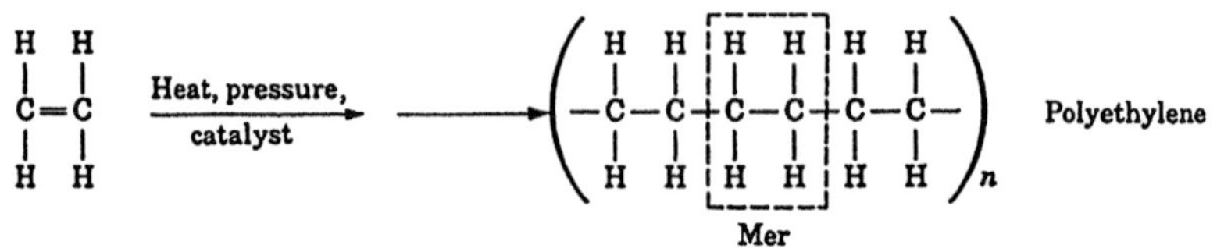

Figure 9.2. Formation of thermoplastic polyethylene by polymerization of ethylene.

Thermosets are plastics that form large molecules by an irreversible chemical cross linking reaction. Figure 9.3 illustrates the formation of a thermoset. In this case, phenol is reacted with formaldehyde to produce the thermoset called bakelite, with water as a by-product. This was the first synthetic plastic developed. It was invented by L. H. Bakeland in 1906. Thermosets cannot be melted and cannot be recycled. Other thermosets are epoxies, alkyds, polyesters, aminos, polyimides, and silicones.

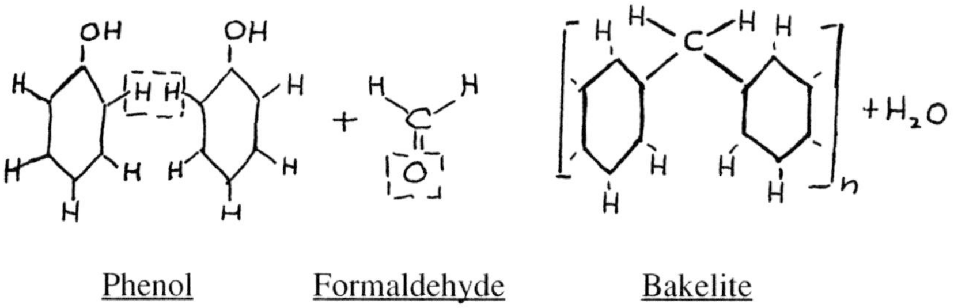

Phenol   Formaldehyde   Bakelite $+ H_2O$

Figure 9.3. Formation of the thermoset bakelite by reaction of phenol and formaldehyde.

Elastomers are high molecular weight polymers that exhibit nonlinearelasticity to large strains (up to 800%). Natural rubber is the oldest and best known example of this class of polymers. The monomer of natural rubber is a form of isoprene (Fig. 9.4). On heating, the two double bonds of the monomer are replaced by a single double bond and growth of a chain. The long polyisoprene chains are cross linked, just as thermosets are, and are tightly coiled. The hardness and stiffness of rubber may be increased by reaction with atomic sulfur which results in additional cross linking between chains. This reaction of S with double bonds is called vulcanization. As stress is applied, the chains between points of cross linking uncoil, but again assume a highly coiled condition when unloaded. Several other elastomers having different chemical compositions and properties are neoprene, polyurethane, and silicones.

Thermoplastics may be produced as a foam. This may be done by mixing the liquid thermoplastic with a material that reacts to form many entrained gas bubbles. On cooling, these bubbles remain in the polymer resulting in a material with many noncommunicating pores.

Polystyrene foam is widely used for floatation and thermal insulation, and this product is produced in a unique way. Polystyrene beads are first produced by co-extruding polystyrene and liquid heptane (C_7H_{16}) in such a way that a small droplet of heptane is trapped at the center of each polystyrene bead. When these beads are placed in a closed mold and heated, the heptane vaporizes and expands producing small polystyrene bubbles. These bubbles expand to fill the mold. The mold pressurizes the mass of bubbles causing them to unite in the form of an approximately hexagonal structure. Figure 9.5 shows the polymerization process for polystyrene and the formation of a porous solid from polystyrene beads. White foamed plastic cups used for hot coffee are produced by expanding beads of polystyrene/heptane in a heated cast iron mold. The boundaries of the individual expanded beads are evident to the unaided eye.

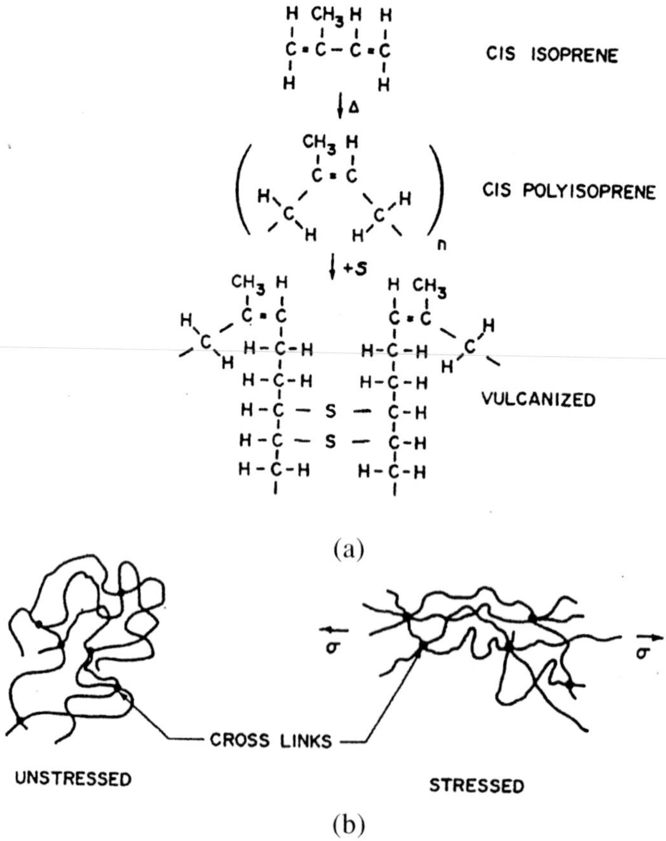

(a)

(b)

Figure 9.4. *(a)* Formation of natural rubber elastomer with vulcanization with S to adjust hardness and stiffness and *(b)* diagrammatic representation of unstressed and stressed elastomer chains with cross linking.

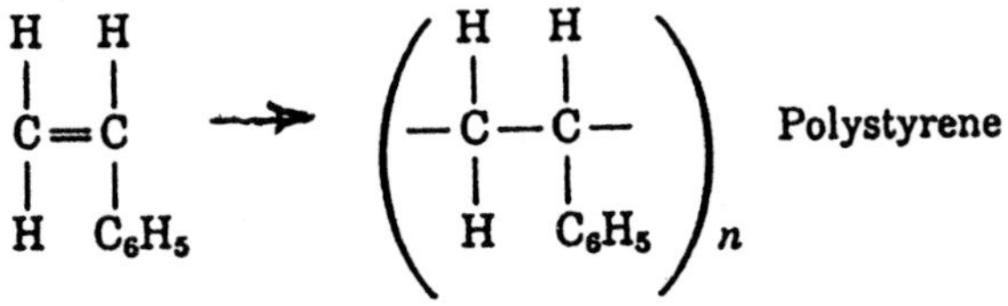

Figure 9.5. *(a)* Polymerization of polystyrene from phenylethylene and *(b)* formation of porous solid by heating polystyrene/heptane beads in a confining mold.

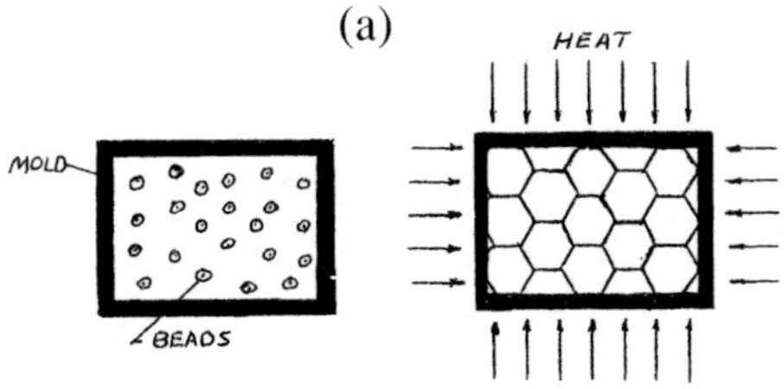

4.0 GLASSES AND CERAMICS

Glass is an inorganic product of fusion that has cooled to a rigid condition without crystallizing. Ceramics are polycrystalline inorganic solids. Both of these are held together primarily by ionic bonds.

The most common constituent of glass is SiO_2. Figure 9.6 shows SiO_2 in the glassy (amorphous state) at (*a*), while the same material is shown in the crystalline (ceramic state) at (*b*). There are large numbers of glass compositions in use. These consist of SiO_2 plus additions of other oxides such as Na_2O, CaO, PbO, B_2O_3, and Al_2O_3. The main influence of these additives is to change the melting temperature, the coefficient of thermal expansion, and the hardness and strength of the glass. Additions of PbO, Na_2O, and CaO lower the melting point; B_2O_3 lowers the coefficient of expansion, while additions of Al_2O_3 increase hardness and strength. The most refractory (highest melting point) glass is fused quartz (100% SiO_2).

Ceramics for the most part are oxides (such as Al_2O_3 and SiO_2), nitrides (Si_3N_4), and carbides (SiC). Both glasses and ceramics are hard, brittle materials that are poor conductors of heat and electricity and are relatively corrosion resistant. Most glasses are transparent while ceramics are opaque because of their grain boundaries. Ceramics are unusually refractive (high melting point) which makes them an attractive choice for structural members that must operate at high temperatures. Glasses do not have a definite melting temperature, like metals, but gradually soften over a range of temperatures. This is a useful characteristic in the production of glass items.

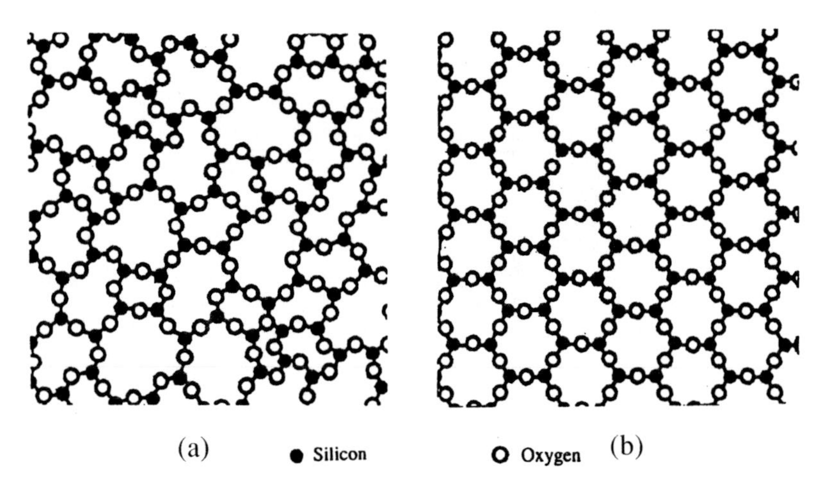

Figure 9.6. *(a)* Structure of SiO_2 glass in amorphous state and *(b)* structure of SiO_2 glass in crystalline state.

Glass may be converted irreversibly to a ceramic by incorporating a nucleating agent that comes out of solution when the glass is cooled to room temperature. On reheating the glass, crystals form on these nuclei. The resulting polycrystalline material, known as pyroceram, is a true ceramic that is opaque and has entirely different properties than the original glass:

- A higher Young's modulus (16×10^6 psi compared with 10×10^6 psi for the glass)
- A lower coefficient of expansion than the glass
- A greater solubility in hydrofluoric acid (about 200 times that for glass)
- Less brittle than glass (only about as brittle as cast iron)

Because of its greater stiffness and toughness, and lower thermal expansion, pyroceram is more useful than glass as a structural material.

By proper choice of additive, the precipitation of nuclei may be controlled photographically. When the additive is a gold salt, it will precipitate only when exposed to ultraviolet light and may then be ceramed by holding it at the softening temperature until oxide crystals form on the nuclei. A photographic negative may be used to obtain selective precipitation which results in selective ceraming when the material is held at the softening temperature. By treating such a partially ceramed body with HF, holes may be produced in the body in the areas that have been ceramed. The remaining glass may then be converted to a ceramic by exposure to ultraviolet light and heating to the softening point. The product in which precipitation is triggered by photon absorption is called photoceram. Many intricate ceramic parts may be made by chemically "machining" photoceram glass.

Still another product that depends on precipitation of a particle in glass is photosensitive glass. In this case, the solubility of an additive in the glass is controlled by the intensity of light. As the intensity increases, more precipitation occurs and the glass becomes more tinted. When the intensity of light is diminished, the additive again goes into solution and the glass becomes clear. Photosensitive glass is presently used for variable intensity sunglasses, but creative engineers will undoubtedly devise many additional novel uses for this interesting product.

5.0 ROCK AND CONCRETE

Rock and concrete are important engineering materials in the construction of buildings and other large civil engineering structures. Rock occurs in nature as many chemical compositions. Three main categories exist: igneous, sedimentary and metamorphic. Igneous rocks are formed by solidification of molten rock (magma). Examples are granite, quartz, and basalt. Sedimentary rocks are formed by deposition of the wear debris of weathered igneous rock or other inorganic material. The debris is transported into land depressions by wind and water. Sandstone, coal, bauxite, and limestone are examples of sedimentary rock. Metamorphic rocks result from alterations to igneous and sedimentary rocks. These alterations are brought about by shifting of the earth's tectonic plates as in earthquakes. Slate, marble, and schist are metamorphic rocks.

Stone is steadily gaining importance as an engineering material in the construction and monument industries. This is because of the emergence of modern abrasive materials and techniques that make stone more economic to cut, size, and polish. Of all the rock categories, granitic igneous rocks offer the best resistance to environmental weathering and air pollution. Granites are available in a wide range of colors and patterns. They can be shaped accurately and polished to a lustrous finish. Hence, they are a popular choice with architects for external cladding of buildings and in the monument industry. An outstanding example of the latter is the Vietnam Veterans Memorial in Washington, D.C.

Concrete is a mixture of portland cement, sand, crushed stone, and water. A typical recipe is (1:2:4):

- 1% portland cement
- 2% sand
- 4% crushed stone

Just enough water is used to produce a consistency suitable for transporting it into the form or mold. The less water that is used, the greater the ultimate strength.

Portland cement is made by mixing 80% $CaCo_3$ with 20% shale and heating (calcining) the mixture in a kiln. The calcined material in the form of lumps is finely ground to give portland cement. Portland cement is also used in mortar (cement and sand) for cementing bricks and other structural shapes together.

Concrete has considerable strength in compression, but is very weak when loaded in tension. When used in beams stressed in bending, steel must be cast in the tensile side of the beam (as discussed in Ch. 4).

6.0 COMPOSITES

Composites are combinations of two or more materials to give a product that has resultant properties different than those of the constituents. Concrete is an example of a composite where sand increases the compressive strength of cement and the aggregate tends to arrest crack growth. Frequently, a strengthening constituent is added to a ductile matrix to produce a composite that has a desired combination of strength and ductility. The strengthening agent may be in the form of particles, short random fibers (whiskers), long continuous oriented threads, or material in sheet form. Examples of composites are:

- Glass, metal or ceramic particles or whiskers in a polymer, metal, or ceramic matrix
- Glass, graphite, SiC, or boron threads wound on a mandrel in a polymer, metal, or ceramic matrix
- Alternate layers of metal or ceramic foil with layers of metal or polymer

An interesting material is a honeycomb composite. This consists of a honeycomb structure of metal or cardboard filled with a polymer or foamed polymer. Figure 9.7 illustrates the method of forming the honeycomb. Thin sheets of metal or cardboard are bonded to form what is called a *hove.* When the hove is extended, a pattern develops that resembles the hexagonal comb in a beehive. The voids are filled with a polymer, foamed polymer, or other material, and continuous sheets are bonded to the upper and lower surfaces. This produces a light, rigid and strong structure. The honeycomb construction is used for aircraft parts in the high technology area and for light weight inexpensive interior doors and panels at the consumer end of the spectrum of products. For doors, the hove may be made of thin cardboard, the voids being filed with a foamed plastic and with upper and lower cover sheets of thin wood veneer.

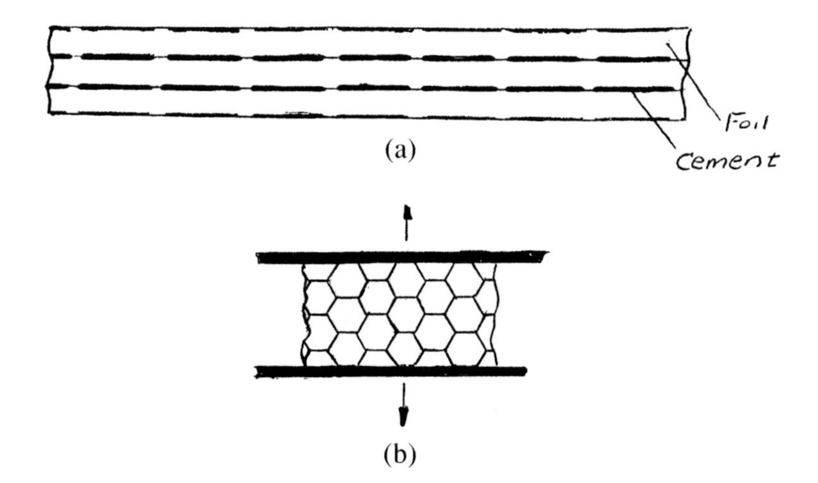

Figure 9.7. *(a)* Magnified representation of hove consisting of flat sheets bonded together as shown and *(b)* hexagonal honeycomb structure obtained by extending hove.

7.0 MATERIALS PROCESSING

7.1 Introduction

Materials processing has to do with the conversion of materials into products of desired shape, integrity, and properties. This involves several well established methods including:

- Casting
- Forming
- Removal of unnecessary material
- Powder metallurgy
- Joining

7.2 Casting

Casting involves the transfer of liquid material into a hollow cavity having the desired shape, where the material solidifies before being extracted from the mold. The cost of a casting operation is directly related to the cost of melting (which depends on the melting temperature) and the cost of the mold. The lower the melting temperature, the lower the energy and furnace costs involved, and the less refractory and costly the mold. Cast iron castings are relatively less expensive than steel castings, primarily due to a difference in melting temperature. Large cast iron castings are usually made using a split sand mold. The upper half of a wooden model (pattern) is first suspended in a box and packed with sand containing a binder. Then the lower half is placed in a second box and packed with sand. When the two boxes are clamped together and passages for transporting liquid in and gas out are provided, the mold is ready for use. After solidification, the mold is broken away and the casting is cleaned by sand blasting. Auxiliary appendages are removed by cut-off grinding.

A great deal of technology is involved in producing high quality castings. This consists of insuring that flow rates are high enough so that the mold is filled before solidification (this involves heat transfer), yet not too high so that the mold material is entrained in the casting due to excessive turbulence (this involves fluid mechanics). The rate of heat extraction from the mold is also important. A rapid cooling rate is desirable from the standpoint of grain size. When a liquid metal is cooled rapidly to a temperature far below the melting point before solidification occurs, there will be a large number of nuclei produced on which crystals grow on further heat extraction, giving rise to a fine grain size. When solidification begins just below the melting point, there will be relatively few nuclei and a coarse grain size results. If heat extraction is rapid, this may cause uneven cooling, hence uneven shrinkage which can lead to internal crack formation.

In addition to sand casting, there are a host of other techniques. These include:

- Permanent mold casting
- Precision casting (lost wax process)
- Evaporative pattern casting (lost foam process)
- Slush casting
- Die casting
- Centrifugal casting

Permanent mold casting is similar to sand casting except that a reusable split metal mold (cast iron, steel, bronze) is used instead of compacted sand. A thin refractory coating (parting agent) of sodium silicate or graphite is often used to ease removal from the mold and to increase mold life.

Precision and evaporative pattern casting are techniques that do not involve a split mold, hence there will be no parting line on the product. In precision casting, a wax part is first produced by casting molten wax into a metal mold. The wax part (or a large number of small parts attached to a central wax column called a tree) is coated by dipping it in a refractory slurry and coating with larger refractory particles to produce a relatively thick layer of stucco. The assembly is heated and the recyclable wax removed. Liquid metal is transported into the cavity, and the refractory material and appendages removed after cooling. Precision casting is mainly used for small parts and parts made from high melting point materials.

Evaporative pattern casting involves producing a foamed polystyrene part for each part cast by the method of Fig. 9.5. The polystyrene part is given a thin refractory coating to improve surface finish, and then packed with refractory sand in a single box. Metal is introduced from the bottom of the mold cavity and converts the polystyrene to a vapor as it rises to fill the cavity. A large central hole must be provided in the polystyrene pattern to enable the gas generated to escape as the pattern burns away. The polystyrene pattern supports the sand until the polystyrene is replaced by metal.

Die casting involves forcing molten metal into a split metal die where it quickly solidifies. The die is opened and the part ejected. This is a high production rate process that is used primarily to produce small intricate parts from low melting materials (usually a zinc alloy) where high strength is not required.

Centrifugal casting involves transporting liquid metal into a split rotating die. Long cast iron water pipes are produced by this technique using rotating steel dies that are only partially filled. A refractory parting agent is usually used after each pipe is cast which enables up to 500 pipes to be cast before the die needs to be reconditioned.

7.3 Forming

There are a number of processes that take advantage of the ductility of metals to change shape by plastic flow. These include:

- Rolling
- Forging
- Extrusion
- Drawing
- Sheet metal forming

Figure 9.8 shows a plate being reduced in thickness by rolling. Both rolls are driven at the same speed. Friction between rolls and plate drive the plate in the direction indicated. A variety of structural shapes (channels, I beams, R.R. rails, etc.) are produced by using shaped rolls. In steel making, cast ingots are hot rolled or hot forged to close shrinkage cracks, and to make the steel more homogeneous (same properties at all points) and more isotropic (same properties in all directions). The surface is improved by grinding or burning away surface defects (called conditioning). The steel may then be cold rolled to improve finish and dimensional accuracy.

Figure 9.9 (*a*) shows a cylinder about to be forged, and Fig. 9.9 (*b*) shows the cylinder after forging with negligible friction. With friction, the forging will be barrel shaped [Fig. 9.9 (*c*)], since the material in contact with the die surfaces will be kept from expanding outward as much as that farther removed from these surfaces. Other more complex forging operations involve shaped die surfaces that may or may not close completely at the end of the stroke.

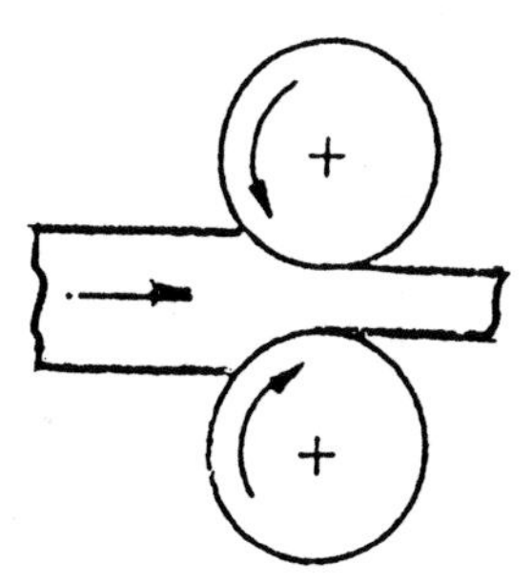

Figure 9.8. Rolling a plate.

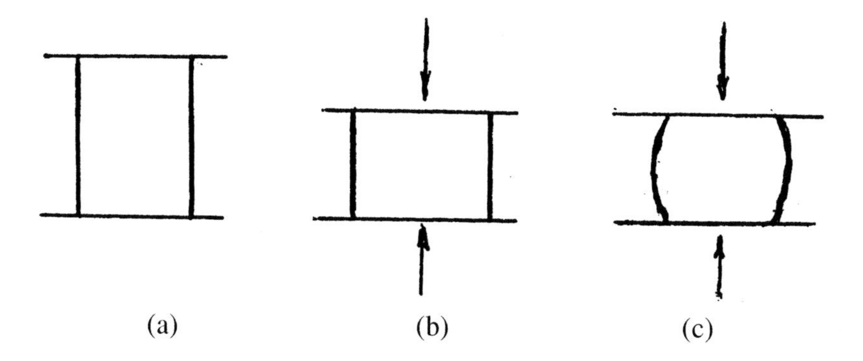

Figure 9.9. Open die forging of cylinder (upsetting); *(a)* Before forging, *(b)* after forging (with no die friction), and *(c)* after forging (with die friction resulting in barrelling).

Figure 9.10 (*a*) shows an extrusion process while Fig. 9.10 (*b*) shows a drawing process. In extrusion, the billet is pushed through the die by a ram, while in drawing, the exiting product is gripped and pulled through the die. Drawing is limited to relatively small reductions in area because otherwise the product may fail in tension. The reduction in area in extrusion may be very much greater than in drawing. Drawing is normally done on a cold billet, but extrusion may be done either hot or cold.

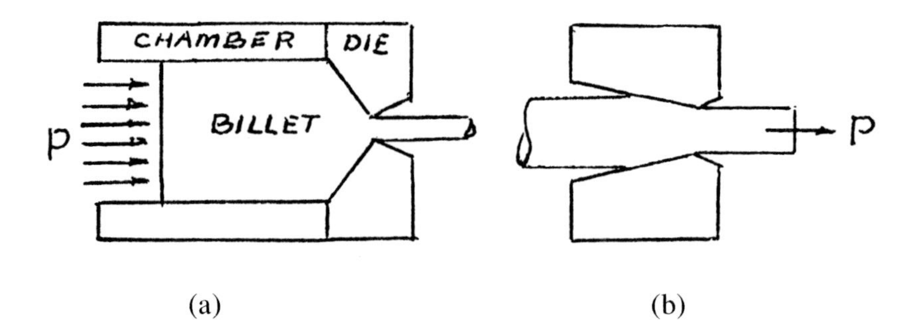

Figure 9.10. *(a)* Extrusion of billet with plunger providing pressure (p) and *(b)* rod drawing with drawing force (P) transmitted through product.

Sheet metal operations involve shearing, bending, and shaping of relatively flat materials in the automotive and appliance industries. Figure 9.11 (*a*) illustrates a shearing operation where the main defect is the formation burrs on the sheared surfaces. Figure 9.11 (*b*) illustrates a bending operation where important considerations are cracking, if the bend angle is too great, and elastic spring back when the bending load is removed. When large sheets are changed in shape by deforming between dies, the major problems are crack formation, if the combination of strains exceed the ductility of the material being used, and surface roughing due to inhomogeneous plastic flow at large plastic strains.

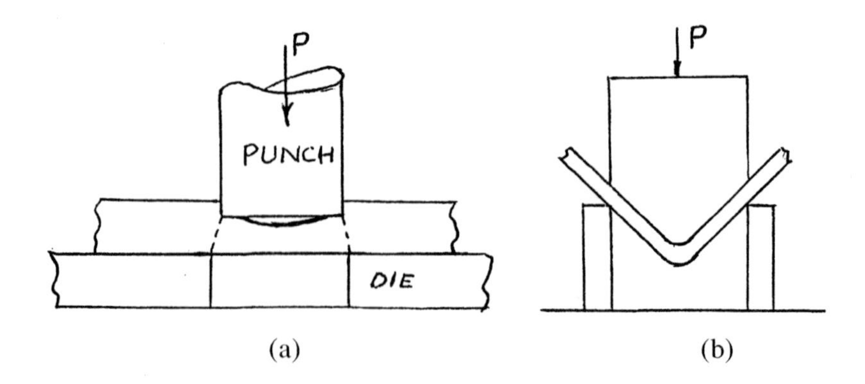

Figure 9.11. *(a)* Sheet metal shearing operation producing blank (disc) and punched sheet and *(b)* production of 90° bend in metal sheet.

7.4 Stock Removal Operations

Many intricate shapes are generated by cutting and grinding which are chip forming operations designed to remove unwanted material. In cutting, a very hard tool with a sharp edge is forced into the workpiece under controlled conditions. Chips are produced by a complex combination of fracture and plastic flow in shear. The machine that controls the motion of the work and tool is called a machine tool. The three most widely used machine tools for cutting are the lathe, the milling machine, and the drill press. There are many more types of machine tools. Figure 9.12 (*a*) shows a typical turning operation, Fig. 9.12 (*b*) is a typical milling operation, while Fig. 9.12 (*c*) shows a twist drill producing a hole.

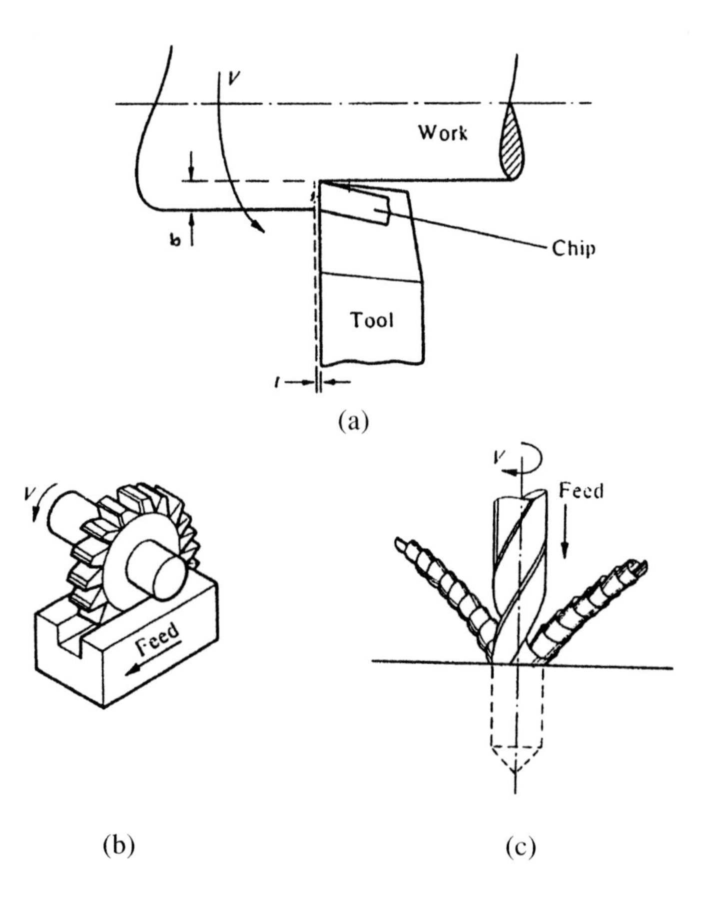

Figure 9.12. *(a)* Turning operation (V = cutting speed, b = depth of cut, t = feed/revolution), *(b)* milling of slot, and *(c)* twist drill producing a hole.

Traditionally, cutting is performed on unhardened material, often as a roughing operation, where the bulk of the unwanted material is removed. This may be followed by a grinding operation after hardening, if greater dimensional accuracy and surface finish are required than can be obtained by cutting alone.

Grinding is performed by hard refractory particles (Al_2O_3 or SiC, called common abrasives, and cubic boron nitride or diamond, called superabrasives) producing many small chips at high speed. The tool normally employed in grinding is a grinding wheel. This consists of abrasive particles bonded together to form a disc with a central hole. The bonding material may be ceramic (vitreous bond), a polyset (resin bond), or metal. Figure 9.13 shows a surface grinding operation at (*a*), an internal grinding operation at (*b*), and an external grinding operation at (*c*). All of these employ grinding wheels that have a cylindrical shape. In addition to these three basic grinding operations, there are many more, some of which involve wheel surfaces of complex shapes that may be reproduced in the work by plunging the wheel into the work (called form grinding).

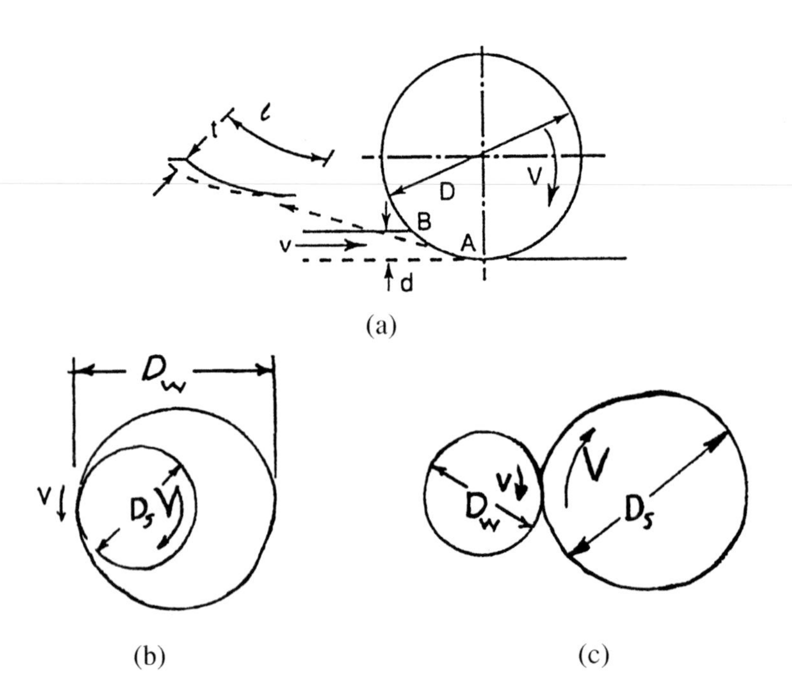

Figure 9.13. Representative grinding operations; *(a)* surface grinding (undeformed chip in inset), *(b)* internal grinding, and *(c)* external grinding. (V = wheel speed, v = work speed, D_s = wheel diameter, D_w = work diameter)

In addition to cutting and grinding, there are other methods of generating a required geometry by stock removal. Not all of these are of a mechanical nature like cutting and grinding. Some are electrical in character while others are thermal and still others are of a chemical nature. An engineer working in this field must be a generalist who has a good acquaintance with all of the engineering sciences, and an ability to solve problems quickly as they develop on the shop floor.

7.5 Powder Metallurgy

This involves parts made by mixing metal powders, compressing them in dies, and holding them just below the lowest melting point of the mixture. The powders join together by diffusion at points of contact to produce a solid. This process of consolidating powders is called *sintering*. Pressures required in the formation of the pre-sintered compact are up to 10^5 psi (690 MPa). Lower values of pressure are used for softer metals such as aluminum, and higher pressures are used for harder metals such as steel. The time required at temperature will vary from a few minutes to several hours depending on the rates of diffusion of the materials involved.

One of the unique characteristics of powder metallurgy is the ability to produce porous parts. This has been used to produce metal filters and porous bearings. In the latter case, the sintered part is vacuum treated to remove all air and then submerged in oil to fill the pores. This produces a bearing capable of operating for the life of a product without further application of lubricant. Nonmetal powders are sometimes mixed with metal powders to produce particle metallurgy parts.

7.6 Joining

Parts may be made by bonding several components together. A large number of bonding techniques are used including:

- Use of adhesives (epoxy or other polymers)
- Soldering (Sn/Pb, Pb/Ag, Cd/Ag)
- Brazing (torch, furnace)
- Welding (gas, arc, resistance, friction, ultrasonic)
- Mechanical (bolts, rivets, screws, seaming, shrink fit)

In *soldering* or *brazing*, two well-fitting surfaces to be joined are heated and molten solder or brazing alloy applied to an edge. Surface tension draws the molten metal between the close fitting surfaces. If the metals to be joined tend to form an oxide film in air, a flux may be applied which reacts with oxide and removes it so that solder or brazing alloy may wet the deoxidized surfaces and form a strong bond. *Welding* involves melting surfaces to be joined, and may or may not employ a filler metal. The large number of welding methods available results from the variety of methods of applying heat.

Ultrasonic welding involves pressing two surfaces together and vibrating one relative to the other at a frequency above that detectable by the human ear (>18,000 Hz). The motion breaks up the oxide present enabling clean surfaces to come together to form strong bonds. Melting is normally not involved in ultrasonic welding.

Seaming, employed in sheet metal work, forms a joint by folding the edges of two very ductile metals together to form a seam consisting of two interpenetrating U bends.

PROBLEMS

9.1 Name three structural arrangements involving carbides in steel.

9.2 Name three elements added to steel to:

a) Increase hardenability

b) Increase wear resistance

c) Increase hot hardness.

d) Decrease defects due to entrained air

e) Improve machinability

9.3 Name three elements that improve steel by:

a) Forming carbides

b) Going into solid solution in ferrite

c) Forming insoluble particles in ferrite

9.4 Should stainless steel have a high or a low carbon content? Why?

9.5 What is the chief role of Ni in stainless steel?

9.6 What are the two most important alloying elements in tool steel?

9.7 Cast iron has a carbon content that is lower than that for steel (T or F)?

9.8 Is grey cast iron harder and more brittle than white cast iron?

9.9 What is precipitation hardening?

9.10 Name five hardening mechanisms.

9.11 Ball bearings are assembled from through-hardened elements while roller bearings use surface hardened parts. Which of these will have the greatest shock resistance?

9.12 Name three ways of softening metals.

9.13 List five characteristics of titanium alloys.

9.14 List the three basic types of polymers.

9.15 What does vulcanization of rubber involve?

9.16 What is the chief difference between a glass and a ceramic?

9.17 What is the main constituent of glass?

9.18 What is the highest melting glass?

9.19 Name three types of rock and give an example of each type.

9.20 What is a typical composition of concrete?

9.21 Is concrete stronger in tension or in compression?

9.22 Name the principle materials processing procedures.

9.23 Name the principle metal forming operations.

9.24 What is the principle difference between extrusion and drawing?

9.25 Name the three most important machining operations in order of importance.

9.26 Three important types of grinding are?

9.27 a) Two important common abrasives are?

b) Two important superabrasives are?

9.28 Three types of grinding wheel bonds are?

9.29 Name and define three joining methods.

10

Electrical Engineering

1.0 INTRODUCTION

Electrical engineering is associated with the application of electrical and magnetic principles to the design, analysis, and performance of devices and services. This covers a very broad set of subtopics. In most technical universities, the largest engineering department is that concerned with electrical-related subjects.

The entire field may be divided into:

- Power generation and distribution
- Electronics and communications
- Systems analysis and control

Subjects that are basic to all of these endeavors include:

- Principles of conduction
- Circuit theory
- Electromagnetic theory
- Performance of electrical components

Since the design and application of electronic computers had their origin in electrical engineering departments, computer engineering subjects are sometimes taught in electrical engineering departments.

However, as computer activity has increased, computer science and engineering are usually covered in separate departments in most technical universities. In this chapter, an overview of a few of the major topics of importance to electrical engineering is given.

2.0 HISTORICAL BACKGROUND

Natural magnets (Herculean stone or lodestone) were known in ancient times, but there was little attempt to explain their behavior. Their properties were described in mystical terms. It was not until 1269 that Peregrine discovered magnetic lines of force and the existence of poles. Further contributions to magnetism were made by W. Gilbert (1540–1603) and Descartes (1596–1650). Boyle demonstrated in 1744 that air plays no role in magnetic attraction by conducting experiments in a vacuum.

Static electricity was known in early times, but also was not understood. The first electrical machine was developed by von Guericke in 1672. This was followed by improved machines in the early 1700s that increased interest in the phenomenon. An outcome of this was the recognition by DuFay (1692–1739) that two types of electricity exist (+ and -).

However, it was not until Franklin (1706–1790) focused his attention on electricity that real progress began to occur. In 1791, Franklin published his book *Observation on Electricity Carried Out at Philadelphia* which had an important influence on research in Europe throughout the next century. Franklin introduced the term "charge" and described electricity as a single fluid consisting of two particles of opposite charge. He also introduced the concept of conservation of electrical charge and explained the sparks that occur when oppositely charged bodies are brought close together. Franklin's best known contribution is the explanation of lightning which led to the invention of the lightning rod—a pointed metal rod attached to a steeple grounded to the earth, designed to draw static electricity out of clouds before a larger damaging strike could occur. In 1752, he performed his famous kite experiment to demonstrate the principle.

Franklin, a very practical person, is a good example of a case where curiosity led to invention. He shared the view of Bacon who is reported to have said that all good science must have practical results.

Electrical concepts and theory are of relatively recent origin compared with other fields of science. While there had been a few important findings mainly concerned with static electricity before 1800, most of the important electrical developments occurred during the 19th and 20th centuries. A significant result was the discovery in 1800 by Alessandro Volta, a professor at the University of Pavia, of the first galvanic cell. This consisted of an alternating pile of zinc, moist cardboard, and copper discs. The zinc was found to become positively charged and the copper negatively charged. After the passage of current, the zinc was oxidized. This was the first example of the conversion of chemical energy into electrical energy, and forms the basis of modern batteries.

Volta's experiments were repeated and verified by a number of people including the chemist Davy. The development of electrical theory and technology was well on its way. For the next forty years, many significant experiments were performed and many discoveries made in the laboratories of Europe. Very important contributions were made by the following:

- Ampere (1775–1836)
- Oersted (1777–1851)
- Ohm (1787–1854)
- Faraday (1791–1867)

Faraday was an extremely talented experimentalist who made many important contributions not only to electrical technology, but also to chemistry and material science. A few of these contributions include:

- A current-carrying conductor in a magnetic field is subjected to a transverse force
- A current is generated in a conductor moving through a magnetic field (electromagnetic induction)
- A current is generated in a conductor in a magnetic field when the strength of the field is charged
- A strong momentary current is produced when a circuit is interrupted (self-induction)
- The discovery of benzene
- The isolation of chlorine and other gases
- The production of high lead glass having a high refractive index for use in optics

The first of these discoveries led to the electric motor and the second and third to two types of electrical dynamos (electrical generators).

After 1840, physicists largely took over and many contributions to the theory of electrical science were made. Important names during the second half of the 19th century include:

- W. Thomson (Lord Kelvin) (1814–1907)
- Stokes (1819–1903)
- Kirchoff (1824–1887)
- Boltzmann (1844–1906)
- Hertz (1857–1894)
- Helmholtz (1821–1894)
- Lorentz (1853–1928)
- J. J. Thomson (1856–1940)

Beginning about 1900, the present theory of the atom was developed. Many important electrical and electronic devices have appeared throughout the 20th century.

3.0 ELECTRICAL CHARGE, CURRENT, AND POTENTIAL

A body may have a net positive or negative electrical charge which will be an integral multiple of the charge on an electron. The unit of charge is the coulomb (C) and the charge on a single electron is 1.6×10^{-19} C (Millikan, 1909). The mass of an electron is 0.911×10^{-27} gms and its radius is about 2.82×10^{-13} cm. Bodies having like charges tend to repel each other, while those with unlike charges attract each other. The forces of attraction and repulsion obey Coulomb's law (1785):

Eq. (10.1) $$F = 8.99 \times 10^{9}(q_1 q_2)/d^2), \text{ Newtons}$$

where q_1 and q_2 are charges on the two bodies in coulombs and d is the distance between them in meters.

Static electricity involves bodies that are positively or negatively charged. When two bodies are rubbed together, the electrons may be removed from one (giving a positively charged body) and transferred to the other (giving a negatively charged body). For example, when glass is rubbed by silk, glass becomes positively charged and silk negatively-charged. Similarly, rubber rubbed with fur gives negatively charged fur and positively charged rubber. The electrons transferred to silk or fur tend to remain static even when one end of the material is in contact with a positively charged body and the other end is in contact with a negatively charged body (i.e., subjected to a potential difference). Fur and silk are termed nonconductors (insulators) since electrons do not tend to flow toward the positive terminal when these materials are subjected to a potential difference. Metals such as copper and aluminum, on the other hand, are termed conductors since their many free electrons will tend to flow from a negative terminal to a positive one. When static charges recombine, sparks may be generated. This is the source of lightening, as Franklin demonstrated many years ago.

The time rate of change of electrical charge through a conductor is called current (I) and the unit of current is the ampere (A). One ampere corresponds to the flow of one coulomb of charge per second. A stroke of lightening involves the flow of a very high current for a very short time (~1,000 A for about 10^{-5} sec). The potential difference responsible for current flow in a conductor is measured in volts (V). A positive flow of current corresponds to flow from a plus to a minus terminal.

A semiconductor is a material having electrical conductivity between a conductor and a nonconductor. Silicon, which is in the fourth column of the periodic table (valence = 4), is normally a nonconductor. It may be converted to a semiconductor by diffusing a small amount (~1 part in 10^6) of boron (valence = 3) or phosphorous (valence 5) throughout its structure. This is called *doping*. When boron is the dopant, it is called a positive type (p-type) semiconductor; but when phosphorous is the dopant it is called an n-type semiconductor. When a potential difference (voltage) is applied across an n-type semiconductor, the unattached electrons where phosphorous atoms are located move toward the positive terminal. When a boron atom is in a p-type semiconductor, there is an unfilled bond site called a *hole*. Holes tend to act as positively charged particles and move toward the negative terminal when a potential difference is applied.

A liquid that conducts current is called an *electrolyte* while one that does not is called a *dielectric*. When a crystal such as copper sulfate ($CuSO_4$) dissolves in water, it yields Cu^{++} and SO_4^{--} ions. When a potential difference is applied to immersed electrodes, positive ions migrate to the negative terminal (cathode) and negative ions to the positive terminal (anode). Upon reaching the cathode, Cu^{++} ions accept two electrons and deposit as a neutral atom of copper. This is called *electroplating*. If a copper part is made the anode, material will be removed (deplated). This is called electrolytic machining. By masking the cathode or anode, different shapes may be generated by differential plating or deplating. When material is differentially deposited to form a shape, this is called electrolytic forming.

The flow of current in a conductor is analogous to the flow of fluid in a pipe where the current is equivalent to the rate of fluid flow, and the potential difference is equivalent to pressure drop along the pipe. Just as work is done when lifting a weight in mechanics, work (dW) is done when voltage (V) causes a displacement of a charge (dq).

Eq. (10.2) $$dW = Vdq$$

where work is in joules when V is in volts and dq is in coulombs.

Power (P) is the rate of doing work, hence under steady state conditions, (constant V and I):

Eq. (10.3) $$P = VI$$

where P is in watts (W) when V is in volts and I is in amps. A watt is one joule per second.

4.0 SOURCES OF EMF

The production of an electrical potential capable of causing a flow of current is generally accomplished by converting another form of energy into electrical energy. This may involve:

- Thermal energy (from fossil fuel, geothermal, or thermonuclear) into mechanical energy and then into electrical energy
- Potential or kinetic energy (hydro, wind, and tidal) into mechanical energy and then into electrical energy
- Solar energy into electrical energy
- Chemical energy into electrical energy (as in batteries)

The first of these is the most important, and involves generation of high-pressure steam or high velocity products of combustion to drive an engine of some sort coupled to an electrical generator. The second involves a turbine coupled to an electrical generator. The third involves use of a solar cell to convert energy in the form of thermal radiation directly into electrical energy. The last of these involves use of non-rechargeable (primary) or chargeable (secondary) batteries.

Two types of primary batteries in common use are the carbon-zinc battery and the amalgamated zinc-mercuric oxide alkaline battery. The first of these has a carbon anode (+) mounted in a zinc containercathode (-) with a manganese-carbon paste electrolyte between the two. Alkaline batteries have an amalgamated zinc anode (+) and a mercuric oxide cathode (-) with a KOH electrolyte between the electrodes. Both of these types of primary cells have a no load emf of about 1.5 volts. An alkaline battery has about four times the life of a carbon-zinc battery, but is correspondingly more expensive. These are termed dry cells since the electrolyte is in the form of a paste, and there is normally no leakage. Most small (flashlight) batteries are of this type. Nickel (+) -cadmium (-) batteries are small dry cell batteries that may be recharged by reversing the direction of current flow through the battery. These batteries contain no free alkaline electrolyte and do not require venting of gases; thus these are well suited for use in power tools and other portable equipment.

The most common rechargeable battery is the one used in automobiles. This has two lead electrodes—an anode coated with PbO_2 and a cathode coated with PbO. The electrolyte is a H_2SO_4 solution. The controlling chemical equation for this battery is:

Eq. (10.4)

$$\underset{\text{(anode)}}{Pb + PbO_2} + \underset{\text{(cathode)}}{Pb + PbO} + H_2SO_4 \underset{\text{charge}}{\overset{\text{discharge}}{\rightleftarrows}} 2PbSO_4 + 2H_2O$$

Current flows through an external circuit from the anode (+) terminal to the cathode (-) of a battery at essentially constant voltage. This is called a direct current (DC).

The conversion of rotational mechanical energy into electrical energy involves a generator that usually produces current that varies from plus to minus sinusoidally at a relatively high frequency. Such a current is called an alternating current (AC). Electrical generators are discussed briefly later in this chapter.

5.0 DIRECT CURRENT

When a direct current flows through a conductor, energy is dissipated in the form of heat, and there is a drop in voltage. The voltage drop (V) in the direction of current flow is proportional to the current (I).

Eq. (10.5) $$V = IR$$

The proportionality constant (R) is called resistance and the unit of resistance is the ohm (Ω). Equation (10.5) is known as Ohm's law (1827). The corresponding loss of power is:

Eq. (10.6) $$P = IV = I^2R$$

where P is in watts (W) when V is in volts, (V), I is in amps (A) and R is in ohms (Ω).

When electric power is transmitted over long distances, it is important that this be done at high voltage and low current to minimize power loss in the transmission line.

An element that absorbs energy, such as a resistor, is termed a passive unit while one that supplies energy is referred to as a source. Figure 10.1 shows a simple flashlight circuit. In this case, the power supply consists of two D size batteries (each of 1.5 volts and 1.25 in. dia. × 2.50 in. length) in series. The load (a bulb) is a fine wire (resistance $R = 10\ \Omega$ in an inert atmosphere) that emits heat and light when current passes through it. From Ohm's law, the current that will flow though the bulb will be

$$I = V/R = (1.5 + 1.5)/10 = 0.3 \text{ A}$$

From Eq. (10.6), the power consumed will be:

$$P = VI = RI^2 = 10(0.3)^2 = 0.90 \text{ W}$$

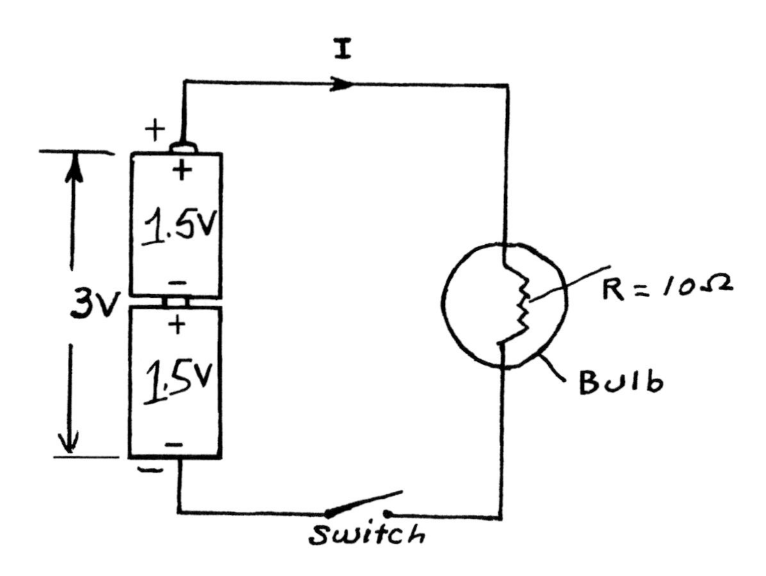

Figure 10.1. Flashlight circuit.

When resistances are connected in series with the same current (I) flowing through them [Fig. 10.2 (*a*)], they may be replaced by an equivalent resistance (R_e) that is the sum of the resistances in the series:

Eq. (10.7) $\quad R_e = R_1 + R_2 + R_3$

When resistances are connected in parallel as shown in Fig. 10.2 (*b*), they may be replaced by an equivalent resistance (R_e) that is the reciprocal of the sum of the reciprocals of the individual resistances:

Eq. (10.8) $\quad 1/R_e = 1/R_1 + 1/R_2 + 1/R_3$

or

Eq. (10.9) $\quad R_e = (R_1 R_2 R_3)/(R_1R_2 + R_2R_3 + R_3R_1)$

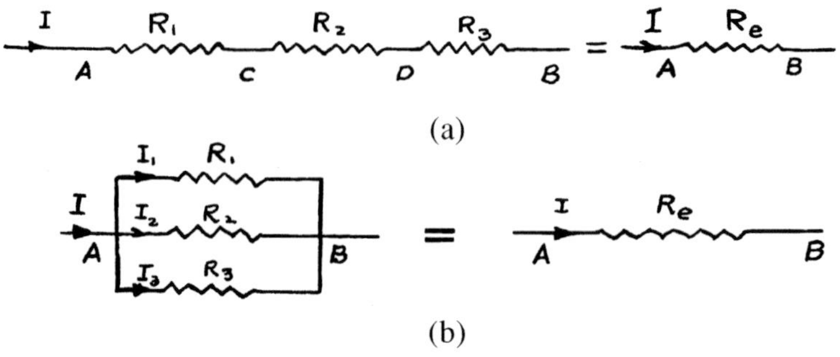

(a)

(b)

Figure 10.2. Resistances *(a)* in series and *(b)* in parallel.

These relations follow from the fact that the potential difference from A to B in Fig. 10.2 (a) and (b) must be the same with several resistances as for the equivalent resistance (R_e).

For resistances in series:

Eq. (10.10) $$E_{AB}/I = R_e = E_{AC}/I + E_{CD}/I + E_{DB}/I = R_1 + R_2 + R_3$$

For resistances in parallel:

Eq. (10.11) $$I_e = E_{AB}/R_e = I_1 + I_2 + I_3 = E_{AB}/R_1 + E_{AB}/R_2 + E_{AB}/R_3$$

or

$$1/R_e = 1/R_1 + 1/R_2 + 1/R_3$$

The resistance of a conductor of area (a) and length (ℓ) will be:

Eq. (10.12) $$R = \rho\ell/a$$

where ρ is a material constant (specific resistance with units $[\Omega/L]$ for a given conductor at a given temperature. For pure Cu at 20°C, ρ = 1.72 μΩcm and for pure Al at 20°C, ρ = 2.83 μΩcm. Resistance of Cu and Al increases about 0.004 Ω per °K.

6.0 DIRECT CURRENT CIRCUIT ANALYSIS

Figure 10.3 shows a simple DC circuit having two sources and two resistors (R_1 and R_2). The no load (0 current flow) potential of the sources are E_1 and E_2 and their internal resistances are r_1 and r_2 respectively. The positive and negative terminals of the sources are indicated (+) and (-) respectively. Assumed directions of current flow in each branch of the circuit are indicated by I_1, I_2, and I_3.

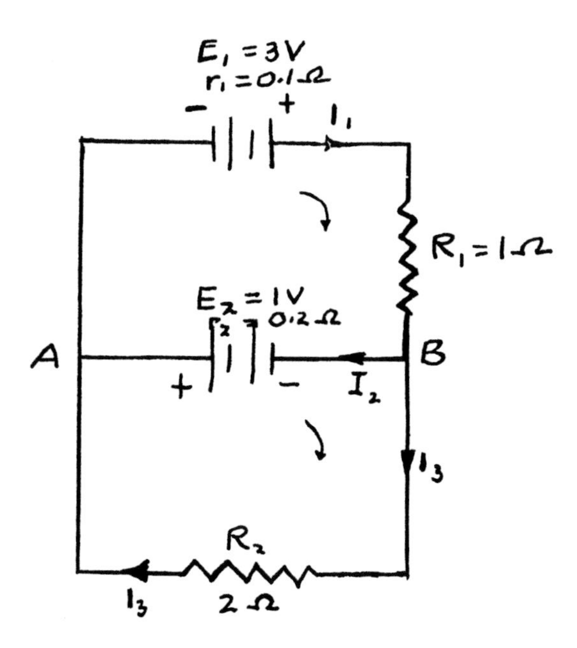

Figure 10.3. Simple DC circuit (r = resistance within a source, and R = an external resistance).

Kirchoff derived the following two laws (1845) based on Ohm's law that are useful in DC circuit analysis:

- In any branching network, the algebraic sum of the currents in all wires that meet at a point (a node) will be zero. This law will be designated KIL. Current flow into a node is considered plus, while current flow from a node is minus.

- The sum of all potential differences around a loop will be zero. This law will be designated KVL. Potential differences for a source are positive when the direction taken around the loop is from minus to plus within a source. Potential differences associated with resistances are negative when the direction taken around a loop opposes the assumed direction of current flow.

The direction of the three currents shown in Fig. 10.3 has been arbitrarily assumed. If any turns out to be negative, this means that the current is actually in the opposite direction to that assumed. The three currents may be found by application of Kirchoff's laws.

Applying KIL to the node at point *B* in Fig. 10.3:

Eq. (10.13) $$+I_1 - I_2 - I_3 = 0$$

Applying KVL to the upper loop in a clockwise direction:

Eq. (10.14) $$3 - 0.1I_1 - I_1 + 1 - 0.2I_2 = 0$$

or

$$4 - 1.1\,I_1 - 0.2I_2 = 0$$

Applying KVL to the lower loop in a clockwise direction:

Eq. (10.15) $$2I_3 - 1 + 0.2I_2 = 0$$

Solving Eqs. (10.13)–(10.15) for I_1, I_2, and I_3:

$$I_1 = 3.05 \text{ A}$$

$$I_2 = 3.23 \text{ A}$$

$$I_3 = -0.18 \text{ A}$$

This means that energy is being provided by the system at source one at the following rate:

$$P = E_1 I_1 = (3)(3.05) = 9.15 \text{ W}$$

Energy is being supplied to the system at source two at the following rate:

$$P = E_2 I_2 = (1)(3.23) = 3.23 \text{ W}$$

or, at a total rate of 12.38 W.

At the same time, energy is being dissipated in the entire system and converted to heat at the following rates:

$$\text{At source one: } I_1^2\, r_1 = (3.05)^2(0.1) = 0.93 \text{ W}$$

$$\text{At source two: } I_2^2\, r_2 = (3.23)^2(0.2) = 2.09 \text{ W}$$

$$\text{At } R_1\text{: } I_1^2(R_1) = (3.05)^2\,(1) = 9.30 \text{ W}$$

$$\text{At } R_2\text{: } I_3^2(R_2) = (0.18)^2(2) = 0.06 \text{ W}$$

$$\text{Total} = 12.38 \text{ W}$$

Many DC circuit problems are much more complex than that of Fig. 10.3, but the principles are the same.

7.0 MAGNETISM

Magnets are metals that have small groups of atoms (called domains or micromagnets) oriented so that collectively they produce a magnetic field. If a wire is wrapped around a cylinder of iron or low carbon steel in the form of a helix and current is passed through the wire as shown in Fig. 10.4, a magnetic field (B) develops. North (N) and south (S) poles will be established as shown for the direction of current flow indicated. Such an iron core and winding is called a *solenoid*. If N poles of two solenoids are brought close together, they will repel each other but a N-S combination will attract each other. When current no longer flows through the helical winding, the iron core ceases to be a magnet. If, however, the core is hardened steel or some special alloy such as Alnico (8% Al, 14% Ni, 4% Co, 3% Cu, 31% Fe), it will retain its magnetism after the current stops flowing, and is termed a permanent magnet.

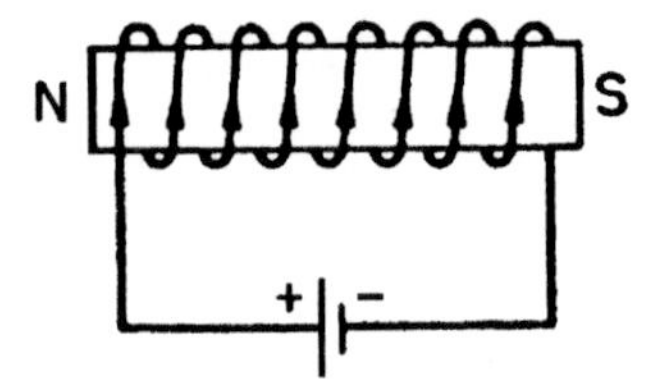

Figure 10.4. Solenoid.

Electromagnetic induction involves the generation of a current when charges move through a magnetic field. If current flows through a conductor into the paper, as shown in Fig. 10.5, a magnetic field will be formed around the conductor in a clockwise direction.

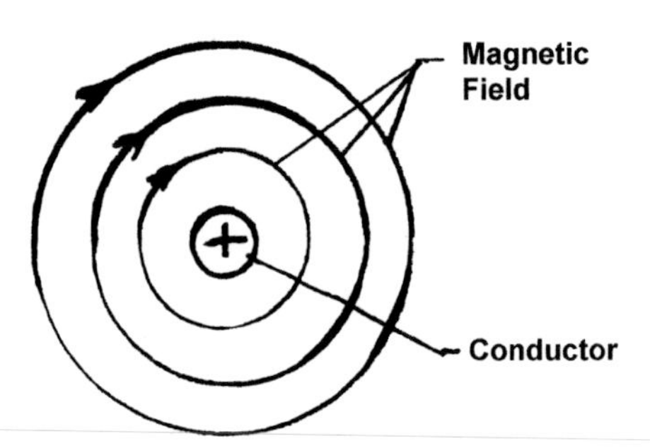

Figure 10.5. Magnetic field around a conductor carrying current into the paper.

8.0 MOTORS AND GENERATORS

If a current carrying conductor is placed in the magnetic field between two solenoids, the two fields will interact to produce an upward force (F) on the conductor as shown in Fig. 10.6. When the direction of current flow is reversed, the direction of the force will be reversed (if the direction of the magnetic flux between the solenoids remains unchanged). This is the basis for the electric motor. A coil of wire carrying current (I) mounted on a rotor (armature) will be caused to rotate, due to the interaction of magnetic fields B_1 and B_2 (Fig. 10.7). If a conductor is caused to move upward in the magnetic field between north and south poles, as in Fig. 10.6, this will cause a current to flow in the wire. This is the basis for an electric generator.

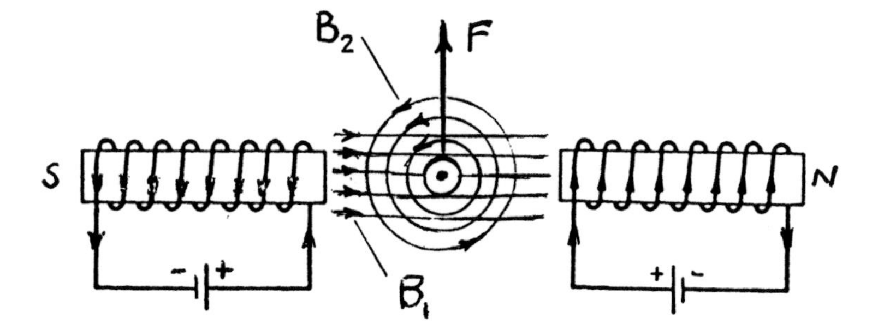

Figure 10.6. Interaction of a magnetic field (B_1) around a conductor carrying current away from paper and magnetic field (B_2) between solenoids, producing a force (F) on conductor.

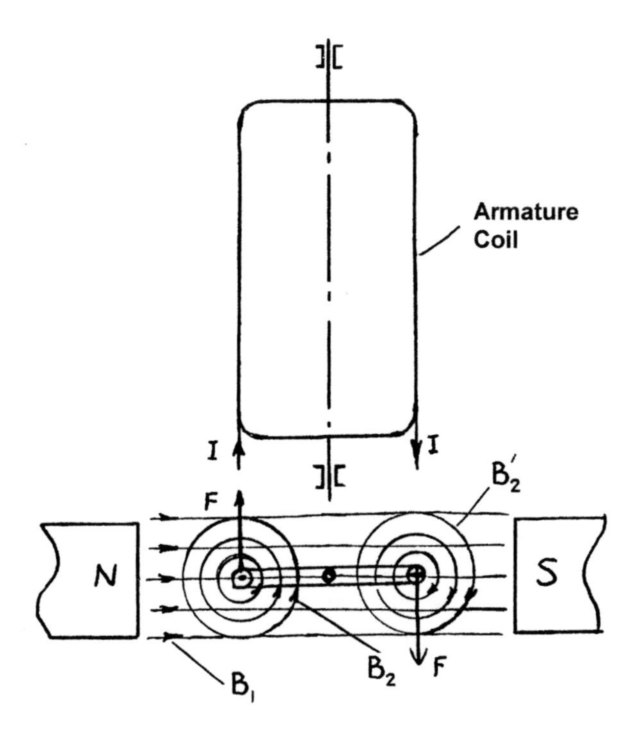

Figure 10.7. Torque to rotate a motor shaft produced by interaction of magnetic fields (B_1) and (B_2).

In the case of both motors and generators, current must be fed or extracted from the rotating coils of wire. This is accomplished by connecting the ends of the coil to insulated collector rings which rotate against stationary brushes (Fig. 10.8). Since it is important that electrical contact resistance and friction be low, collector rings are usually made of copper and brushes of graphite.

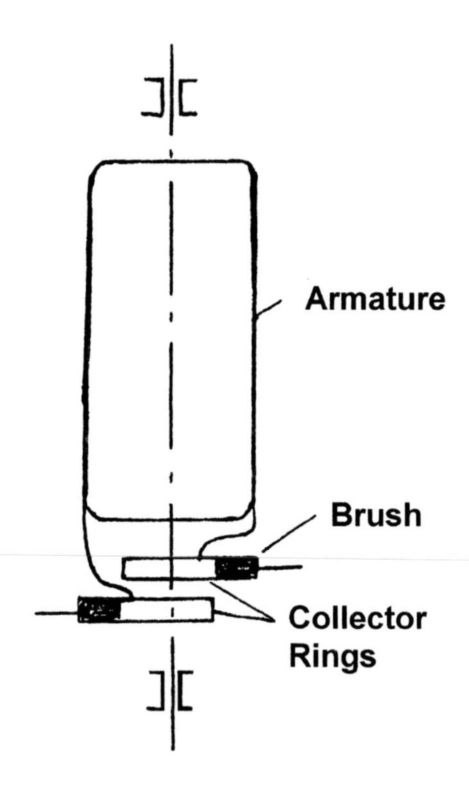

Figure 10.8. Collector rings and brushes to conduct current to or from a rotating armature.

As the coil shown in Fig. 10.7 rotates, forces (F) will vary sinusoidally from a maximum, in the position shown, to zero when the coil has rotated 90° in the motor mode. In the generator mode, the current will vary sinusoidally from a maximum value (I_m) to zero:

Eq. (10.16) $$i - I_m \sin \omega t$$

where ω is the angular velocity in radians per second and t is time in seconds. This means that the current will change direction from plus to minus once during each revolution for a two-pole machine such as that shown in Fig. 10.8. This is called alternating current (AC). For a machine having (P) poles, the AC frequency (f) will be as follows for a rotational speed of N rpm:

Eq. (10.17) $f = NP/120$ cycles/s (Hertz)

In an AC generator, the current induced in the rotating conductors alternates from plus to minus. However, this may be rectified to DC by means of commutator and brushes as illustrated in Fig. 10.9 for a single coil of wire. In Fig. 10.9 (*a*), the wires in the coil are moving at right angles to the magnetic field (B_2). This will cause a current to flow in the direction shown for a counterclockwise rotation. The ends of the wires are connected to a split ring that rotates with the coil. Figure 10.9 (*b*) shows the coil after rotating 90°. The wires in the coil are now moving parallel to field B_2 and no current will be generated. The split ring will also have rotated 90° and the stationary brushes are now opposite the gap. Figure 10.9 (*c*) shows the situation after another 90° rotation of the coil and split ring. The current will again be at maximum. Thus, by use of the rotating split ring (commutator), the current collected by the brushes will vary sinusoidally in magnitude, but will be rectified to have a single direction. In practice, a number of coils are used, as shown in Fig. 10.10, each one connected to a split ring. The collection of split rings constitutes the commutator, and the output current will be not only unidirectional, but also essentially constant in magnitude.

9.0 ALTERNATING CURRENT CIRCUITS

Most large generators produce AC and the electrical power produced is used without rectification. In AC circuits, in addition to resistance, there are two other types of passive units of importance (capacitance and inductance). These are active only when there is a time rate of change of current associated with the unit. The first of these involves the storage of energy due to the build up of charge between conducting plates separated by an insulating material (called a condenser or a capacitor). The second of these AC passive units is due to a voltage drop associated with the magnetic flux induced whenever there is a time rate of change of current in a circuit. The voltage drop due to inductance goes to zero when there is no longer a change of current with time.

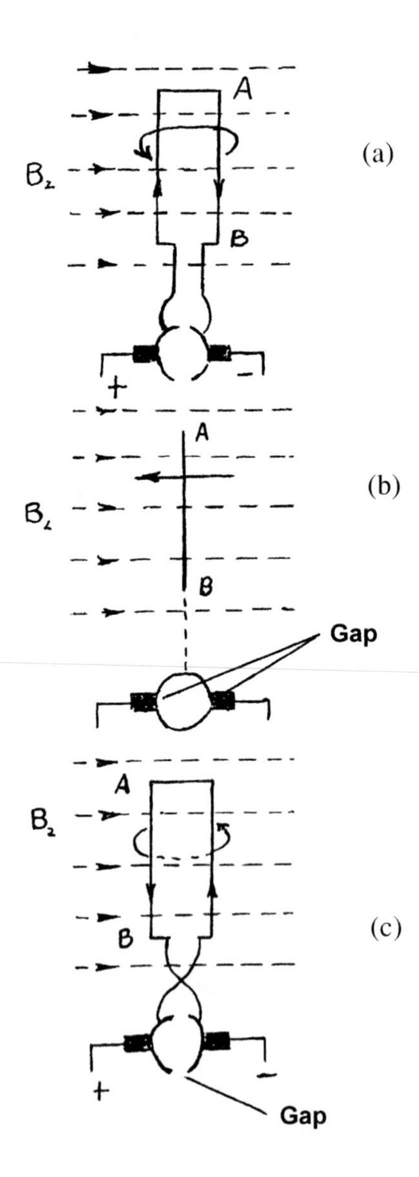

Figure 10.9. Schematic to illustrate principle of split ring commutator for rectifying AC to DC.

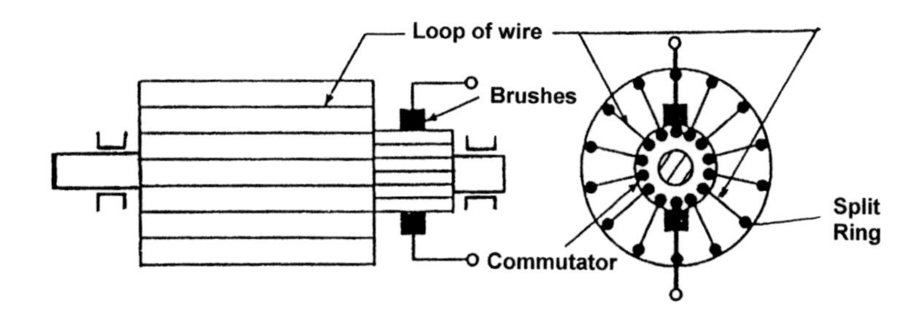

Figure 10.10. Schematic representation of DC generator.

Figure 10.11 shows simple circuits involving the three types of passive resistance to the flow of current: (*a*) resistance, (*b*) inductance, and (*c*) capacitance. In each case, the voltage drop is in the direction of current flow.

When an alternating current flows through a resistance as shown in Fig. 10.11 (*a*), there is no change in phase between current and voltage. However, in the case of Fig. 10.11 (*b*) (inductance only), the current lags the voltage by 90°, while for Fig. 10.11 (*c*) (capacitance only), the current leads the voltage by 90°. In the general case when all three passive entities are present, the current may lag, lead, or be in phase with the voltage depending on the values of resistance, induction, and capacitance.

For an AC generator, the current (i) and emf (e) will vary with time (t) sinusoidally as follows:

Eq. (10.18) $$i = I_m \sin(\omega t)$$

Eq. (10.19) $$e = E_m \sin(\omega t + \theta)$$

where: ω is the angular velocity of the armature in radians/s
I_m and E_m are the maximum values of current and emf
θ is the phase difference between e and i (Fig. 10.12)

The effective value of current (I_e) responsible for power loss is the root mean square value for a half cycle ($\omega t = 0$ to π)

Eq. (10.20) $$I_e = \int_0^{\pi} [id(\omega t)]^{0.5} = 0.707 I_m$$

and the effective value of emf E_e will be:

Eq. (10.21) $$E_e = \int_0^{\pi} [ed(\omega t)]^{0.5} = 0.707 E_m$$

The power (P) for an AC situation will be:

Eq. (10.22) $$P = E_e I_e \cos\theta = [(E_m I_m)/2]\cos\theta$$

where θ is the phase angle between e and i (Fig. 10.12).

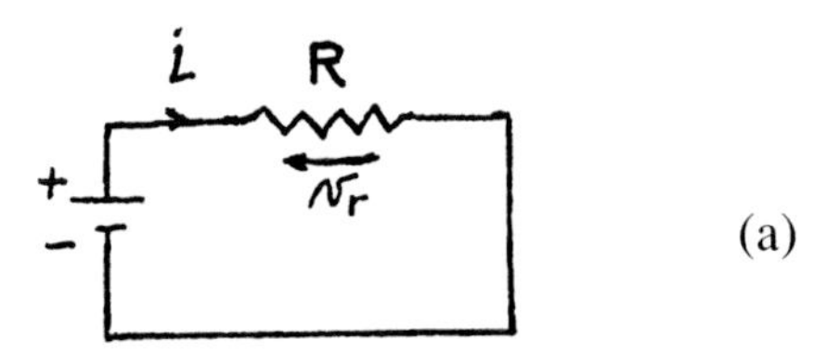

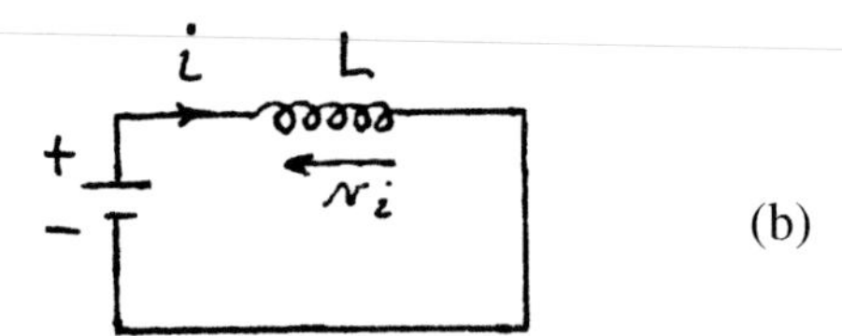

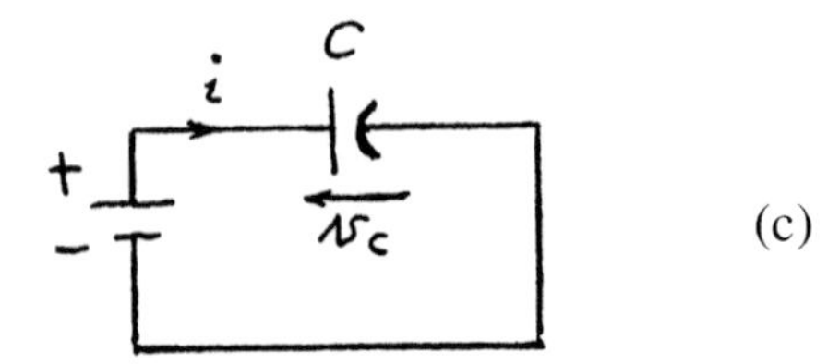

Figure 10.11. Simple circuits involving single impedances due to: *(a)* resistance, *(b)* capacitance, and *(c)* induction.

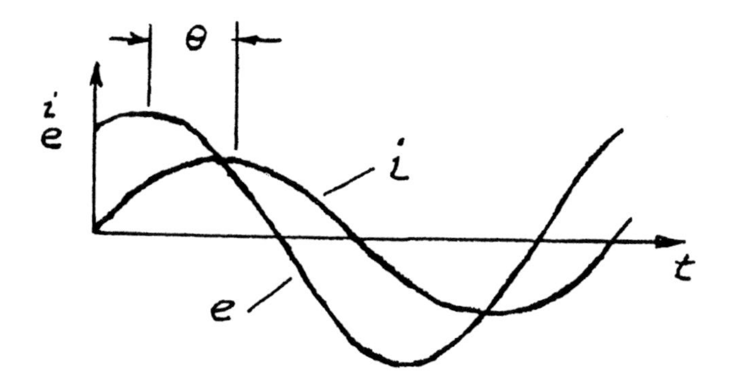

Figure 10.12. Sinusoidal variation of current (i) and emf (e) for an AC generator. Here (i) leads (v) by θ radians (positive θ).

The quantity $\cos\theta$ is called the *power factor*. For high efficiency, it is desirable that the power factor be as large as possible. In most industrial situations, the power factor is negative (i lags v), but it may be improved (raised) by use of a bank of capacitors in parallel with the load.

While analysis of AC circuits is mathematically more complex than analysis of DC circuits, the basic principles are the same involving corresponding applications of Kirchoff's two laws.

10.0 TRANSFORMERS

A transformer transfers AC energy from one circuit to another by means of a coupled magnetic field usually to achieve a change of voltage. Figure 10.13 is a schematic sketch of a transformer. This consists of a steel loop with a primary winding (N_1) turns on the left and a secondary winding of N_2 turns on the right. The main characteristics of such a unit may be approximated by considering what is termed an ideal transformer. This is a unit where resistance in both primary and secondary windings may be ignored, and there is no loss of flux ϕ in the magnetic circuit. For such an ideal transformer:

Eq. (10.23) $\quad V_1/V_2 = N_1/N_2$

Eq. (10.24) $\quad V_1 i_1 = V_2\, i_2$ (no power loss)

hence,

Eq. (10.25) $\quad I_1/I_2 = N_2/N_1$

In actual practice, there will be losses associated with both the electric and magnetic circuits, and the technology of transformer design is far more complex than the simple picture presented above.

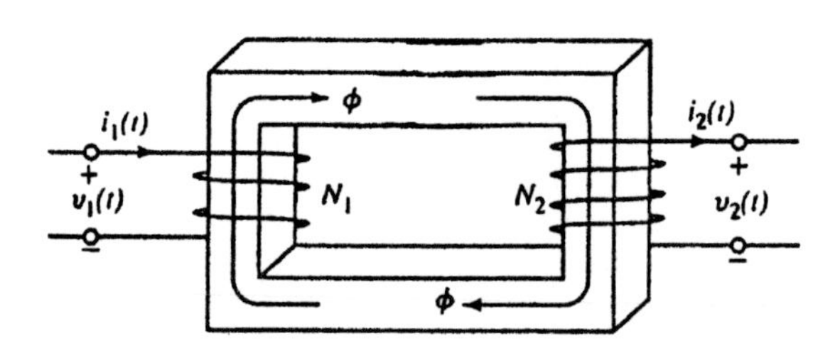

Figure 10.13. Schematic of electrical transformer.

Electronic transformers are small units used in radio and television sets, calculators, computers, and small electrical appliances where batteries are less convenient or too costly. Power transformers are used to convert moderate voltages at which power is generated to higher voltages for transmission and back down to much lower values for distribution and consumption. For example, a 220 kW generator might generate at 1,000 A and 220 V, then by use of a power transformer convert this to 1 A at 220 kV for transmission. When the power reaches the distribution point, it would be transformed back down to 1,000 A at 220 V. Power transformers are usually immersed in a dielectric oil to prevent arcing. Also, the steel core usually consists of many thin sheets of high silicon iron in order to reduce power loss and noise.

11.0 INSTRUMENTS AND MEASUREMENTS

The D'Arsonval galvanometer plays a very important role in electrical measurements. A coil of wire located in a magnetic field produced by permanent magnets is mounted on low friction bearings (Fig. 10.14). Rotation is constrained by coil springs that also conduct current to and from the coil. A pointer attached to the coil moves across a scale that indicates rotary displacement proportional to the current (i), flowing through the coil.

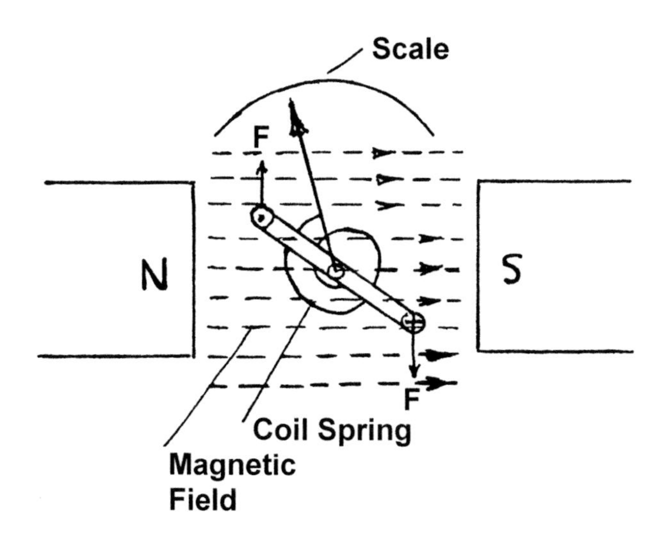

Figure 10.14. D'Arsonval galvanometer.

Galvanometers may be used to measure voltage or current. Figure 10.15 (*a*) shows the arrangement used to measure voltage and Fig. 10.15 (*b*) shows the arrangement for current. For the measurement of voltage, a large resistance (R) is placed in series with the small internal resistance of the galvanometer (r) in order to limit the voltage across the galvanometer coil. The voltage read on the galvanometer scale should then be multiplied by $(R + r)/r \cong R/r$ to obtain the voltage across AB in Fig. 10.15 (*a*).

In the case of Fig. 10.15 (*b*) a shunt of large resistance (*R*) is in parallel with the small internal resistance of the galvanometer (*r*). The current in the external circuit from *A* to *B* in Fig. 10.15 (*b*) will then be:

Eq. (10.26) $I_e = [(R + r)/R]I_g \cong I_g$

By use of a wide range of resistances, *R* in Fig. 10.15, a very sensitive galvanometer may be capable of measuring a wide range of voltages and currents.

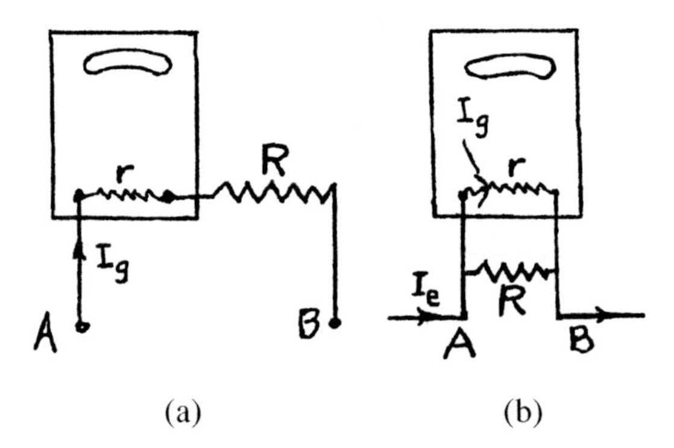

Figure 10.15. Use of a galvanometer for measuring: *(a)* voltage across *AB*, *(b)* current flowing from *A* to *B* (I_e), *R* equaling external resistance for changing peak current through the galvanometer and *r* standing for internal galvanometer resistance.

The Wheatstone bridge (Fig. 10.16) is a convenient means for measuring small values of resistance. No current will flow through the galvanometer (G) if:

Eq. (10.27) $R_1/R_4 = R_2/R_3$

Let R_2 and R_3 be fixed precision resistances, R_1 a variable resistance, and R_4 an unknown resistance. Then the variable resistance may be adjusted until no current flows through the galvanometer. The three known resistances may be substituted into Eq. (10.27), and the unknown resistance (R_4) determined. The resistance may then be read from a scale on a self-balancing potentiometer, an instrument that balances automatically.

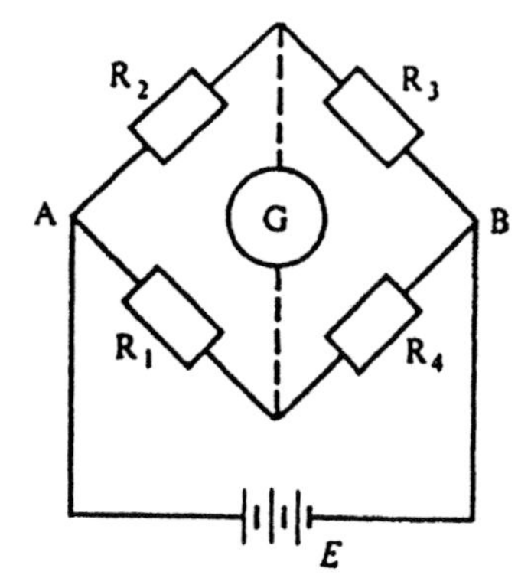

Figure 10.16. The Wheatstone bridge.

Bonded wire resistance strain gages are widely used for measuring the elastic strain in a loaded structure. Such a gage consists of a flat coil of very fine wire cemented between two thin insulating sheets of paper or plastic (Fig. 10.17). The gage is cemented to an unloaded structure with the long dimension of the wire in the direction in which the strain will be measured. If a tensile strain develops at this point, the wires will be stretched an amount $\Delta\ell$. At the same time, the diameter will be reduced in accordance to Poisson's ratio. Both of these effects will cause the resistance of the gage to increase in accordance with Eq. (10.12). Since the collective length perpendicular to the ℓ direction will be negligible relative to the collective length of wire in the ℓ direction, the change in resistance perpendicular to the ℓ direction will be negligible compared with that in the ℓ direction.

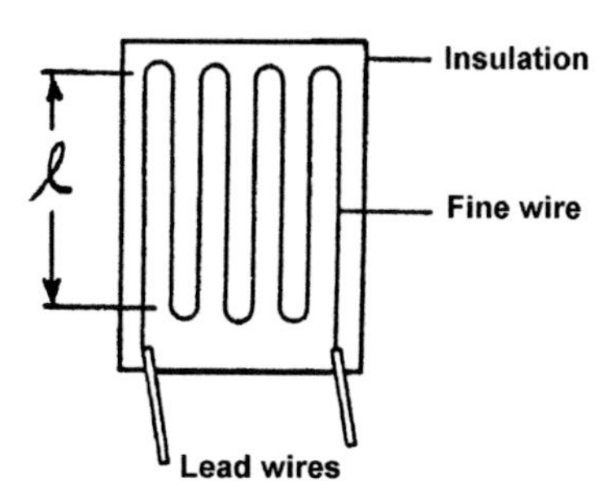

Figure 10.17. Schematic of wire resistance strain gage.

The wire resistance strain gage is a very useful transducer for converting an electrical change (ΔR) into a mechanical signal (Δ strain). It has been used extensively in the design of dynamometers for measuring metal cutting forces. Figure 10.18 shows a very simple arrangement for measuring forces (F_P) and (F_Q) in a two-dimensional turning operation. A gage placed at (A) with its long dimension in the radial direction of the work will measure force (F_P). Since the tensile strain at (A) will depend not only on force (F_P) but also on moment arm (r), it is important that r be the same when the dynamometer is used as it was when calibrated by applying known loads in the F_P direction at the tool tip. The sensitivity of the dynamometer may be increased by a factor of two by attaching a second gage on the surface below A (call this gage A_1). Gage A_1 will be subjected to a compressive stress when a force (F_P) is applied. Gages A and A_1 should be placed at R_4 and R_1 respectively in the bridge circuit of Fig. 10.16. Gages at B and B_1 in Fig. 10.18 will respond to force (F_Q) and these should be placed in position R_1 and R_4 in a separate Wheatstone bridge circuit.

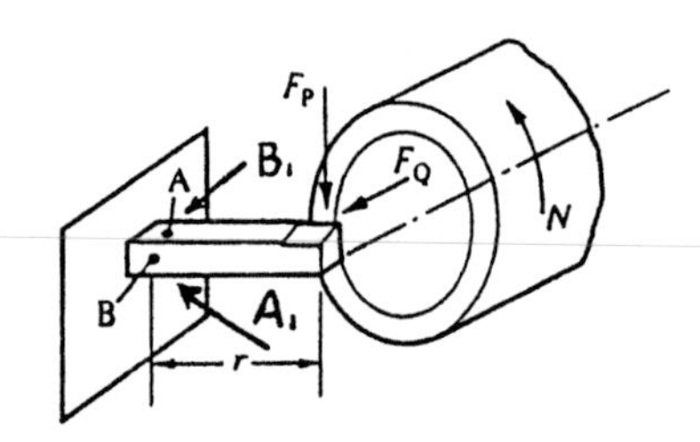

Figure 10.18. Schematic of strain gage dynamometer for measuring forces (F_P) and (F_Q) in a two-dimensional metal cutting turning operation.

12.0 ELECTRONICS

Electronics involves devices and circuits in the design of components such as rectifiers, amplifiers, control circuits, etc., used in radio, sound systems, television, computers, radar, and telecommunications equipment of all sorts. This is a huge area of activity, which is changing rapidly, and today many electrical engineers work in this area. In the space available, only two of the many components in use today are discussed—electronic rectifiers for converting AC to DC, and amplifiers for intensifying a signal.

A diode is a two terminal circuit element that offers high resistance to current flow in one direction, but low resistance to current flow in the opposite direction. It is used mainly to convert AC to DC. Vacuum tubes were introduced in the early 1900s. A vacuum diode is shown in Fig. 10.19. This consists of a hot (~2,000 K) cathode (C) and a cool anode (A) contained in an evacuated ($p \cong 10^{-6}$ mm Hg) sealed glass envelope. The cathode (usually tungsten) is heated by current flow through an internal resistance beneath the cathode. The AC to be rectified is connected across the positive and negative terminals. Electrons will be emitted from the hot cathode, drawn to, and enter the plate when the AC is plus, and current will flow. However, when the AC is minus, electrons will be repelled from the plate, and no current will flow. Thus, the diode acts as a switch in a circuit being closed, and allows current to flow when the AC is plus, but opens and prevents current flow when the AC is minus. This allows a half sine wave of positive current to flow for each cycle. By use of a special bridge circuit, full wave rectification may be achieved, and the sinusoidal fluctuation reduced to essentially a DC of constant amplitude by means of a capacitance filter circuit.

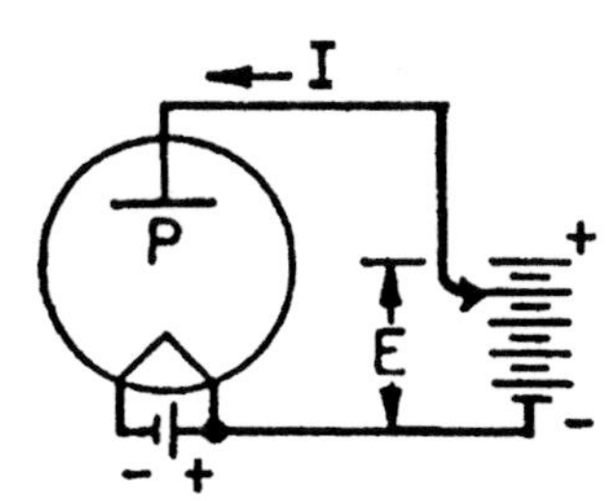

Figure 10.19. Vacuum tube diode.

In the early 1950s solid state units were introduced; gradually they replaced vacuum tubes. Figure 10.20 shows a solid state diode which consists of a combination of n- and p-type semiconductors described earlier in this chapter. The solid dots represent free electrons in the n-type semiconductor while the open dots represent holes in the p-type material. The flow of current in such a unit will be due to flow of free electrons (-) in one direction and of holes (+) in the other direction. When an electric potential is connected with the negative side at *a* and the positive at *b*, electrons (-) will be repelled to the right and holes (+) to the left, and current will flow through the diode. When terminal *a* is connected to the negative side of an external potential and terminal *b* to the plus side, this is referred to as a forward bias. However, when the positive side is connected to *a* and the negative side to *b* (reverse bias), no current will flow. This is because free electrons will tend to pile up on the left and holes on the right, leaving a central nonconducting region without either free electrons or holes, and no current will flow.

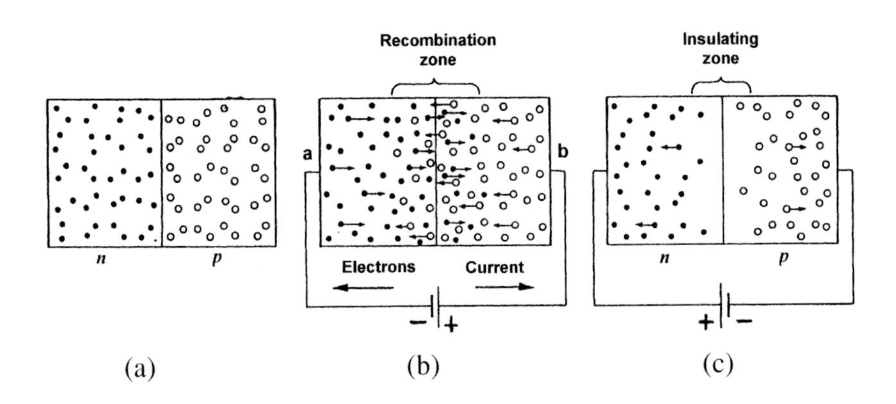

Figure 10.20. *(a)* Solid state diode (solid dots are free electrons, open dots are holes), *(b)* current flow with forward bias (plus terminal attached to p-type semiconductor), and *(c)* no current flow with reverse bias (*Courtesy of L. H. VanVlack*).

When an AC is connected across terminals *a* and *b*, current will flow only during the positive half of a sine wave (forward bias) but not during the negative half sine wave (reverse bias). Solid state diodes have replaced vacuum diodes because they are less expensive, more compact, have a longer life, and require less power and less environmental cooling.

Formerly, devices used to amplify current in a signal were vacuum tubes called triodes. Figure 10.21 shows a vacuum triode. This is like the diode except that a grid (g) is inserted between the cathode and the anode. A small change in voltage between grid and cathode has a much larger effect on current flow than when the same voltage is applied between cathode and anode.

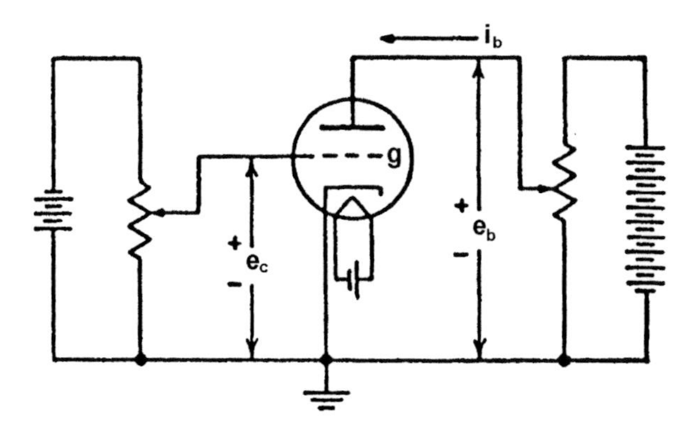

Figure 10.21. Vacuum tube amplifier.

The corresponding solid state amplifier is called a transistor (developed in 1947 by Bardeen, Brittain, and Shockley of Bell Labs). Figure 10.22 shows a transistor which consists of two n-type semiconductors separated by a relatively thin region of p-type semiconductor. When a varying emf is connected across *ab* with the negative at *a* and positive at *b* (called negative bias), electrons will flow to the right, and for the most part will continue across the very thin p section into the collector section. This produces a potential difference between *c* and *b* that is very much (exponentially) greater than the potential difference between *b* and *c*. The thin section (base) could be n-type material and the more extensive emitter and collector regions p-type material. However, then the sign of the *a* and *c* terminals should be reversed. The rectifier and amplifier elements discussed above are two of the many units used in a wide variety of electronic circuits and devices.

Figure 10.23 shows a constant source of emf connected through its fixed internal resistance (r) to an external load of resistance (R). It is desired to know how large R should be relative to r such that the greatest energy possible is dissipated at R. The current flowing in the circuit will be:

Eq. (10.28) $\quad i = E/(R + r)$

and the rate energy dissipated at R will be:

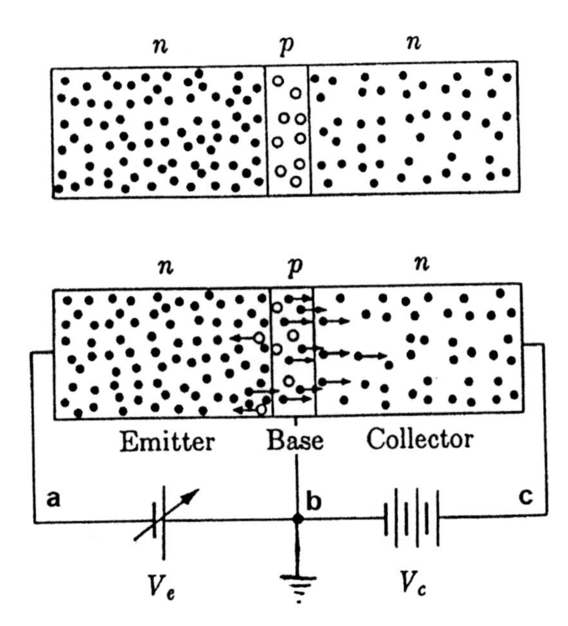

Figure 10.22. *(a)* Solid state transistor and *(b)* circuit arrangement for amplification of voltage (V_e) to (V_c) where the gain is V_c/V_e *(Courtesy, L.H.VanVlack)*.

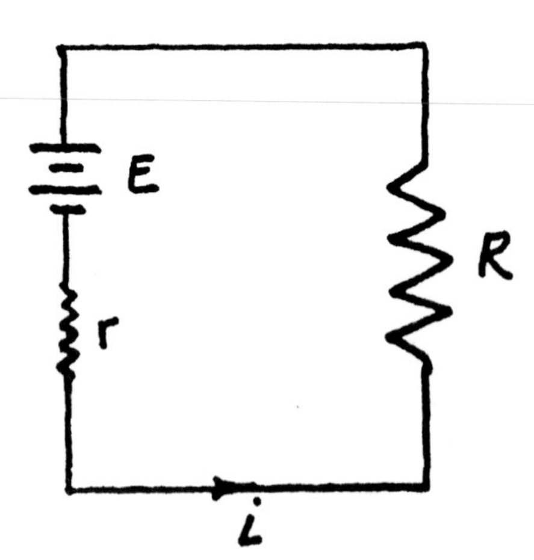

Figure 10.23. Electrical circuit where r is the internal resistance and R equals external resistance.

Eq. (10.29) $i^2R = [E^2/(R + r)^2]\ R = E^2/(R + r)$; if $R>>r$

This will be a maximum when (see App. A):

Eq. (10.30) $(i^2R)/R = [(R + r)^2E^2 - 2E^2R(R + r)]/(R + r)^4 = 0$

or when

Eq. (10.31) $R = r$

Thus, maximum power will be available to external load (R) when the external impedance is equal to the internal impedance (which is referred to as impedance matching). Impedance matching is the technique used to select a speaker to obtain the greatest output from a given HiFi system.

13.0 MEASUREMENT OF TIME

Water clocks used in Egypt measured the time of outflow or inflow of water through an orifice. These were used as well as sundials for many centuries to divide relatively long periods of times into shorter intervals. The hour (sand) glass used primarily for measuring short periods of time did not appear until about the 8th century AD, since it depended on the development of glass working skills. In the Orient, time was measured by progression of the zone of combustion along a path of powdered incense. It was not until the Middle Ages, that the first mechanical clocks appeared which were driven by falling weights. Galileo's observation in 1583 that the period of a swinging pendulum depends upon its length, but not its width of swing suggested a new method of regulating clocks. Before 1700, the accuracy of clocks regulated by a simple pendulum was reduced to ten seconds per day. To make clocks more portable, falling weights were replaced by coiled springs.

The present, most widely used, method of regulating modern time-pieces is a quartz crystal. A properly oriented quartz single crystal will expand and contract at a very precise frequency when subjected to an oscillating voltage produced by an inductive-capacitive (L-C) circuit oscillating at about the same frequency as the crystal. The crystal fine-tunes the oscillating frequency of the current [$f = 2\pi(LC)^{0.5}$] where L is the inductance and C the capacitance of the circuit. Then, a frequency reducing circuit is employed, and the output fed to a high frequency motor that drives the clock. The quartz clock is the electronic equivalent of the pendulum clock, but the maximum accuracy of a quartz clock is about 10^3 times that of the best pendulum clock (about 3 s/yr). Thus, the best quartz clock has an accuracy of about 0.003 s/yr.

The phenomenon utilized in the quartz clock is called the *piezoelectric effect*. When a crystal is subjected to alternate compressive and tensile strains, oscillating (+, -) electric charges appear on opposing faces of the crystal. Similarly, when a crystal is subjected to oscillating charges, it undergoes expansion and contraction.

A still more accurate timepiece is the atomic clock. In this case, regulation depends upon the frequency of molecular vibration. The maximum accuracy of a gaseous ammonia clock is about 1,000 times that of the best quartz clock. Thus, the best atomic clock has an accuracy of about 3 μs/yr.

14.0 ELECTRONIC SENSORS

Electronic sensors are elements used to measure and record a host of quantities of importance to engineering systems. These include sensors to measure force, torque, acceleration, pressure, strain, temperature, humidity, etc. In addition to wire resistance strain gages used to measure static strains, the piezoelectric effect is used to measure dynamic strains.

15.0 ELECTROMAGNETIC WAVES

Electromagnetic waves are cyclic (sinusoidal) disturbances that are generated in a variety of ways. They are capable of traveling long distances in the atmosphere and completely penetrating materials of considerable density to a substantial depth. Electromagnetic waves travel in straight lines at the speed of light in air, which is approximately 3×10^5 m/s (180,000 mi/s). They are capable of being reflected, combined, diffracted, and absorbed. The main characteristics that define their character are frequency (f) and wavelength (λ). Table 10.1 lists a number of electromagnetic wave types together with the frequency range involved (f) in Hz

(Hertz equals cycles per sec), wave lengths (λ) in meters and a typical method of generation. Electrical engineers in communications technology, thermal engineers concerned with radiation from hot bodies, optical engineers concerned with the behavior of light, and materials engineers concerned with x-ray photography and diffraction are all involved with electromagnetic waves and their behavior.

Table 10.1. Electrodynamic Waves

Type	f(Hz)	λ(m)	How Excited
Radio	10^5–10^7	10^2–1	oscillating antenna
TV	10^8–10^9	10^{-1}–10^{-2}	oscillating antenna
Microwaves	10^{11}	10^{-3}	oscillating circuit
Infra red	10^{12}–10^{14}	10^{-4}–10^{-6}	warm body radiation
Light-red	4.3×10^{13}	7×10^{-7}	hot body radiation
Light-blue	7.5×10^{13}	4×10^{-7}	hot body radiation
Ultraviolet	$10^{15} \times 10^{16}$	$10^{-7} \times 10^{-8}$	very hot radiation
X-rays	$10^{17} \times 10^{20}$	$10^{-10} \times 10^{-12}$	electron bombardment
Gamma rays	$10^{18} \times 10^{21}$	$10^{-9} \times 10^{-12}$	ion bombardment

$$f\lambda \cong 3 \times 10^7 \text{ m/s}$$

Electromagnetic waves consist of two magnetic waves at right angles to each other and perpendicular to the direction of propagation. Both of these subwaves have the same frequency. In the case of radio and TV, electromagnetic waves in the frequency range of 10^8 to 10^{10} Hz they act as a carrier for the information being transmitted. This information is contained in rapid variations of amplitude, frequency, or phase angle of the carrier wave. Conversion of a voice or visual signal to an altered carrier wave at the transmitter is called *modulation* while conversion from the altered carrier wave back to a voice or visual signal is called *demodulation*. When the alteration of the carrier wave is in the form of a variable amplitude, this is called amplitude modulation (AM). Similarly when frequency is modulated, this is called frequency modulation (FM), and when the phase angle is coded, this is called phase modulation (PM). Radio signals, first demonstrated by Hertz in 1888, employ either AM or FM while color television employs all three. Telecommunication engineers are concerned with the design of special circuits to perform these modulations as well as the design of the antennas (aerials) required for the transmission and reception of the microwaves involved. Radio broadcasting was developed in the early 1920s.

The microwave oven, developed in 1947, is used to heat food by energizing water molecules that generate heat. Nonconductors (glass, ceramics, and polymers) are unaffected, but metallic conductors are energized to a destructive level; they must not be used in a microwave oven.

Radar (radio detection and ranging, 1930) measures the total time for a radio wave to travel to an object and for the reflected wave to return to the source. The transmitted wave is usually pulsed so that the same electronics may be used for transmission and reception. By rotating the transmitting/receiving antenna through 360°, the angular orientation as well as the distance to the target is obtained.

The *laser* (light amplification by stimulated emission of radiation) was first developed by Maiman in 1960. Energy is absorbed by electrons as they are reflected back and forth in a tube. Upon reaching a semitransparent area of one of the mirrors, the electrons are emitted as a high intensity light (laser) beam having a fixed frequency (monochromatic) with all electrons in step (coherent). A laser beam can travel long distances in a straight line in air without a change in intensity. When brought to rest, energy is released as the electrons return extremely rapidly (in $\sim 10^{-8}$s) to their normal energy state. Lasers are used for producing holes precisely in metals and nonmetals. They are also used in eye surgery, in bar code scanners, and in surveying, and machine alignment operations.

Electromagnetic waves are not to be confused with sound waves that are generated by explosions, thunder, and musical instruments. Sound waves are compression waves of relatively low frequency (256 Hz for middle C on the musical scale) that travel at much lower speeds than electromagnetic waves (about 331 m/s $\cong$ 12 mi/min in air).

16.0 GALILEO

Galileo has nothing to say about magnetism or electromagnetic waves. The latter were not discovered until the 19th century. However, he does discuss sound waves and their relatively slow speed of movement compared with the speed of light. In passage 87 of his text, Galileo points out that the speed of light is very much greater than the speed of sound, and describes an experiment designed to estimate the speed of light. He proposed that two people each with a lantern and means for uncovering and covering (occulting) the lanterns stand a few feet apart. When assistant A sees B's light, he would uncover his own. After practice to reduce any time lag, the experiment was to be repeated at a spacing of three miles, and at 10 miles with the aid of telescopes. After a large number of cycles, Galileo suggests that the time to complete a certain number of cycles should be different for different spacing, if the speed of light is not instantaneous.

Galileo indicates he actually performed the experiment, but detected no evidence that the speed of light is not instantaneous. However, based on observations of lightning, he concluded that the speed of light is not instantaneous but exceedingly high. The fact that Galileo's experiment was not conclusive is understandable. For a lantern spacing of 10 miles, it would take light only $20/180{,}000 = 10^{-4}$ seconds to make the 20 mile round trip per cycle. For 10,000 cycles, the difference in the total time elapsed at a 10 mile spacing would be only one second. Galileo had no way of measuring time to this precision. It is also inconceivable that cycle time could be reproducible to the small fraction of 10^{-4} seconds that would be required by Galileo's experiment.

In 1849, Fizeau performed a mechanical version of Galileo's experiment to obtain the speed of light. Figure 10.24 shows the arrangement employed. A wheel with 720 teeth was rotated as light was projected through the teeth using a semi-silvered mirror. The light traveled between Montmartre and Suresne in Paris a distance (L) of 8,633 meters (5.63 miles). The speed of the wheel was such that the light going out through a tooth spacing was blocked out on the way back by the adjacent tooth. This speed was found to be 12.68 rps. Equating the total transit time of the light to the time for the periphery of the wheel to travel distance ℓ in Fig. 10.24 (*b*):

Eq. (10.32) $(2L)/V_L = \ell/(\pi DN')$

or

$$V_L = [(2L)(\pi DN')]/[(\pi D)/(2t)] = 4LtN'$$

where:
- V_L = speed of light, m/s
- $2L$ = total distance light travels, m
- D = diameter of wheel, m
- t = number of teeth
- N' = rps for complete blocking of returning light

Substituting into Eq. (10.32):

$$V_L = 4(8{,}633)(720)(12.68) = 315{,}263 \text{ km/s}$$

In 1879, Cornu repeated the experiment using a greater distance of travel and a higher wheel speed. He obtained 300,030 km/s for the speed of light. The generally accepted rounded value is 300,000 km/s (180,000 mi/s).

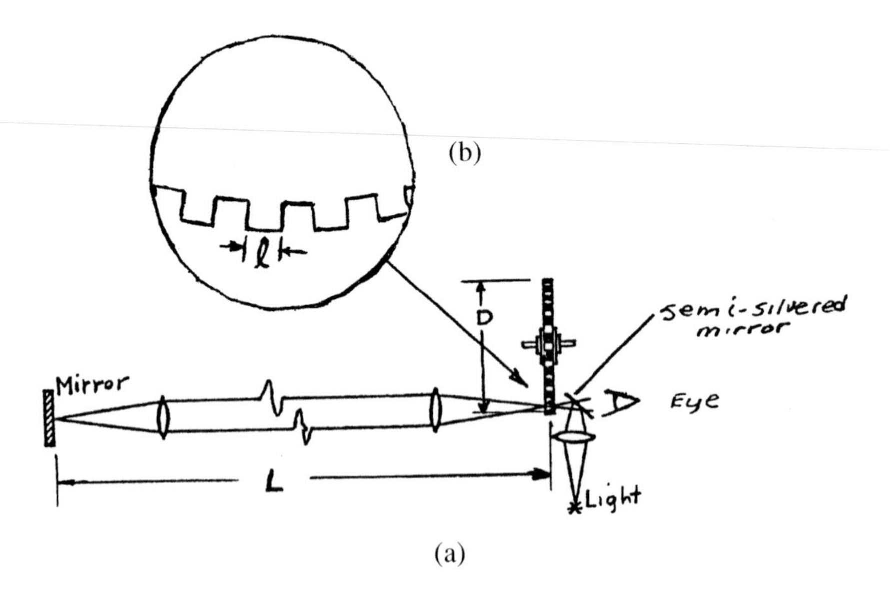

Figure 10.24. *(a)* Arrangement used by Fizeau to measure the speed of light and *(b)* magnified view of edge of wheel.

PROBLEMS

10.1 Two bodies each with a charge of 10^{-5} coulombs of opposite sign are 10 cm apart. What is the force acting between these two bodies?

10.2 a) If a charge of one coulomb is displaced by a voltage of 100 V, how much work is done?

b) If this takes place in one second, what power is involved?

10.3 List four types of energy conversion.

10.4 Describe the least expensive primary dry cell battery used in flash-lights and other electrical devices.

10.5 a) For the parallel circuit shown in Fig.10.2 (*b*), $R_1 = 1\ \Omega$, $R_2 = 2\ \Omega$ and $R_3 = 3\ \Omega$. What is the equivalent resistance from *A* to *B*?

b) If a current of 1 A flows from *A* to *B*, what is the power involved?

10.6 a) What is the resistance of a Cu wire 0.01 in. dia. and 1 in. long? at 20°C?

b) What is the resistance of an Al wire of the same dimensions at 20°C?

10.7 Find the equivalent resistance for the circuit of Fig. P10.7.

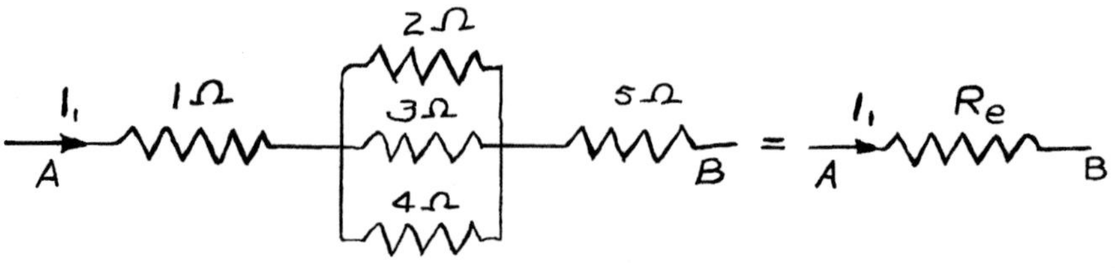

Figure P10.7.

10.8 a) Find the equivalent resistance for the circuit of Fig. P10.8.

b) Find current I_1.

c) Find currents I_2 and I_3.

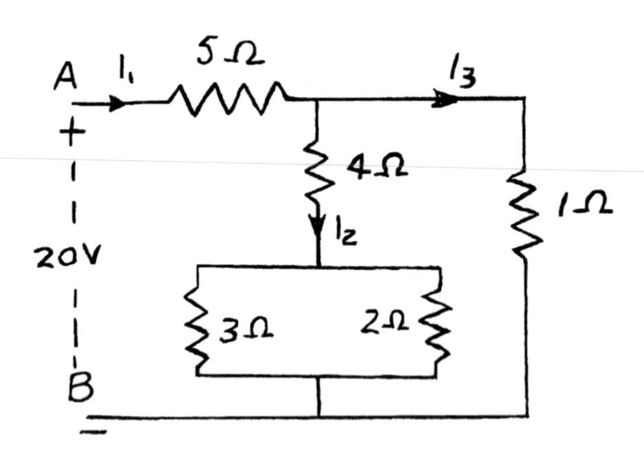

Figure P10.8.

10.9 Find I_1, I_2, and I_3 for the circuit of Fig. P10.9.

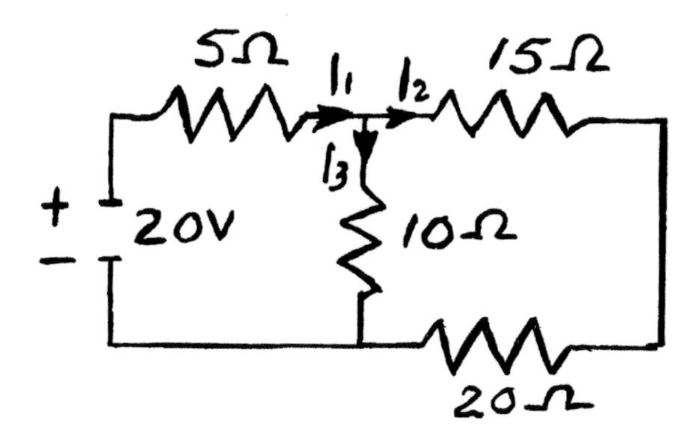

Figure P10.9.

10.10 Show that the equivalent resistance (R_e) of R_1 and R_2 in parallel is $R_e = R_1R_2/(R_1 + R_2)$.

10.11 a) What is the equivalent resistance (R_e) across AB in Fig. P10.11?

b) What are the currents I_1 and I_2?

c) What is the voltage across AB?

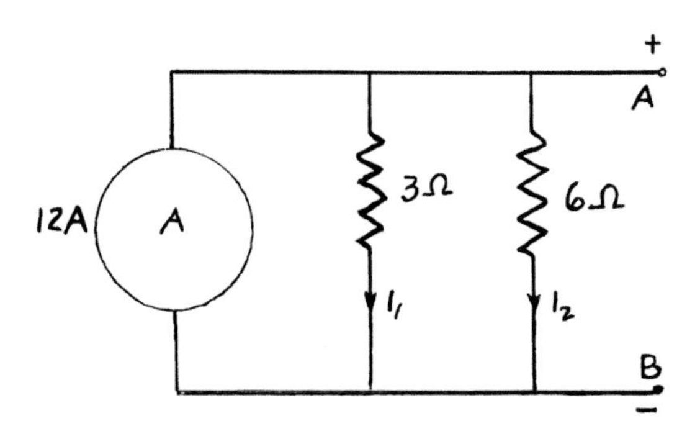

Figure P10.11.

10.12 What is the origin of 120 in the denominator of Eq. (10.17)?

10.13 When the right hand encircles a wire as shown in Fig. P10.13 and the fingers represent a coil carrying current in the direction of the finger tips, will the thumb point in the direction of the north or south pole?

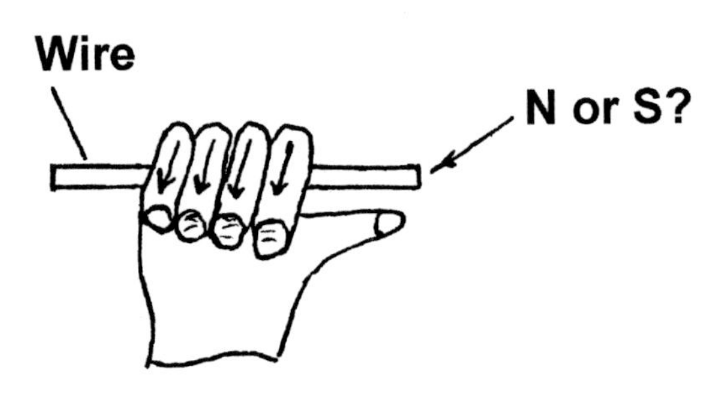

Figure P10.13.

10.14 If a current flows through a wire in the direction of the thumb (Fig. P10.13), will the magnetic field surrounding the wire, as viewed from the thumb, run clockwise or counterclockwise?

10.15 Is the magnetic field between two oppositely charged poles directed from N to S or S to N?

10.16 If a conductor moves upward through a magnetic field going from left to right, will the direction of current flow induced in the moving wire be as shown in Fig. P10.16 or in the opposite direction?

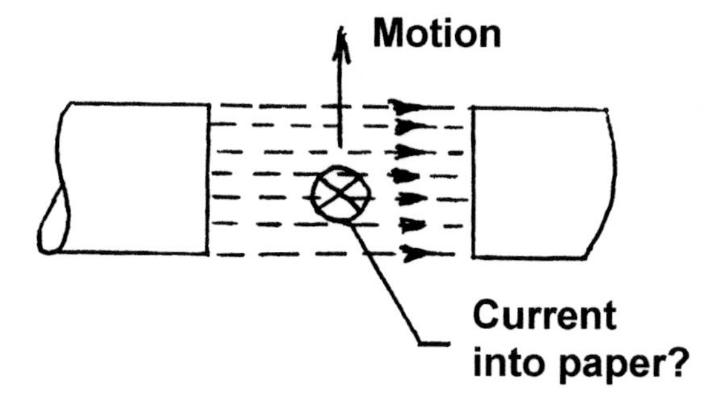

Figure P10.16.

10.17 Would the situation of Fig. P10.16 correspond to that pertaining for a motor or for a generator?

10.18 Figure P10.18 shows a light emitting diode (LED) that will light when the current flowing through it is 20 mA, and when lit, the drop across its 1 Ω resistance is 2 V. Find the value of *R* when the current through the LED is 20 mA and the voltage of the source is 5 V.

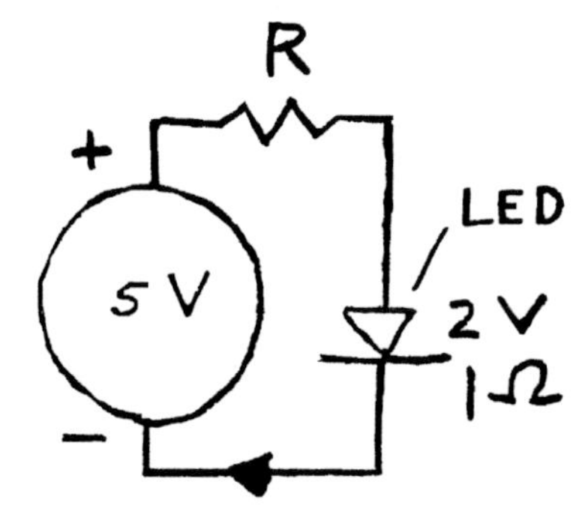

Figure P10.18.

10.19 A transformer reduces the voltage from 110 V to 1.5 V.

a) Ignoring losses in the transformer, estimate the number of turns (N) at the 110 volt side if N at the 1.5 V side is 25.

b) If the current at the 1.5 V side is 1 A what is the current in the 110 V winding?

10.20 A large generator produces energy at 220 V and 1,000 A. This is to be converted to 220 kV at 1 A for transmission. If this power transformer has 100 turns on the 220 V side, estimate the number of turns required on the 220 kV side.

10.21 For the AC voltage-current situation shown in Fig. 10.12, what is the power factor if $\theta = 60°$?

10.22 A DC ammeter gives a full scale reading of 2.5 mA when subjected to a voltage of 100 mV.

a) What is the internal resistance for this ammeter (r_m)?

b) This meter can be used with a full scale reading of 20 mA when subjected to a voltage of 100 mV by employing a resistance in parallel (R_p) (called a shunt) as shownin Fig P10.22. What should R_p be?.

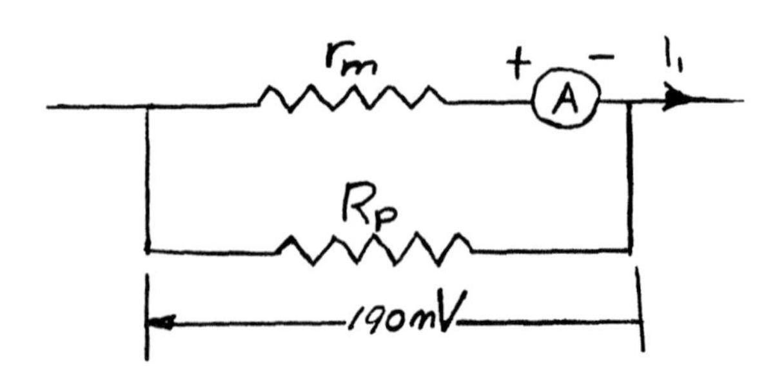

Figure P10.22.

10.23 A DC voltmeter gives a full scale reading of 20 mV when a current of 100 mA flows through the meter.

a) What is the internal resistance of this voltmeter (r_m)?

b) If the range of this meter is extended by use of resistance (R_s) in series (called a multiplier) as shown in Fig. P10.23, so that its full scale range is 200 mV. What is the value of R_s required?

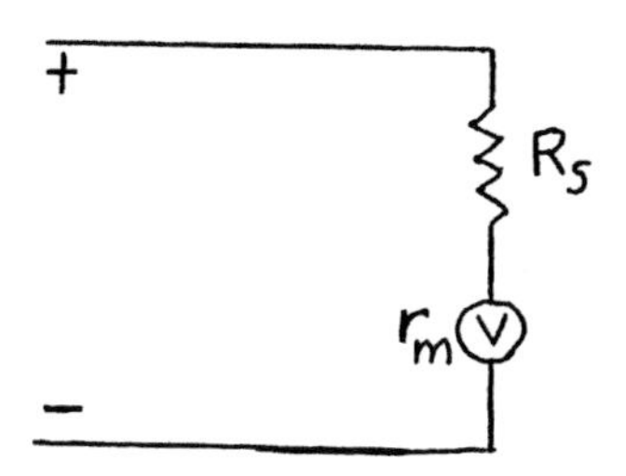

Figure P10.23.

10.24 a) The resistance ($R_?$) may be obtained as shown in Fig. P10.24. If V = 100 V when A = 1 amp, what is the value of $R_?$?

b) An ohmmeter is a more convenient instrument for measuring resistance than use of an ammeter and voltmeter as shown in Fig. P10.24, however, a Wheatstone bridge (Fig. 10.16) is a far more accurate method. If $R_1/R_4 = 5$ in Fig. 10.16, and R_2 = 50 mΩ, what is the resistance (R_3) when no current flows through a galvanometer (G)?

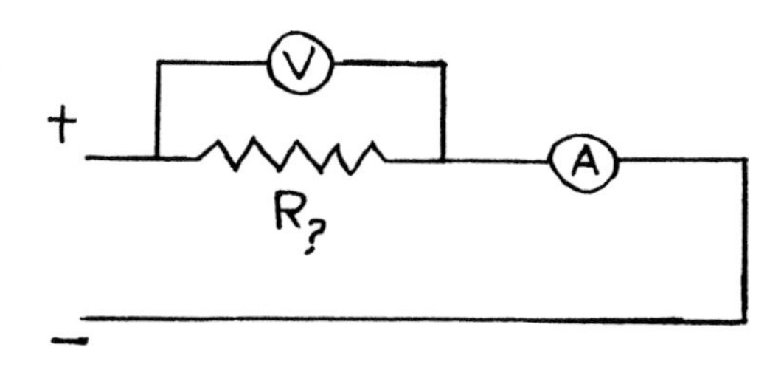

Figure P10.24.

11

Thermal Engineering

1.0 INTRODUCTION

Important topics to be considered in this chapter are thermodynamics, thermal transformation systems, and heat transfer. Thermodynamics involves fundamental relationships between heat, work, and the properties of a system. It is concerned with the transformation of one form of energy into another and the basic laws that control such transformation. Of particular importance is the transformation of thermal energy into mechanical energy, which is the first step in the conversion of the energy associated with fossil fuels into electrical energy as discussed in Ch. 10. Thermal transformation systems are systems that transform thermal energy into mechanical energy. This includes steam power plants, steam engines, steam turbines, gas turbines, and internal combustion engines. Heat transfer is concerned with the transfer of thermal energy from one medium to another by:

- Radiation
- Conduction
- Convection

With radiation, heat is transferred by electromagnetic waves emanating from a hot body to a cold body where radiation waves are absorbed resulting in a temperature rise. Conductive heat transfer involves the passage of heat through a solid from a region of high temperature to one of lower temperature. Convective heat transfer involves the transport of thermal energy from a hot body to a fluid flowing across the hot body.

2.0 HISTORICAL BACKGROUND

Before the 17th century, little attention was given to thermal energy. The Phlogiston Theory of heat championed by Stahl (1660–1734) was the first generally accepted. This proposed that all combustible materials contain a massless material (phlogiston) that escapes on combustion. Some materials like sulfur were considered rich in phlogiston while others contained very little. When this theory failed to explain all experimental results, considerable ingenuity was displayed in attempts to keep it alive.

The Caloric Theory of heat that persisted well into the 18th century suggested that caloric was a special form of fluid whose content increased with the temperature of a body. This theory led to the concept of specific heat (Black, 1766) which was defined as the caloric content needed to raise the unit mass of a substance 1°C. It also led to the concept of heat conduction. Ingendhousz (1789) and later Rumford (1792) measured the relative conductivities of metals by placing wax coated metal rods in hot water and noting the extent of melting along a rod in a given time. However, the Caloric Theory failed to explain the fact that friction releases heat. These theories eventually gave way to the concept that thermal energy is just another form of energy like mechanical energy, and that there exists a mechanical equivalent of heat.

One of the first measurements of the mechanical equivalent of heat was made by Benjamin Thomson (better known as Count Rumford). Rumford measured the heat evolved during the boring of brass cannon in Bavaria (Germany). He immersed the work, tool, and chips in a known quantity of water and measured the temperature rise corresponding to a measured input of mechanical energy. Rumford published his results in England in 1798. Although his value was 25% too high, it was the accepted value for several decades.

From 1840–1849, Joule made many calorimetric measurements to determine the mechanical equivalent of heat, and came within 1% of the presently accepted value (1 gm calorie = 4.1868 Joules, or 1 Btu = 778.2 ft lbs). The *British thermal unit* (Btu) is the thermal energy needed to increase the temperature of one pound of water 1°F, while the calorie is the thermal energy needed to raise the temperature of one gram of water 1°C. Joule's method involved measuring the temperature rise of water in an insulated container after churning the water by a paddle rotated by a falling weight.

Pepin (1681) demonstrated the principle on which the steam engine is based, but it was not until 1763 that James Watt perfected the first expansion steam engine. By 1800, Watt's engines were in wide use pumping water from mines and in cotton mills throughout England. Watt's steam engine played a very important role in the Industrial Revolution that changed manufacturing from a manual cottage operation into a mechanized factory operation. The automobile and refrigeration systems were invented at the close of the 19th century.

3.0 HEAT, WORK, AND TEMPERATURE

Heat is thermal energy in transit. It cannot be accumulated or stored. When thermal energy flows from a hot body to a cold one (the reverse never occurs without the expenditure of work), it should be called heat. However, when flow ceases as a result of temperature equilibrium, the thermal energy gained by the cold body should not be referred to as heat.

Work (W) is also a quantity in transit. It is a force acting through a distance [equivalent to lifting a weight a distance (ℓ) vertically: $W = F\ell$]. *Static energy* is the capacity for doing work.

Temperature expresses the thermal energy level of a body. The molecules of a body vibrate at very high frequency, but with an amplitude that increases with the thermal energy level of the material. The structural configurations called solid, liquid, and gas are states of matter. The two temperature scales in common use today are defined in terms of the transition of one of the mostcommon substances (water) from solid to liquid (freezing point) and from liquid to vapor (boiling point) at one atmosphere pressure. Celsius (1741) took the freezing point of water as 100° and the boiling point as 0°, and divided the scale between into 100 equal parts. Later the scale was inverted with freezing at 0° and boiling at 100°. Mean body temperature on this scale is about 37° Celsius (37°C).

Earlier, Fahrenheit had introduced a similar temperature scale in which the freezing point of water was 32°, and the boiling point of water was 212°. The scale between was divided into 180 equal intervals called degrees Fahrenheit (F). Mean body temperature on the Fahrenheit scale is about 98.6°F.

The conversion for °C to °F is as follows:

Eq. (11.1) $\quad °F = (°C)(1.8) + 32°$

Kelvin (1848) introduced a temperature scale having the same size degree as the Celsius degree, but with zero at the point where all vibrational motion ceases and the pressure of a gas is zero. Zero temperature on the Kelvin (absolute) scale is -273°C = -460°F. This is a more useful scale than the Celsius scale for scientific purposes since the amplitude of molecular vibration is approximately proportional to the absolute temperature (°K). Conversion from °C to °K is as follows:

Eq. (11.2) $\quad °K = °C + 273°$

The Fahrenheit equivalent to the Kelvin scale is the Rankin (R) scale:

Eq. (11.3) $\quad °R = °F + 460°$

The nondimensional homologous temperature (T_H) discussed in Ch. 8, Sec 7.0 is particularly valuable with regard to materials behavior:

Eq. (11.4) $\quad T_{H.} = T/T_{mp}$

where T is the absolute temperature of the body and T_{mp} is the absolute temperature of the melting point of the material. Different materials with the same T_H tend to have similar behavior.

Temperatures are measured in many ways:

a) Change of volume of liquid in a sealed tube (mercury thermometers)

b) Change of length of a material due to thermal expansion

c) Change of electrical resistance

d) Change in color of oxide on steel (blue—yellow—orange—red—white, with increased temperature)

e) Emf developed in a thermocouple

f) Radiation from a hot surface

g) Change in color of special paints or waxes

h) Melting or softening points of various materials

Mercury thermometers are useful from -32°F to 900°F. Those that are used above 600°F must have an inert gas (N_2, vacuum, etc.) above the mercury column instead of air to prevent oxidation of the mercury.

Bimetallic materials (thin strips of materials having widely different coefficients of expansion that are bonded together) are used in the form of helical or coil springs that expand or contract with a change in temperature. This change in length is amplified and used to move a pointer over a fixed calibrated temperature scale. Bimetallic strips are also used to open and close contacts with a change of temperature in control devices.

Certain materials called thermistors undergo an appreciable change in electrical resistance with temperature and are used to measure particularly low temperatures and small temperature differences (as small as 10^{-6}°C).

When certain materials are bonded together, electrons tend to transfer from one to the other. This is called the *Volta effect*. If two such materials are joined together with two junctions at the same temperature, the plus Volta emf at one junction will be balanced by a minus emf at the other and no current will flow. However, if the two junctions have different temperatures, a current will flow from one junction to the other. This is called the *Seebeck effect*, and is the basis of the thermocouple. Figure 11.1 (*a*) shows a thermocouple with a small voltmeter in series with two thermocouple wires (iron and 60 Cu-40 Ni constantan). The emf will be proportional to (T_2 - T_1). Figure 11.1 (*b*) gives the calibration curve for an iron-constantan couple. Thermocouples are used to measure very high (furnace) temperatures, and when the upper range of an iron-constantan thermocouple is reached, a platinum-rhodium couple [also shown in Fig. 11.1(*b*)] may be employed.

Another method of measuring high temperatures is an optical pyrometer. This matches the thermal radiation from a hot surface with that from a hot wire. The current through the wire required for a match is related to the temperature of the body on a nonlinear scale.

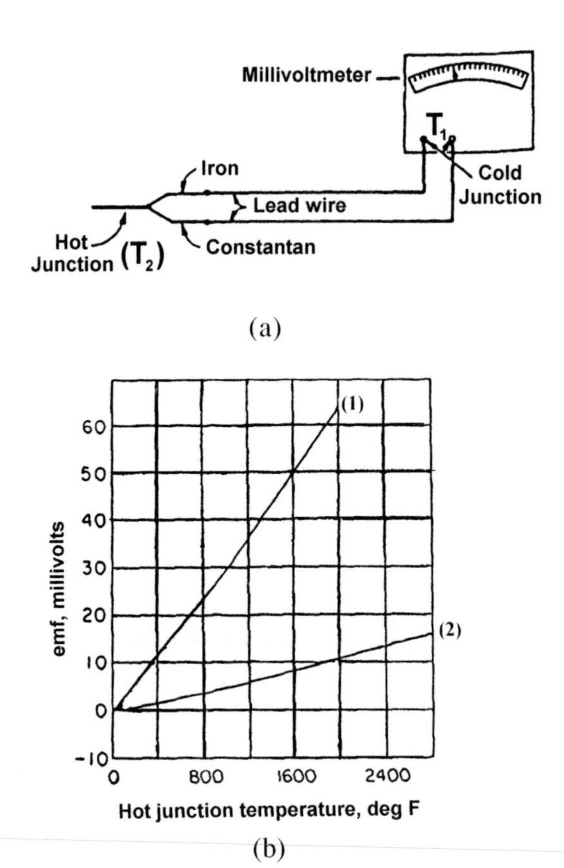

Figure 11.1. *(a)* Thermocouple for measuring temperature in terms of emf generated and *(b)* calibration curves for (1) iron-constantan and (2) for platinum-platinum 10% rhodium.

Commercial products are available that employ the last two methods of temperature measurement called "Tempilsticks" in the former case and "Seger cones" in the latter case.

4.0 THERMODYNAMICS

There are two laws of thermodynamics that have been found to hold experimentally without exception. The first law involves the conservation of energy (energy cannot be created or destroyed). For a steady state process, this may be expressed as follows for a system:

Eq. (11.5) $Q - W = \Delta E$

where Q is heat transfer (+ if to the system), W is work involved (+ if done by the system) and ΔE is the change in energy of the system (potential, kinetic, and internal). In fluid mechanics, applications potential and kinetic energy are usually significant, but in thermal applications they are not.

The first step in an application of the first law is to identify the system. This is conveniently done by drawing a line around the system, which is called the control volume. The quantities Q, W, and ΔE are then introduced into Eq. (11.5). The change in thermal energy (ΔE) in thermal problems usually involves changes in temperature (T) and pressure (P). Tables of relative values of specific energy per unit mass (u) in kJ/Kg for different materials such as air, water, steam, refrigerant, etc. are available. These tables also give values of other properties at different combinations of P and T, such as volume per unit mass. The Bernoulli equation [Eq. (5.15)] is a special application of the first law for applications where there are no losses, no heat transfer, and no work done.

Problem. As an example, the first law may be used to solve the following problem:

Find the heat (Q) that must be added to three kg of steam in a rigid container if the initial absolute pressure is 3 MPa, the initial temperature is 300°C and the final absolute pressure is 6 MPa.

Solution. Take the walls of the rigid container as the control volume that defines the system, and $W = 0$, since there is no change in volume. From a steam table for $P_1 = 3$ MPa and $T_1 = 300$°C, the initial specific energy (u_1) is found to be 2,750 kJ/kg and the specific volume (v_1) is 0.081 m^3/kg. The final specific volume will be the same ($v_2 = 0.081$ m^3/kg). From a standard thermodynamic property table, this is found to correspond to $T_2 = 800$°C and $u_2 = 3{,}643$ kJ/kg. Applying the first law:

$$Q - W = m(u_2 - u_1)$$

$$Q - 0 = 3(3{,}643 - 2{,}750) = 2{,}679 \text{ kJ}$$

In Imperial units, the answer would be in British thermal units (1 Btu equals the thermal energy required to raise one pound of water 1°F) and would be 2,679/1.054 = 2,542 Btu.

5.0 SECOND LAW OF THERMODYNAMICS

The second law of thermodynamics states that it is impossible to have an engine with a cycle that produces work while exchanging heat with a single reservoir. Another statement is that heat will not transfer from a cooler body to a warmer one without work being added. When heat is added to a body the mean amplitude of vibration of its molecules increases as its degree of structural order decreases. The maximum amplitude of vibration of the molecules in a body cannot be reduced without work being done on the body.

6.0 THE CARNOT CYCLE

A heat power cycle is a fluid cycle that converts thermal energy into mechanical energy returning periodically to its initial state. Carnot (1824) visualized an ideal engine cycle without energy losses in order to determine the maximum effectiveness possible for an engine operating between two reservoirs—one at a low absolute temperature (T_1) and a second at a higher absolute temperature (T_2). Figure 11.2 shows a pressure (P) - volume (V_0) plot for the Carnot cycle.

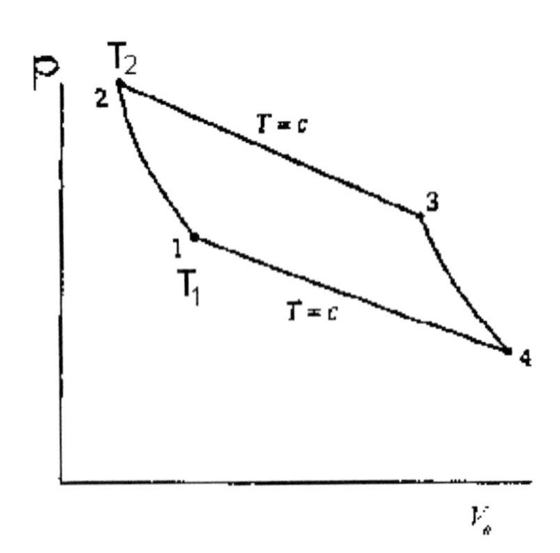

Figure 11.2. The Carnot P-V_0 diagram.

Starting at point (1), the fluid is compressed without heat loss (adiabatically) or mechanical loss to point (2). The absolute temperature rises from T_1 to T_2 during this compression. The fluid expands at constant temperature (T_2) without losses to point (3) as it takes heat (Q_2) from a reservoir at temperature (T_2). It then expands without heat or mechanical loss to point (4) as the temperature of the fluid drops to T_1. The fluid is compressed adiabatically back to point (1) at constant temperature as it rejects heat (Q_1) to a second reservoir having a constant temperature (T_1). From points (2) to (3) and (3) to (4), work equal to Q_2 is delivered to an external system, but from (4) to (1) and (1) to (2), work equal to Q_1 is taken from an external system. The net work done is (Q_2 - Q_1) and the efficiency of the process (e_c) is:

Eq. (11.6) $$e_c = (Q_2 - Q_1)/Q_2 = 1 - Q_1/Q_2$$

However, the Kelvin temperature scale is based on the fact that:

Eq. (11.7) $$Q_1/Q_2 = T_1/T_2$$

hence the efficiency of the ideal Carnot cycle is:

Eq. (11.8) $$e_c = 1 - (T_1/T_2)$$

This is the maximum efficiency possible for any cycle operating between absolute temperatures (T_1) and (T_2).

7.0 THE PERFECT GAS LAW

It was found experimentally by Boyle (1662) that the volume of a given gas varies inversely with absolute pressure (P), and by Charles (1800) that volume varies directly with absolute temperature (T). It was subsequently found that these two results could be combined as follows:

Eq. (11.9) $$PV_0/T = R$$

where: P is the absolute pressure $[FL^{-2}]$
V_0 is the volume $[L^3]$
T is the absolute temperature $[\theta]$
R is a dimensional constant for a given gas $[FL/\theta]$

All dimensions in brackets are expressed in terms of the fundamental set: force (F), length (L), and temperature (θ).

It was later found that for all gases, R equals a universal constant (R_U) where:

Eq. (11.10) $$R = R_U/M$$

and R_U is 1,545 for Imperial units (lbs, ft, °R), and M is the molecular weight in pounds. Thus, for air having a molecular weight (M) of 29:

$$R_{air} = 1{,}545/29 = 53.3 \text{ ft lbs/}(M \text{ in lbs})(°\text{R}) \quad (0.28 \text{ kJ/}(M \text{ in kg})(°\text{K})$$

Similarly for oxygen ($M = 32$):

$$R_{Oxy.} = 1{,}545/32 = 48.3 \text{ ft lbs/}(M \text{ in lbs})(°\text{R}) \quad (0.25 \text{ kJ/}(M \text{ in kg})(°\text{K})$$

Equation (11.9) is known as the ideal gas law, and is found to hold over a wide range of conditions. For example, for air Eq. (11.9) is good to an accuracy of 2% for pressures below 3,200 psia (100 atmospheres), and for temperatures from 0°F to 750°K.

8.0 THERMAL TRANSFORMATION SYSTEMS

8.1 Steam Power Plants

Figure 11.3 is a schematic of a steam power plant and Fig. 11.4 is the P-V_0 diagram for the widely used Rankin engine cycle. The pump takes water exiting the condenser at temperature (T_1) and pressure (P_1) and raises the pressure to (P_2) at, essentially, a constant temperature [from point (1) to point (2) in Fig. 11.4]. The water is vaporized and, possibly, superheated to temperature (T_2) at, essentially, a constant pressure (P_2) [from point (2) to point (3) in Fig. 11.4] in the boiler and superheater

(Fig. 11.3). The steam then enters the engine (reciprocating or turbine) where it expands (essentially adiabatically) to pressure (P_1) [from point (3) to point (4) in Fig. 11.4]. Finally, the exhaust steam is condensed and returned to the initial state of (T_1) and (P_1) [from point (4) to point (1) in Fig. 11.4]. The work done by the engine is the crosshatched area in Fig. 11.4.

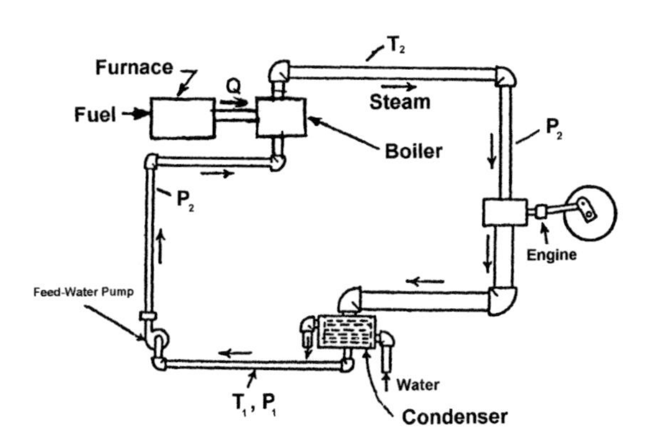

Figure 11.3. Schematic of steam power plant.

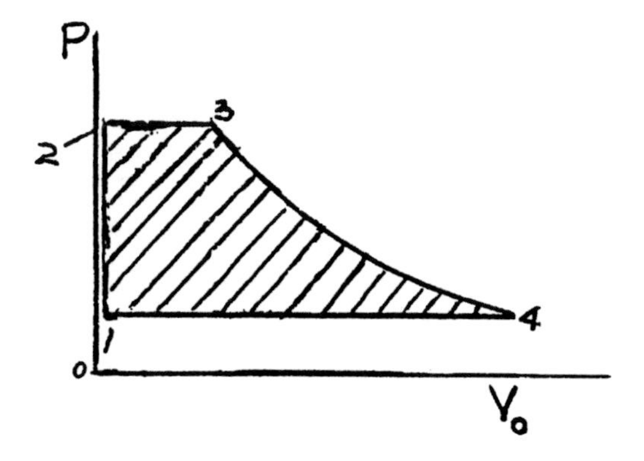

Figure 11.4. Rankin cycle P-V_0 diagram.

The ideal Carnot efficiency of a steam engine operating with an inlet steam temperature of 420°F (488°K) and an outlet temperature of 230°F (383°K) is from Eq. (11.8); (1-383/488) × 100 = 21.5%. The actual corresponding Rankin thermal efficiency would be about 19.5%.

A great deal of technology is involved in the design and construction of the numerous components associated with a steam power plant. These include furnaces and accessories for burning fuel, feed water pumps, boilers for transferring heat from combustion gases to feed water to produce steam, engines for converting the energy in steam at high temperature and pressure into shaft work (reciprocating engines or turbines), condensers for converting low pressure steam into feed water, and feed water treatment to name a few.

8.2 Internal Combustion Engines

Internal combustion (IC) engines convert chemical energy in a mixture of air and liquid or gaseous fuel into thermal energy in the engine cylinder. This is done in four-stroke cycle engines, two-stroke cycle engines, or in diesel engines. Figure 11.5 shows the basic mechanism employed in the three types of engines. This crank mechanism converts reciprocating motion into rotary motion. Figure 11.6 is the P-V_0 diagram for a four-stroke IC engine where the four strokes are numbered in sequence. Starting with the piston at the top of the stroke, the fuel-air mixture, previously proportioned and atomized in a carburetor, is drawn into the cylinder during the first downward stroke. The vapor is compressed on the next upward stroke. When the piston nears the top of this stroke, a spark plug fires igniting the gas. The pressure developed during combustion drives the piston downward. At the bottom of the stroke, the exhaust valve opens, and the combustion gases are ejected during the next upward stroke. Thus, a complete cycle consisting of four strokes of the piston, 1) intake, 2) compression and combustion, 3) expansion, and 4) exhaust is repeated over and over.

Figure 11.7 shows the P-V_0 diagram for a two-stroke cycle engine. In this case, the exhaust valve opens before the piston reaches its bottom position and the remainder of the downward power stroke is used to expel the combustion products. Introduction of the new charge starts just before the piston reaches bottom and continues for a short time after the piston starts up. This is made possible by use of a blower to increase the fuel intake pressure.

The compression ratio (V_1/V_2) has an important influence on IC engine efficiency. The efficiency increases with an increase in (V_1/V_2), but so does the cost of the antiknock fuel required with an increase in compression ratio.

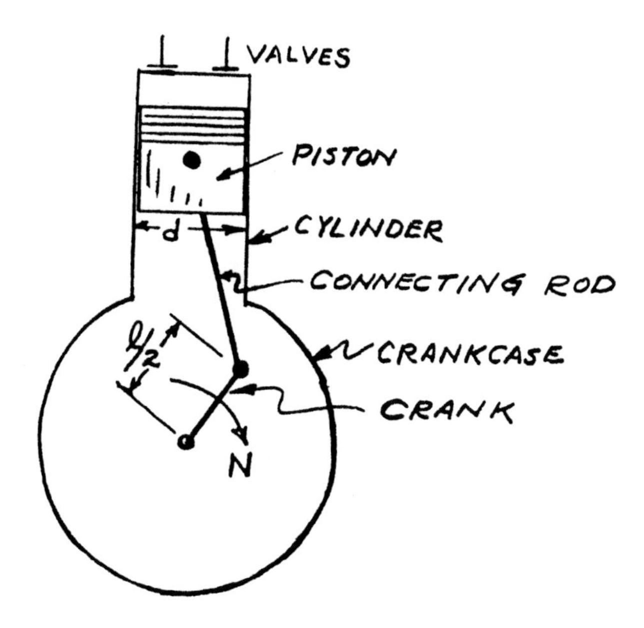

Figure 11.5. Internal combustion engine mechanism (single cylinder).

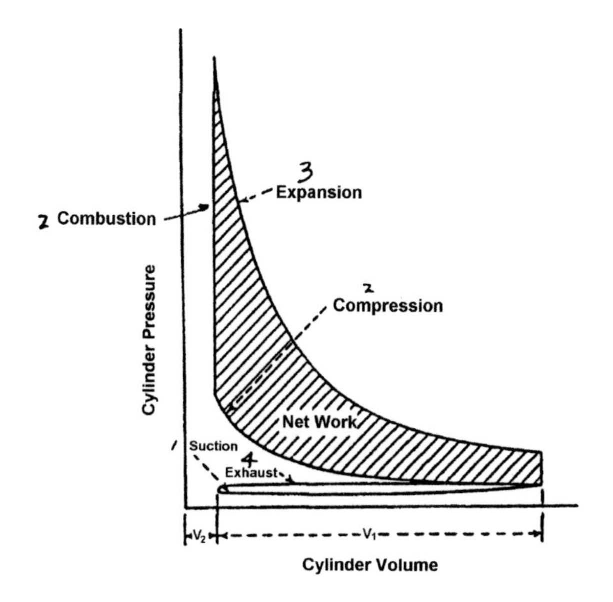

Figure 11.6. P-V_0 diagram for four-stroke cycle IC engine (Otto cycle).

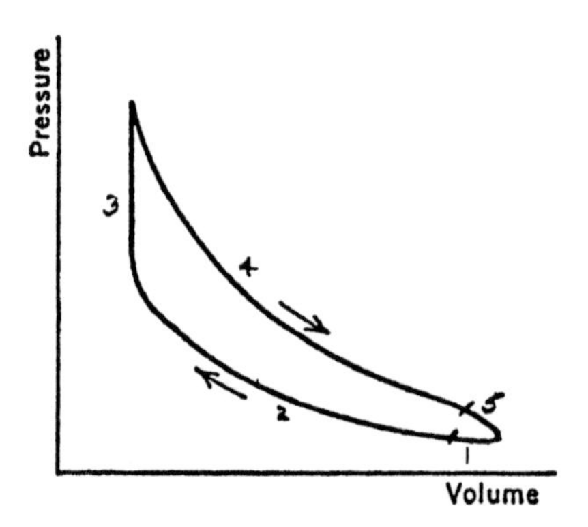

Figure 11.7. P-V_0 diagram for two-stroke cycle IC engine.

A P-V_0 diagram for a four-stroke diesel cycle is shown in Fig. 11.8. In this case, air is compressed to a high pressure. Fuel is then injected under pressure and burns with extreme rapidity. The pressure remains constant during the fuel injection phase [from (a) to (b) in Fig. 11.8], and then falls as the gas expands and as the piston does work.

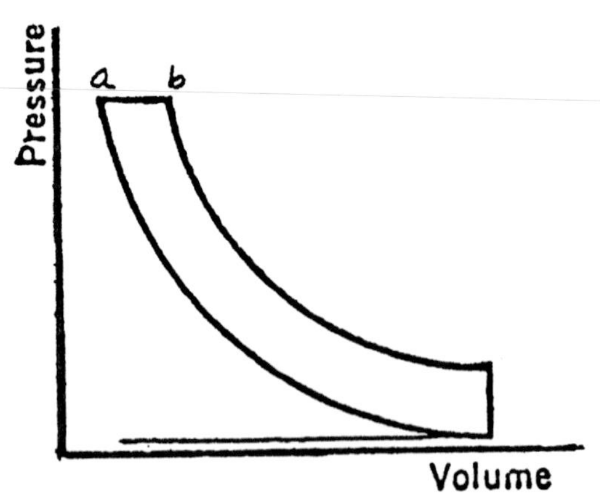

Figure 11.8. P-V_0 diagram for IC engine (diesel cycle).

In most cases, IC engines have more than one cylinder. These are usually arranged in line or in a V formation, and the crankshaft is designed so that the several pistons are subjected to maximum pressure in a definite sequence to provide good engine balance.

Some of the characteristics of an IC engine depend strongly upon the size of the unit (the extensive properties), while others remain more or less independent of size (the intensive properties). Representative values of each type are presented in Table 11.1 for very large and very small engines in commercial production. The intensive characteristics are obviously useful in comparing engines of different size, just as the breaking stress of a beam is a more convenient quantity to use than the breaking load when comparing large and small beams.

Table 11.1. Comparison of Dimensions for Large and Small IC Engines

Extensive Characteristics	**Model Airplane Engine**	**Large Diesel Engine**	**Ratio**
Bore (d), in.	0.495	29	0.017
Stroke (ℓ), in.	0.516	40	0.013
Displacement volume, cu. in.	0.10	26,500	3.8×10^{-6}
Hp per cylinder	0.136	710	1.9×10^{-4}
Rotational speed (N), rpm	11,400	164	69.5
Total engine weight per cylinder, lb	0.26	78,300	3.3×10^{-6}
Intensive Characteristics	**Model Airplane Engine**	**Large Diesel Engine**	**Ratio**
Mean cylinder pressure, psi	47	66	0.71
Mean piston speed, fpm	980	1,100	0.89
Specific power hp/in.^{2}	0.71	1.075	0.66
Weight/displacement lb/cu.in.	2.6	2.94	1.29
Weight/hp, lb/hp	1.91	110	

The IC engine is a very complex device. Carefully planned experiments must be coupled with appropriate analysis in order to design equipment of high efficiency, favorable specific fuel consumption, long life, and smooth, stable operating characteristics.

8.3 Dimensional Analysis

While there are a number of important dimensional analyses that may be made for the IC engine, the following one for inertia- and pressure-induced stresses is representative. Here, we assume that components of stress due to weight are negligible compared with inertia-induced components of stress. The pertinent variables for this problem are listed in Table 11.2.

Table 11.2. Quantities Involved in Dimensional Analysis for Stress Induced in an IC Engine

Quantity	Symbol	Dimensions
Inertia- and pressure-induced stress	σ	$[FL^{-2}]$
Stroke*	ℓ	$[L]$
Density of metal	ρ	$[FL^{-4}T^2]$
Young's modulus	E	$[FL^{-2}]$
Speed	N	$[T^{-1}]$
Crank angle	θ	—
Viscosity of lubricant	μ	$[FTL^{-2}]$
Mean piston pressure	$\bar{p}$	$[FL^{-2}]$

*Any other characteristic size factor may be used such as bore, etc.

Before dimensional analysis:

Eq. (11.11) $\sigma = \psi_1(\ell, \rho, E, N, \theta, \mu, \bar{p})$

After dimensional analysis:

Eq. (11.12)

$$\sigma/(\rho\ell^2N^2) = \psi_2[\text{shape}, E/(\rho\ell^2N^2), \mu/(\rho\ell^2N), \bar{p}/(\rho\ell^2N^2), \theta]$$

where ψ_2 is some function of the five nondimensional quantities involved.

It should be noted that (ℓ) is associated with (N) as a product in all cases. This indicates that mean piston speed $U = (\pi \ell N)$ would have been a better choice to express speed. If U is used in place of ℓN, Eq.(11.12) becomes:

Eq. (11.13)

$$\sigma/(\rho U^2) = \psi_3[\text{shape}, E/(\rho U^2), \mu/(\rho \ell U), \bar{p}/(\rho U^2), \theta]$$

For:

- Geometrically similar engines (same shape)
- Identical materials (same ρ and E)
- Identical mean piston speed (same U)

The stress will be the same for model (m) and prototype (p) at the same crank angle (θ) if:

Eq. (11.14) $\quad (\mu/\rho \ell U)_p = (\mu/\rho \ell U)_m$

or

Eq. (11.15) $\quad \mu_p/\mu_m = \ell_p/\ell_m$

and if

Eq. (11.16) $\quad (\bar{p}/\rho U^2)_p = (\bar{p}/\rho U^2)_m$

or

Eq. (11.17) $\quad \bar{p}_p = \bar{p}_m$

Thus, stresses determined by means of a strain gage attached to a model connecting rod will be the same as those for the prototype, if the viscosities stand in a ratio equal to the scale factor ℓ_p/ℓ_m, the mean cylinder pressures are equal, and the three conditions enumerated below Eq. (11.13) are fulfilled.

9.0 HEAT TRANSFER

9.1 Radiation Heat Transfer

The quantity of thermal energy leaving a surface by radiation varies with the fourth power of the absolute temperature of the surface and certain characteristics of the surface. A perfect emitter of thermal radiation (called a black body) emits energy at a rate (Q) according to the Stefan-Boltzmann (1884) equation:

Eq. (11.18) $$Q = \sigma A T^4$$

where, A = the radiation area, m^2
T = absolute temperature of surface, °K
σ = Stefan-Boltzmann constant
= 5.72×10^{-5} erg/s/cm²/°K⁴, or 5.72×10^{-11} kW/m²/°K⁴

The net exchange of radiant energy between two black bodies, one at temperature (T_1) and area (A_1) and the other at temperature (T_2) completely surrounding the first will be:

Eq. (11.19) $$Q = \sigma A_1 \left(T_1^4 - T_2^4\right)$$

A real body (called a gray body) will emit less than a black body by a factor (ε_1), which is called the *emissivity* of body 1. Table 11.3 gives emissivities for a variety of materials at two different temperatures. The emissivity of a gray body of area (A_1) and temperature (T_1) to an extensive black body having a temperature (T_2) will be:

Eq. (11.20) $$Q = \sigma A_1 \varepsilon_1 \left(T_1^4 - T_2^4\right)$$

If neither body is a perfect radiator (both gray), then the net radiation heat transfer between the two bodies will be:

Eq. (11.21) $$Q = \sigma A_1 F \left(T_1^4 - T_2^4\right)$$

where F is a view factor depending on the two emissivities, the relative geometries of the two bodies, and the angles of directions of radiation to the surfaces. Calculations for factor (F) are further complicated when a radiation absorbing gas or soot is present between source and sink. Radiation heat transfer is an important consideration in furnace and combustion chamber design, and can be quite complex. Thermal radiation is only one of a number of electromagnetic phenomena. Others have been mentioned at the end of Ch. 10.

Table 11.3. Emissivities (ε) for Different Materials

Material	ε	
	100°F	**1,000°F**
Polished aluminum	0.04	0.08
Anodized aluminum	0.94	0.60
Polished chromium	0.08	0.26
Polished iron	0.06	0.13
Oxidized iron	0.96	0.85
Polished silver	0.01	0.03
Asbestos paper	0.93	—
White enamel	0.90	—
White paper	0.95	0.82
Paints (various colors)	0.95	0.70
Ice	0.97	—
Water	0.96	—
Wood	0.93	—
Glass	0.90	—

All electromagnetic waves travel through space in a straight line until they strike another body, at which time part of their energy is absorbed by the body, resulting in an increase in its temperature and internal energy. The sun, which has a surface temperature of about 10,000°R, radiates in the wavelength range of 0.1–3 μm.

Thermal radiation may be transmitted or absorbed by a solid body, the relative amounts of these actions depending on the wavelength of the radiation. Some bodies are transparent to thermal radiations, while others are opaque. An opaque surface of medium roughness may exhibit specular reflection like a mirror for heat waves of relatively long wavelength, but exhibit diffuse reflection for radiation of shorter wavelength. The effective sky temperature on a clear cold night, or over the desert, is about (-46°C) which can cause water to freeze by radiation cooling. This was used by Egyptians to make ice on the desert.

Solar energy may be utilized by use of a collector or a solar thermoelectric generator. The former uses a working substance that is heated and activates a heat engine, while the latter is an example of direct energy conversion in which thermal radiation is converted directly to electricity. Under ideal weather conditions, solar collectors can generate about 75 kW/acre of collecting surface over a year (1 acre = 44,000 ft^2 = 4,000 m^2). However, this calls for such a high capital investment that the cost of power generated in this way is several times that currently produced from fossil fuel. In order to reduce the collection area required and to increase the operating temperature, solar concentrators in the form of large parabolic mirrors have been used. Concentrators can be used to produce temperatures as high as 6,000°R (>3,000°C) for use in high temperature research.

9.2 Conductive Heat Transfer—Steady State

Conduction is a process where heat flows within a body (solid, liquid, or gas) from a region of high temperature to one of lower temperature. While the basic mechanism of heat conduction is different for metals, non-metals, and fluids, the result is the same—an increase in the vibrational amplitude as heat flows into a body. The basic law of conductive heat transfer is Fourier's equation (1822):

Eq. (11.22) $$Q = kA(\Delta T/\Delta X)$$

where: Q = the rate of conductive heat flow, Btu/hr (J/hr)
A = the area across which heat flows, ft^2 (m^2)

k = a proportionality constant called the coefficient of thermal conductivity, Btu hr^{-1}ft^{-2} (°F/ft)$^{-1}$ (W m^{-1}·K^{-1})

$\Delta T/\Delta X$ = the temperature gradient in the x direction °F/ft (°C/m)

In general, values of k vary with the temperature, but in most problems, this variation may be neglected by using a value corresponding to the mean temperature. The Btu is a unit of thermal energy, and in terms of the fundamental set of dimensions (F, L, T, θ) it may be expressed as $[FL]$. Therefore, the dimensions of k become:

Eq. (11.23) $$k = [FL/TL^2(L/\theta)] = [FT^{-1}\theta^{-1}]$$

Representative values of k for different materials are given in Table 11.4 together with other thermal properties.

For a steady state (no change with time), and assuming a constant value of k with temperature, Eq. (11.22) may be written as follows for a plane slab where the temperature of the hot and cold walls are T_2 and T_1 respectively, and b is the slab thickness.

Eq. (11.24) $$Q = Ak[(T_2 - T_1)/b]$$

9.3 Convective Heat Transfer

Heat transfer is by convection when fluid flows across a surface and carries thermal energy away. If the flow of fluid over the body is due to density differences caused by temperature gradients, we speak of free or natural convection. If the fluid is pumped, we speak of forced convection. Fluid mechanics plays a very important role in convective heat transfer.

Newton's law of cooling (1701) relates the rate of heat transfer $\dot{Q}$ to the temperature difference ($T_{surface} - T_{fluid}$), the surface area of the solid body (A), and a proportionately constant called the coefficient of heat transfer at the interface (h):

Eq. (11.25) $$\dot{Q} = hA(T_s - T_f)$$

The dimensions of h are $[FL^{-1}T^{-1}\theta^{-1}]$, and this quantity is usually determined experimentally.

Table 11.4. Thermal Coefficients for Various Materials

Material	Density ($g\,cm^{-3}$)	Melting point (°C)	Thermal conductivity, k ($Wm^{-1}K^{-1}$)	Volume specific heat, ρc ($Jm^{-3}K^{-1}$)	$\beta = \sqrt{k\rho c}$ ($Jm^{-2}s^{-0.5}K^{-1}$)	$\alpha = k/\rho c$ (m^2s^{-1})
Cu	9.0	1,082	390	3.5×10^6	36.8×10^3	111.4×10^{-6}
Al	2.7	660	220	2.4×10^6	23.0×10^3	91.7×10^{-6}
1020 st.	7.9	1,520	70	3.5×10^6	15.7×10^3	20.0×10^{-6}
430 s.s.	7.8	1,510	26	3.6×10^6	9.7×10^3	7.2×10^{-6}
303 s.s.	7.8	1,420	16	3.9×10^6	7.9×10^3	4.1×10^{-6}
Ni	8.9	1,453	92	3.9×10^6	18.9×10^3	23.6×10^{-6}
Ti	4.5	1,668	15	2.3×10^6	5.9×10^3	6.5×10^{-6}
W	14.7	3,410	170	2.7×10^6	21.4×10^3	63.0×10^{-6}
D	3.5	>3,500	2,000	1.8×10^6	60.0×10^3	11.11×10^{-6}
CBN	3.5	3,300	1,300	1.8×10^6	48.4×10^3	722×10^{-6}
SiC	3.2	>2,300	100	2.3×10^6	15.2×10^3	43.5×10^{-6}
Al_2O_3	4.0	2,050	50	3.1×10^6	12.5×10^3	16.1×10^{-6}
Si_3N_4	3.2	>1,900	33	2.3×10^6	8.7×10^3	14.3×10^{-6}
PSZ†	6.1	——	2	3.8×10^6	2.8×10^3	0.5×10^{-6}
Average Ceramic	3.9	——	7	3.0×10^6	4.6×10^3	2.3×10^{-6}
Average Glass	2.4	1,000	1.0	1.8×10^6	1.3×10^3	0.6×10^{-6}
Water	1.0	32	0.6	4.2×10^6	1.6×10^3	0.14×10^{-6}

† Partially stabilized zirconia.
st. = steel, s.s. = stainless steel

For free convective heat transfer, heat is transferred from a heated body to a fluid. The change in density of the fluid gives rise to a vertical buoyant effect that causes the fluid to rise relative to the body. Thus, fluid flow due to buoyancy replaces forced fluid flow due to a pressure difference. Values of h for free convection are much lower than values for forced convection. The subject of free convection heat transfer is very complex and highly empirical.

Free convective heat transfer often involves boiling. When this is the case, heat transfer increases when bubbles of vapor form and are free to rise to the surface. This is called nucleate boiling. When bubbles become very numerous, they coalesce to form an insulating steam blanket, and heat transfer decreases as the temperature of the surface increases leading to surface damage. This leads to what is called *burnout*.

9.4 Nonsteady State Conduction

An important nonsteady state heat transfer problem is to determine the temperature (T) of a small body initially at a temperature (T_0), a time (t) after it is plunged into a liquid whose temperature is T_f (Fig. 11.9). In solving this problem, it is assumed that the body is small enough so that its temperature is always uniformly equal to (T). This is equivalent to assuming that the conductivity of the solid body is infinite, or that the controlling heat transfer process is convective and not conductive. It is further assumed that the volume of fluid and degree of agitation are sufficient for T_f to remain constant.

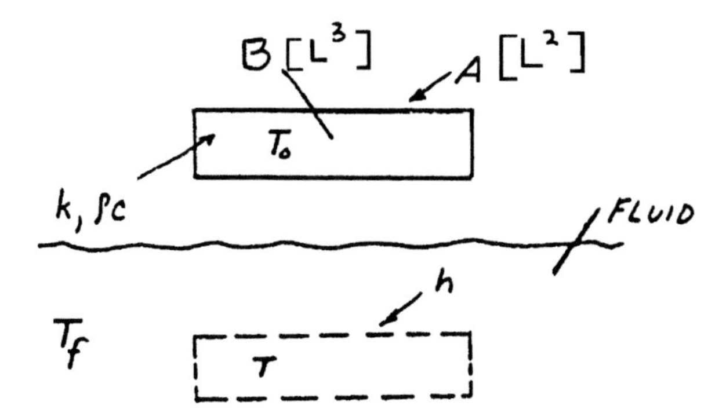

Figure 11.9. Hot body (temperature T_0) plunged at time (t) = 0 into fluid having a constant temperature (T_f).

Before dimensional analysis:

Eq. (11.26) $$\Delta T_1 = \psi_1[\underset{x}{\Delta T_2}, \underset{x}{t}, h, \underset{x}{B/A}, \underset{x}{k_S}, (\rho c)_S]$$

where the quantities involved are defined in Table 11.5, and the quantities indicated by x are the dimensionally independent set.

Table 11.5. Dimensional Analysis for Change of Temperature with Time of Hot Body Plunged into Cooler Fluid

Quantity	Dimensions
$\Delta T_1 = T - T_f$	$[\theta]$
$\Delta T_2 = T_0 - T_f$	$[\theta]$
t = time	$[T]$
h = convective heat transfer coefficient	$[FL^{-1}T^{-1}\theta^{-1}]$
B/A = volume to surface ratio for solid	$[L]$
k_S = thermal conductivity of solid	$[F\theta^{-1}T^{-1}]$
$(\rho c)_S$ = volume specific heat of solid	$[FL^{-2}\theta^{-1}]$

After dimensional analysis:

Eq. (11.27) $$\underset{\pi_1}{\Delta T_1/\Delta T_2} = \psi_2\,[\underset{\pi_2}{(hB)/(Ak_S)}, \underset{\pi_3}{\{(k_S t)/[(\rho c)_S(B/A)^2]\}}$$

The quantity (π_2) is called the Nusselt Number (N), while (π_3) is called the Fourier Number (F_0). It is found experimentally that the temperature of the solid body may be considered constant throughout with time as long as $N < 0.1$. It may be shown by more complete analysis that:

Eq. (11.28) $$\Delta T_1/\Delta T_2 = e^{-NF_0}$$

where e is the Napierian log base. Equation (11.28) is useful in the heat treatment of small items involving quenching.

Another nonsteady state conductive heat transfer problem involves a semi-infinite body (Fig. 11.10) where not enough time has elapsed for thermal equilibrium to be established. A typical problem of this sort is where the top surface of the semi-infinite body is heated to a constant temperature (T_S), and the problem is to find the temperature at an interior point (A) a distance (y) beneath the surface (Fig. 11.10), after a certain elapsed time (t). Before the temperature of the surface is brought to temperature (T_S) at (t) equals 0, the body is at uniform temperature (T_0). This is similar to the problem of heating one end of a long rod whose sides are perfectly insulated so that heat can flow only along the axis of the rod.

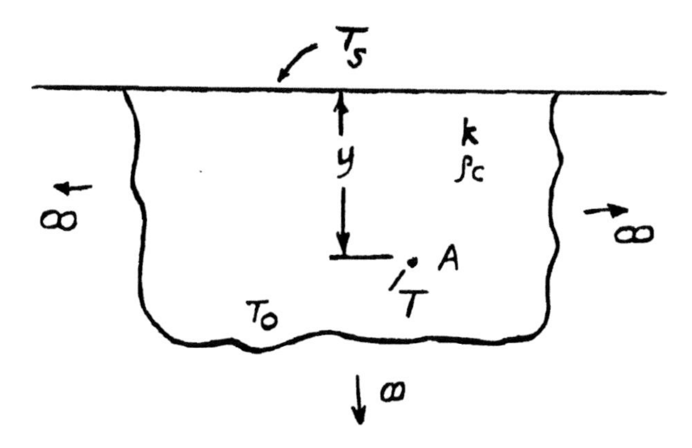

Figure 11.10. Body with boundary condition of first kind (constant surface temperature). The problem is to find temperature at (A), a distance (y) below the surface, time (t) after the surface temperature equals (T_S).

The important variables in this problem with their dimensions in terms of (F, L, T, θ) are listed in Table 11.6.

Before dimensional analysis:

Eq. (11.29) $\quad \Delta T_1 = \psi_1(\Delta T_2, y, k, \rho c, t)$

After dimensional analysis:

Eq. (11.30) $\quad \Delta T_1/\Delta T_2 = \psi_2 \{y/[(k/\rho c)t]^{0.5}\}$

Table 11.6. Dimensional Analysis for Temperature Rise in Semi-Infinite Body with Time when Top Surface is at Constant Temperature after Time Zero

Quantity	Dimensions
$\Delta T_1 = T - T_S$	$[\theta]$
$\Delta T_2 = T_0 - T_S$	$[\theta]$
y = distance below surface	$[L]$
k = conductivity of solid	$[F\theta^{-1}T^{-1}]$
ρc = volume specific heat of solid	$[FL^{-2}\theta^{-1}]$
t = time	$[T]$

The thermal quantity $(k/\rho c)$ that appears in this problem is called thermal diffusivity, and is frequently assigned the symbol (α). The dimensions of thermal diffusivity are:

Eq. (11.31) $$\alpha = k/\rho c = [L^2/T]$$

A complete analysis yields the result shown in Fig. 11.11 where $\psi_2 = \Delta T_1/\Delta T_2$ is shown plotted against the nondimensional quantity $y/(\alpha t)^{0.5}$.

Example. How deep should a water pipe be buried so as not to freeze if the soil temperature is initially at 40°F and the temperature of the surface drops to 0°F for 8 hours? The thermal diffusivity of the soil (α) may be taken to be 0.012 ft²/hr.

$$\Delta T_1/\Delta T_2 = (32 - 0)/(40 - 0) = 0.80$$

From Fig. 11.11, the corresponding value of $y/(\alpha t)^{0.5} = 1.85$, hence

$$y = 1.85\,(\alpha t)^{0.5} = 1.85\,[(0.012)(8)]^{0.5} = 0.573 \text{ ft} = 6.9 \text{ in.}$$

Freezing will not occur if the pipe is at least 7 inches below the surface. It should be noted that Fig. 11.11 is a nondimensional plot; any units may be used in determining the value of the ordinate and abscissa.

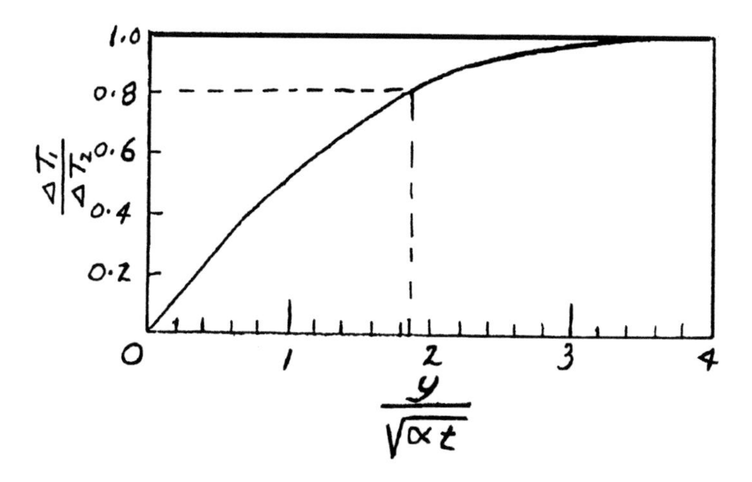

Figure 11.11. Variation of $\Delta T_1/\Delta T_2$ with $y/(\alpha t)^{0.5}$.

The problem just considered is an example of a class of problems having boundary conditions at the surface that are designated as being of the first kind. These are problems where the surface temperature remains constant from time zero, and the temperature at a given point a time (t) later is required. Both thermal conductivity (k) and volume specific heat (ρc) are of importance in such problems. However, for this class of problems only the ratio $k/\rho c = \alpha =$ thermal diffusivity, $[F/L^{-2}]$ is involved. This is a very important observation since it reduces the number of variables in a dimensional analysis for this type of boundary condition.

9.5 Moving Heat Source

Heat transfers with boundary conditions of the second kind are another important class of problems. These are problems where the energy dissipated at the surface remains constant rather than the temperature. For this class of problems, only the geometric mean of k and ρc is involved [that is $(k\rho c)^{0.5}$]. This will be called *heat diffusivity* and the symbol (β) will be used for this quantity. The dimensions of β are:

Eq. (11.32)

$$\beta = (k\rho c)^{0.5} = [(FT^{-1}\,\theta^{-1})(FL^{-2}\,\theta^{-1})]^{0.5} = [FL^{-1}T^{-0.5}\theta^{-1}]$$

The following problem is a very important example of nonsteady state heat transfer having a boundary condition of the second kind. Figure 11.12 shows a perfect insulator sliding across a stationary surface having thermal properties k = thermal conductivity and ρc = volume specific heat. Thermal energy (q) per unit area per unit time is being dissipated at the surface. The problem is to estimate the mean surface temperature (θ). Since this is a problem with a boundary condition of the second kind (constant q), the surface temperature will be a function of $\beta = (k\rho c)^{0.5}$ only.

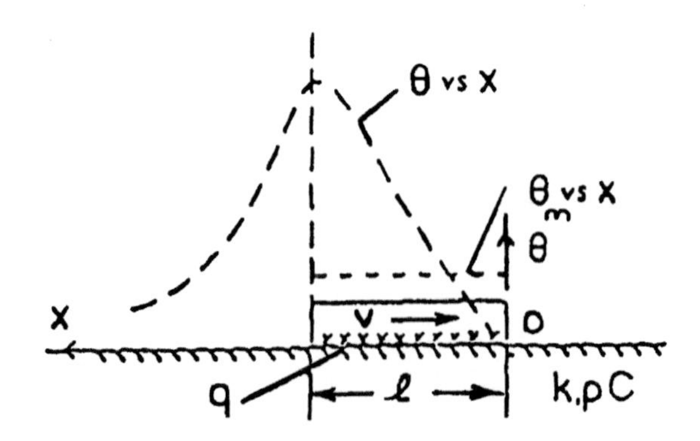

Figure 11.12. Moving heat source with boundary condition of the second kind [constant thermal flux (q) at the surface].

Table 11.7 lists the dimensions of importance for this problem. The width of the slider is not important since this is a two dimensional problem, and each element of width will have the same temperature.

Table 11.7. Dimensional Analysis for Mean Surface Temperature for a Moving Heat Source

Quantity	Dimensions
θ = mean interface temperature	$[\theta]$
q = heat flux	$[FL^{-1}T^{-1}]$
V = slider speed	$[LT^{-1}]$
ℓ = slider length	$[L]$
β = heat diffusivity = $(k\rho c)^{0.5}$	$[FL^{-1}T^{-0.5}\theta^{-1}]$

Before dimensional analysis:

Eq. (11.33) $\quad \theta = \psi_1 (q, V, \beta, \ell)$

Only the following nondimensional quantity involving these five variables may be formed:

$$\theta\ [(\beta/q)(V/\ell)^{0.5}]$$

This means that in accordance with the principle of dimensional homogeneity (every term of any dimensionally correct equation must have the same dimensions), this combination of variables must equal a nondimensional constant which leads to the following proportionality:

Eq. (11.34) $\quad \theta \sim (q/\beta)(\ell/V)^{0.5}$

This is as far as we may proceed by dimensional analysis. To determine the proportionality constant requires at least one experiment or a much more complex heat transfer analysis. More complex analysis reveals that the nondimensional constant in Eq. (11.34) is about 0.75.

Example. For a friction slider:

Eq. (11.35) $q = \mu WV/A$

where: W is the normal load
V is the sliding speed
μ is the coefficient of friction
A is the area of the slider

Substituting Eq. (11.35) into Eq. (11.34):

Eq. (11.36) $\theta \sim (\mu WV/A\beta)(V\ell)^{0.5}$

From this, it is evident that the temperature rise of a slider varies:

- Directly with μ and W/A
- Inversely with $\beta = (k\rho c)^{0.5}$
- Directly with $(V\ell)^{0.5}$

The surface temperature for a slider may be reduced by making the following changes given in order of decreased effectiveness:

- Reduce μ and W/A
- Increase β
- Reduce V and ℓ

For most cases, the slider will not be a perfect insulator and some heat will flow into the slider. This may be taken into account by multiplying the right side of Eq. (11.35) by a fraction (R), where $1 - R$ is the fraction of the total energy dissipated at the surface going to the slider.

PROBLEMS

11.1 Give six methods of measuring temperature.

11.2 How would you calibrate a thermocouple of two metals for which a calibration curve is not available?

11.3 Figure P11.3 (*a*) is a plan view of a bimetal strip of initial length (ℓ_0) and thickness (t) for each element. These were bonded together at temperature (θ_0) and rest on a smooth, flat surface. Figure P11.3 (*b*) shows the composite after a temperature rise of $\Delta\theta$. The coefficient of linear expansion is the change in length per unit length per degree temperature rise, and has the dimension [θ^{-1}], where θ is a fundamental dimension like L, F, and T in mechanics problems. The coefficient of expansion for the upper metal is e_1 and that for the lower is e_2 ($e_1 > e_2$). The angle α increases with temperature rise $\Delta\theta$ and with Δe (e_1 - e_2). Before dimensional analysis:

$$\alpha = \psi(\ell_0, t, \Delta e, \Delta\theta)$$

a) Assuming the friction between the composite and the surface is negligible and $\ell_0 >> t$, perform a dimensional analysis for α.

b) Assuming t is the same in Figs. P11.3(*a*) and (*b*) (very good approximation if $\ell_0 >> t$), perform a dimensional analysis for R.

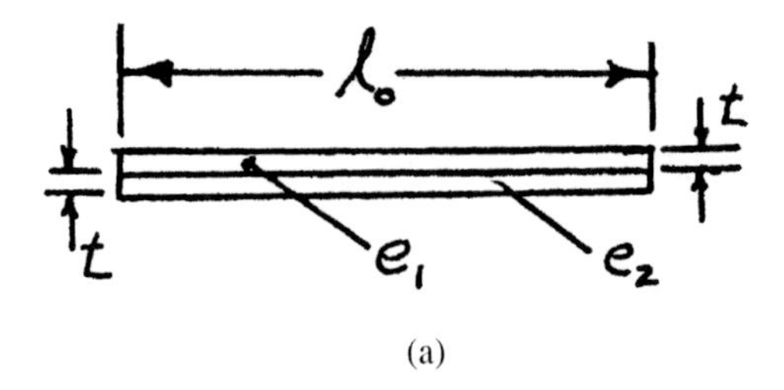

(a)

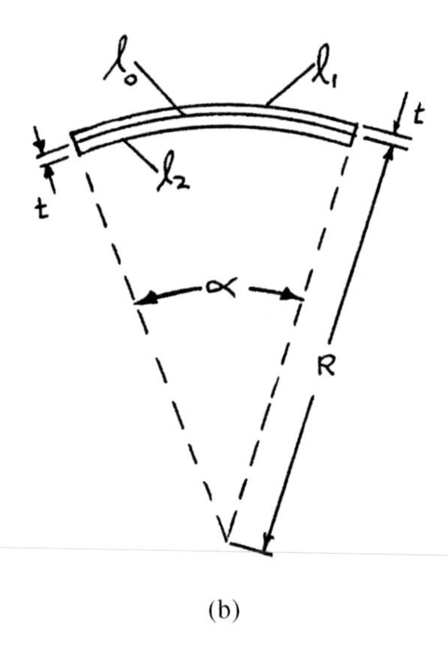

(b)

Figure P11.3

11.4 Assuming ℓ_0 is the same in Figs. P11.3 (*a*) and (*b*) (very good approximation if $\ell_0 >> t$), perform a complete analysis for α and for R. Show that these results are in agreement with the results of Problem 11.3.

11.5 A given gas has its pressure doubled, and its temperature halved. What is the ratio of the final volume to the initial volume?

11.6 The temperature of the air in a tire rises from 25°C to 50°C. If the volume remains constant, by what percentage does the pressure change?

11.7 Charles' law for a perfect gas states that V/T = constant.

a) According to this, at $T = 0$°K, what would the volume be?

b) What is the relation of this to the absolute temperature scale?

11.8 Estimate:

a) The volume of air at an absolute pressure of 14.7 psi and a temperature of 68°F.

b) The volume of air at an absolute pressure of 100 psi and a temperature of 212°F.

c) The pressure of O_2 at a volume of 13.3 ft^3 and a temperature of 68°F.

11.9 If steam enters a turbine at a temperature of 500°C and leaves at a temperature of 200°C, what is the ideal Carnot efficiency?

11.10 The power (P) of an IC engine may be assumed to depend upon the following variables: $P = \psi(\text{shape}, \ell, N, \bar{p}, e)$, where:

ℓ = stroke or other characteristic size
N = rpm
$\bar{p}$ = mean cylinder pressure
e = efficiency

a) Perform a dimensional analysis.

b) Simplify this for two geometrically similar engines.

11.11 Find the rate of heat flow (Q) across:

a) A 1 m^2 steel plate that is one half in. thick, if the temperature on one side is 30°C and that on the other is 15°C.

b) Repeat a) if the metal is Al.

c) Repeat a) if the material is glass.

11.12 a) After sundown on the desert, the temperature of the desertfloor will soon reach a temperature of 20°C. Estimate the rate of energy radiated to the sky in kW by a km^2 if the temperature of the night sky is -50°C. Assume the emissivity of the desert is 0.95.

b) Repeat part a) when the temperature has fallen to -10°C.

11.13 If the temperature of a small body that is initially 500°C is plunged into water having a temperature of 20°C after one second of 100°C, estimate the temperature of the body after two seconds, assuming there is sufficient agitation for the temperature of the liquid to remain constant.

11.14 Estimate how deep a water pipe should be buried to avoid freezing if the mean soil temperature is 40°F and the surface temperature drops to -40°F for 4 days. Thermal diffusivity for the soil is 0.29 ft^2/day.

11.15 A friction slider (Fig. 11.12) operates under the following conditions:

W = normal load = 2 N

V = sliding speed = 100 m/min

μ = coefficient of friction = 0.1

ℓ = length of slider = 2 cm

b = width of slider = 2 cm

β = see Table 11.4 for 1020 steel

R = percent of energy to slider = 0.1

a) Estimate the mean surface temperature.

b) If the material is aluminum and all other conditions are the same, estimate the mean surface temperature.

12

Engineering Design

1.0 INTRODUCTION

Design has many meanings to different people. These include the conception of a new process, a new product, a new use of a physical effect, the preparation of detailed drawings or tools for the workshop, a manufacturing plan, or even a marketing strategy. Design is involved in essentially all engineering activities. Also, the elements of creativity and innovation are involved in all types and levels of design whether it is design for improved reliability and functional life, improved esthetics, reduced cost, improved ergonomics, or manufacturing methods.

Design is a very important engineering activity. It involves meeting some need by applying the laws of physics and chemistry, using mathematics, and the computer where appropriate. Steps in the design process involve some or all of the following:

1. Identifying requirements of the device to be designed.
2. Devising as many general solutions as possible.
3. Comparing possible approaches and selecting the one which appears to be most promising.
4. Performing detailed design based on the selected approach.
5. Specifying shapes, materials, tolerances, standard parts, methods of manufacture, and methods of testing and evaluation.

6. Building a prototype and testing.
7. Redesigning to take care of problems uncovered, optimizing.
8. Adjusting the design for manufacture, taking into account cost, availability of materials and processes, ease of assembly and maintenance, useful life, appearance (esthetics), service life, safety and consequences of failure, energy consumption, and impact on the environment.
9. Sales and application procedures.
10. Updating the design based on performance in the field.

While all steps in the design process involve creativity, it is particularly true for the initial steps where the basic concept of the design is established. Although it is not possible to make all people creative, there are techniques that enhance latent creative ability. The importance of having a clear understanding of design objectives at the outset cannot be over emphasized. Each year, untold millions of dollars are wasted by designers who begin before having a clear idea of what needs to be accomplished.

Detailed design is usually done by specialists who are expert draftsmen. In sophisticated design projects, the design process is usually in the charge of a design engineer who will have a number of engineering analysts (stress analysis, fluid flow, heat transfer, etc.) at his disposal for analytic support. Wherever possible, design alternatives should be based on sound analytical investigation since it is usually far less expensive in time and money to explore alternatives on paper than with hardware.

Tolerances that are specified in design should be given particular attention. The costs of over specifying a part are great. There is a strong temptation for design engineers to play it safe and specify dimensional accuracy or a surface finish that is better than actually required. The skillful design engineer will distinguish between critical and noncritical areas and specify close tolerances only for the former. Often, tolerances are relaxed as a design matures and operating experience is accumulated.

Materials selection is rapidly becoming a design function of great importance. Formerly, only cost, strength, ease of manufacture, and wear and corrosion resistance in specifying a material needed to be considered. In the future, it will become more and more important to consider such socio-technical factors as material availability, safety and product liability, service life, and the possibility of recycling. It is no longer acceptable to waste materials by practicing planned obsolescence of a product. It will also be necessary for design engineers to be aware of substitutes that may be used for materials that happen to be in short supply. A shortage of capital is another important problem that makes the job of the design engineer more challenging by narrowing manufacturing options.

At every step in the design process, the design engineer is faced with a comparison of alternatives. Where possible, this should be done on a quantitative basis using a rating system to measure the achievement of the particular physical attribute under consideration.

There has not always been a good relationship between those involved in design and those concerned with the implementation of designs. One group has often been jealous of the other. Each would tend to blame the other if a given project was unsuccessful. Those in manufacturing might delight in encountering an interference of parts in the workshop that the designer had failed to notice. They would then be tempted to continue the construction instead of bringing the problem to the attention of the designer so it could be corrected immediately. At the same time, design engineers have delighted in embarrassing those in the workshops by specifying items beyond current workshop capability.

It should be noted on the first page of his text, passage number 49, that Galileo had considerable respect for the skills of those in the workshops of the Arsenal of Venice. At the same time, he mentions in passage 50 the importance of true understanding and the danger of accepting proverbial explanations of things. Today, attempts are being made to encourage cooperation between those who design and those who implement. This is being done by acknowledging that both groups have equally important talents and by encouraging designers to consult with implementers and to work as a team before a design is finalized.

Standardization is a very important consideration in design and manufacture. Where possible, the design engineer should take a modular approach to design. When a family of products of different size and capacity is involved, as many of the parts as possible should consist of subunits that are identical, or at least similar. This not only reflects in reduced manufacturing cost, but in ease of maintenance.

One of the important functions of the design engineer involves synthesis of standard components that are mass-produced by another manufacturer. It is unthinkable for a designer to specify a special ball bearing that is not available from the catalog of an existing manufacturer. The cost of such a special bearing would be at least an order of magnitude greater than the standard one, and the quality and life expectancy would probably be far lower than for the standard product. Where possible, the designer should specify materials that are available in a semi-finished form. For example, tubing or ground and finished bars should be specified instead of solid hot rolled steel when the corresponding differences in form and finish result in savings in cost and manufacturing capacity.

Step four in the design sequence outlined above requires the designer responsible for detailed design to be thoroughly familiar with mechanical components (gears, bearings, shafting, tubing, fasteners, springs, etc.) that are available from manufacturers' catalogs.

Steps six and seven might be classified as development. In these steps, the design engineer must work closely with the test engineer involved in developing the new design.

In step eight, the design engineer must work in close cooperation with industrial engineers who are responsible for specifying the selection of processes and equipment, shop layout, production scheduling, quality control and inspection, and packaging and shipping of the finished product.

Sales engineering is a very important function that ranges from something that consists of almost pure merchandising for a simple product used in large quantity, to a highly sophisticated technical function involved in the application of complex products such as the selection and specification of stationary power plants, machine tools or other products that are not mass produced, but are made strictly to order.

The main job of the applications engineer in the latter situation is to be sure that the customer gets the proper piece of equipment for the intended use. The applications engineer will also usually be responsible for making sure the equipment selected is properly installed and that it meets performance specifications in initial trials. The applications engineer also plays an important role in reporting on field performance of equipment (although in many cases this is the function of the service engineer). Most manufacturers keep careful records of service failures and their cause so that these results may be incorporated into future design improvements of a product.

Engineering design is thus seen to embrace a wide spectrum of activities ranging from conceiving new products to sales and service engineering. As we proceed down the above list of design and development activities, work becomes more disciplined and the number of considerations involved increases in scope. It should proceed in an orderly manner from research to application.

Design is not an exact science, but an art. As a consequence, it is not possible to reduce this activity to a few well-chosen theorems or principles as in the case of an engineering science. An important consideration in design is that there is not one correct solution, but, in general, a number of acceptable solutions. A challenging job of the design engineer is to attempt to consider all alternatives and to arrive at the optimum solution for the particular requirements involved. Expertise in design is best acquired by practice and by studying the solutions of seasoned designers having extensive experience.

Success in design involves a good deal of common sense, the ability to plan and organize, a preference for simple solutions as opposed to complex ones, dissatisfaction with the status quo, attention to detail, and an appreciation for constraints of time, material properties, and cost. In what follows, creativity will be considered first followed by description of a number of design solutions.

2.0 CREATIVITY

2.1 Introduction

The main activity of the engineer is to put science to work. This frequently calls for new ideas and concepts and an approach that is novel and untried. Increasing one's creativity involves the development of a personality in harmony with innovation. At the same time, it requires the adoption of a few habits of action, which are particularly useful in this endeavor. Attributes that enhance and discourage creativity will first be discussed, and some useful habits of action which enhance creativity will then be identified. Finally, two examples illustrating the points being made will be given.

It is not difficult to identify creative engineers since they are the ones who come forward with new and unusual solutions to problems. When the characteristics and attributes of more creative people are identified, they are found to correspond to Fig. 12.1. Blocks to creativity are given in Fig. 12.2. A personality favorable toward creativity may be developed by moving in the direction of Fig. 12.1 and away from Fig. 12.2. Of course, the earlier this is done, the more effective will be the result as in all character molding activities.

Inquisitive
Seeks alternatives
Likes to explain ideas to others
Visionary—enjoys the unusual ideas or solutions
Unusual memory
Likes change
Readily accepts failure
Optimistic
Takes risks
Persistent
Energetic and hard working
Does not value security and status symbols
Has good senses of humor
Independent and outspoken
Insensitive to surroundings

Figure 12.1. Characteristics of a creative person.

Overly protective of own ideas
Unwilling to accept ideas of others
Desire to conform
Seeking political rather than functional solutions
Cynicism and pessimistic attitude
Fear of making a mistake
Fear of ridicule
Timidity—lack of self confidence
Competitive jealousy
Thinking too small or too conservatively

Figure 12.2. Blocks to creativity.

A creative person is inquisitive and always on the lookout for a new idea. If this attribute is accompanied by an equally strong urge to understand what is observed and then to apply the new knowledge in design, we have a useful pattern of action. This pattern of action may involve observing a relatively trivial occurrence in nature, understanding the performance of some clever product, or transferring knowledge gained in work performed in one field to that of another. Thus, an effective pattern of behavior involves:

- Identifying something of interest
- Describing its behavior in fundamental terms, even though only qualitatively
- Considering ways of applying these results to the design of as many useful devices or systems as possible

If such a procedure is repeated, a large number of times even when there is no assigned problem, it is bound to lead to enhanced creative ability. A good memory and the ability to explain what is observed in fundamental terms are very important assets in taking full advantage of the results of this pattern of behavior.

2.2 Falling Paper

If a piece of paper falls to the ground, an alert person will observe the following:

- The paper rotates rhythmically with approximately constant velocity
- It glides horizontally with an approximately constant linear velocity instead of falling straight down

When a few other experiments are performed, it is observed that samples of greater width rotate with lower rpm than those of smaller width. The inquisitive person will then question why the paper rotates and why the paper moves along a glide path instead of falling straight down. Answers to these questions are evident from elementary fluid mechanics.

Figure 12.3 shows a piece of rotating paper as a free body and its glide path making an angle α with the horizontal. Since the linear and angular velocities very quickly reach terminal (constant) values, the following three forces acting on the paper must be in static equilibrium:

W = force due to gravity

L = a lift force perpendicular to linear velocity (V)

D = a drag force parallel to linear velocity (V)

Rotation is caused by the alternate peeling of vertices (Karman Vortex Street) from the edges of the paper when oriented perpendicular to linear velocity vector (V) as in Fig. 12.4 (*a*). As each vortex leaves, it subjects the edge of the paper to a force opposite to (V) due to a momentum change similar to that for a rotary lawn sprinkler or reaction turbine. For a non-rotating paper, there would be no net torque, hence no cause for rotation. However, if the paper shown in Fig. 12.4 (*a*) rotates counterclockwise, the vortices leaving the lower edge at (*B*) will be weaker than those leaving the upper edge at (*A*) due to a lower relative velocity between paper and air at (*B*).

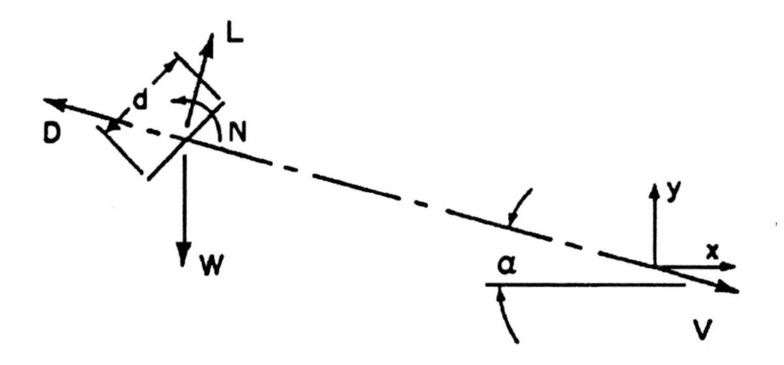

Figure 12.3. Forces acting on paper falling freely in air.

The rotating paper will cause a circulation of air just as though it were a solid cylinder of diameter (*d*). The resultant air velocity will be greater above the paper (at *A*) than below (at *B*) for counterclockwise rotation [Fig. 12.4 (*b*)]. This will give rise to a greater pressure below than above the paper, hence an upward lift force (*L*) (Magnus effect). At the same time, the swept cylinder will be subjected to a drag force (*D*) parallel to and opposing the direction of (*V*).

The frequency (f) with which vortices peel alternately from the top and bottom edges of a stationary plate of width (*d*) oriented perpendicular to an airstream of velocity (V) [Fig. 12.4 (*a*)] is given by the nondimensional Strouhal Number [Eq. (6.17)]. The rotational frequency of the paper (f) will quickly adjust so that:

Eq. (12.1) $$f = 2N$$

From Fig. 12.3, it is evident that there are five unknowns: N, V, L, D, α. Hence, five independent relations are required for a complete solution:

Strouhal equation [Eq. (6.17)]

Equation (12.1)

Equation for lift on a cylinder rotating in air [Eq. (6.8)]

Drag for airflow past a stationary cylinder [Eq. (6.2)]

Equation of statics $\Sigma F_y = 0$ for free body (Fig. 12.3)

Eq. (12.2) $\qquad L = W \cos \alpha$

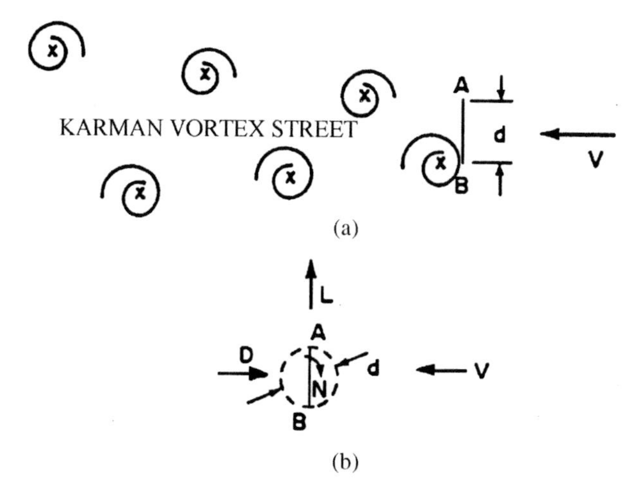

Figure 12.4. (*a*) Source of rotation, (*b*) lift (*L*), and drag (*D*) on paper that rotates as it falls.

If the foregoing experiment is performed in water instead of air by dropping a flat piece of sheet metal in a swimming pool, it is found that there is no rotation, lift, or glide. This is because a moment sufficient to induce rotation is not obtained when a flat rectangle falls in water. However, it is readily observed that two additional turning moments are obtained if the object is "S" shaped instead of flat [Fig. 12.5 (*a*)]. When a S-shaped rotor is oriented in an airstream as in Fig. 12.5 (*b*), there will be more drag on the upper concave half than on the lower convex half, and this will induce rotation as shown. This is the principle of the wind velocity-measuring anemometer [Fig. 12.5 (*c*)]. When an S-shaped rotor is oriented as shown in Fig. 12.5 (*d*), streamlines are closely spaced on the two convex sides giving rise to low pressure relative to that on the concave sides. This, in turn, gives rise to a turning moment (M) tending to cause rotation in the same direction as the peeling of vortices and the anemometer effect.

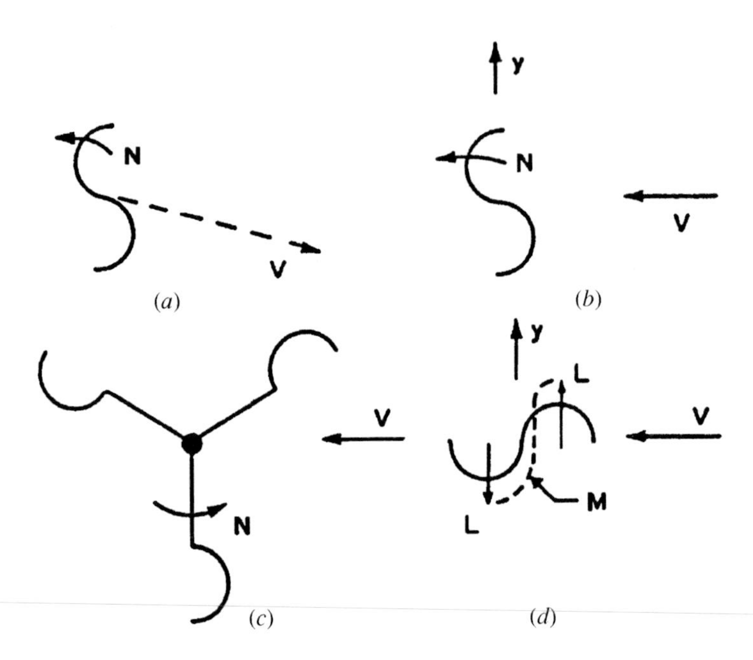

Figure 12.5. The "S" shaped rotor (*a*) direction of rotation and translation when falling, (*b*) source of rotation due to anemometer action, (*c*) plan view of anemometer for measuring wind velocity, and (*d*) source of rotation due to lift forces acting toward convex sides of rotor oriented parallel to the velocity vector.

Having achieved at least a qualitative understanding of the performance of a falling flat paper or S-shaped rotor, the next step is to find some applications. A little thought reveals the following list, which may be readily extended:

- Dynamic kite [Fig. 12.6 (*a*)]
- Reentry radar target [Fig. 12.6 (*b*)]
- Cloud seeding means for rain making [Fig. 12.6 (*c*)]
- Flettner rotor ship [Fig. 12.6 (*d*)]
- Savonius rotor ship [Fig. 12.6 (*e*)]

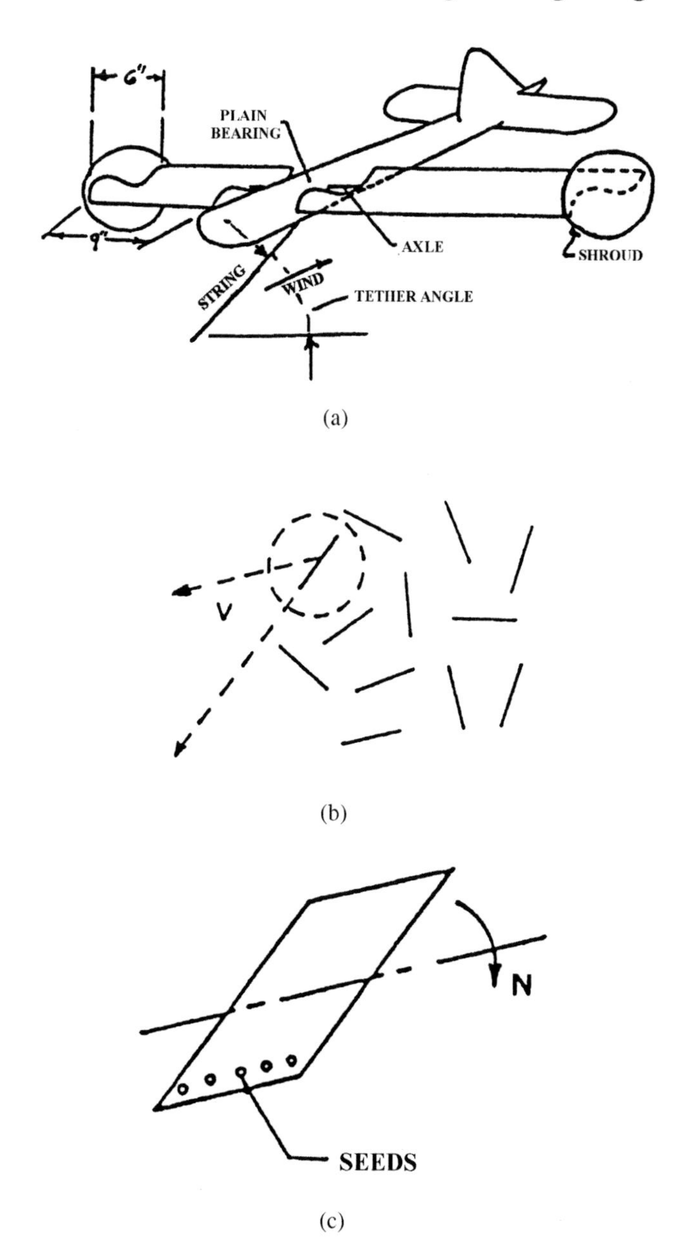

Figure 12.6. Possible applications of falling paper principle. *(a)* Dynamic kite made from expanded polystyrene bead material with S-shaped rotor wing. *(b)* Radar target consisting of pieces of aluminum foil rotating and translating in random directions. *(c)* Means for economic use of nucleating agent in a cloud seeding operation. *(d)* Auxiliary power driven Flettner rotor ship. *(e)* Axial view of Savonius wind driven rotor. *(f)* Sailboat with two collapsible nylon sails.

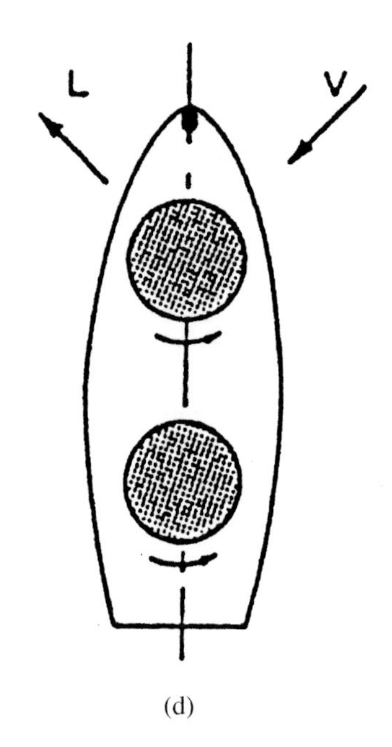

(d)

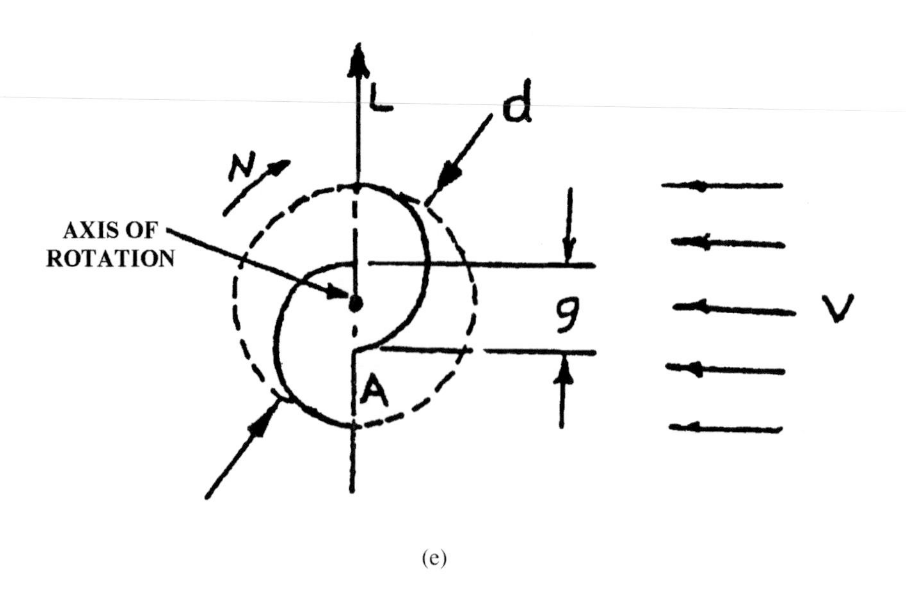

(e)

Figure 12.6. *(Cont'd.)*

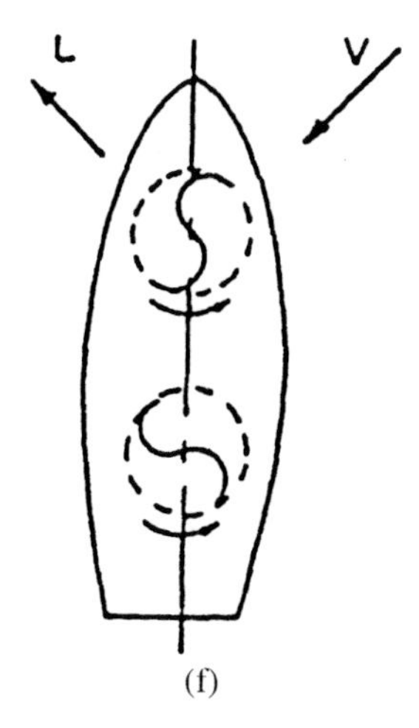

Figure 12.6. *(Cont'd.)*

The dynamic kite is made from expanded polystyrene beads and has a rotating S-shaped wing that gives sufficient lift in a light breeze to cause the kite to fly with a tether angle of about 45°. The radar target consists of a large number of pieces of aluminum foil that glide out in all directions when released, to generate a huge target for radar detection. The cloud seeding elements have a row of dots of the seeding chemical (silver iodide) printed on the outer edges of light cards to provide good interaction between the cloud and the seed material. The Flettner rotor ship [Fig. 12.6 (*d*)] was a sailing vessel built in 1920 that converted energy from the wind into a lift force via the Magnus Effect. The 550,000 kg vessel had two large cylinders (2.8 m dia × 18.6 m high) each driven by a 15 hp diesel engine. The energy taken from the wind by the two cylinders produced a driving force equivalent to 1,000 hp at a speed of 13 knots. (1 knot = 1 nautical mile of 6,080.2 ft/hr.)

Savonius (1924), a Finnish sea captain, improved on Flettner's concept by making it self-acting instead of requiring a drive for the Magnus effect cylinders. He employed a special S-shaped rotor that swept out a cylinder when it rotated [Fig. 12.6 (*f*)]. The Savonius rotor was in two parts with an air gap between so that air entering at (*A*) [Fig. 12.6 (*e*)] could be channeled over the inside surface of the second half of the rotor. This configuration was found to give a greater efficiency (η) than a simple S-shaped rotor. Rotors such as those of Figs.12.6 (*a*) or 12.6 (*e*) are usually fitted with shrouds (discs) located at the ends of the rotor. This reduces end leakage from the rotor, thus increasing efficiency.

The foregoing example is meant to illustrate the inquisitive approach to creative design that may lead to new products. Identification of the physical actions pertaining, and even a qualitative theory of the behavior involved, are obviously important in assessing feasibility of any use proposed for a new concept.

2.3 Micro-Explosive Action

The following discussion is a second example of the inquisitive approach to creativity. For many years, light for taking indoor photographs was produced by the rapid burning of fine wire in a glass tube in an atmosphere of pure oxygen. Originally, the wire was ignited by a spark produced by the rapid discharge of a condenser previously charged slowly by a small battery. Such flash tubes had two wires sealed in the base of the tube to enable ignition by spark discharge. Frequently, the required battery was dead when an important picture was to be taken. Therefore, a new type of flash tube was invented that required no battery. The new unit was in the form of a cube having a flash tube mounted at the center of each of four vertical sides of the cube [Fig. 12.7 (*a*)]. The new unit was called a magic cube because it did not require a battery, and it had only a single electrode sealed into the base of each tube.

When something as interesting as this is encountered, it is usually profitable to take it apart to see how it works, and then to explore how any new ideas uncovered might be used for other purposes. When this was done, each tube was found to be constructed as shown schematically in Fig. 12.7 (*b*). The single electrode is actually a thin walled tube that contains a solid pin on which a small amount of explosive material is deposited. A cocked torsion spring restrained by post (*P*) is released when a pin (*A*) in the camera lifts the end of the spring clear of post (*P*). The end of the released spring strikes the hollow tube, crushing it and detonating the explosive material on the solid pin. The flash from the explosion is spread into the main region of the tube igniting 15 g of fine pure zirconium wire. The glass tube is coated with a film of plastic dyed blue that acts as an optical filter and a safety shield to prevent flying glass in the event the glass tube fractures on firing.

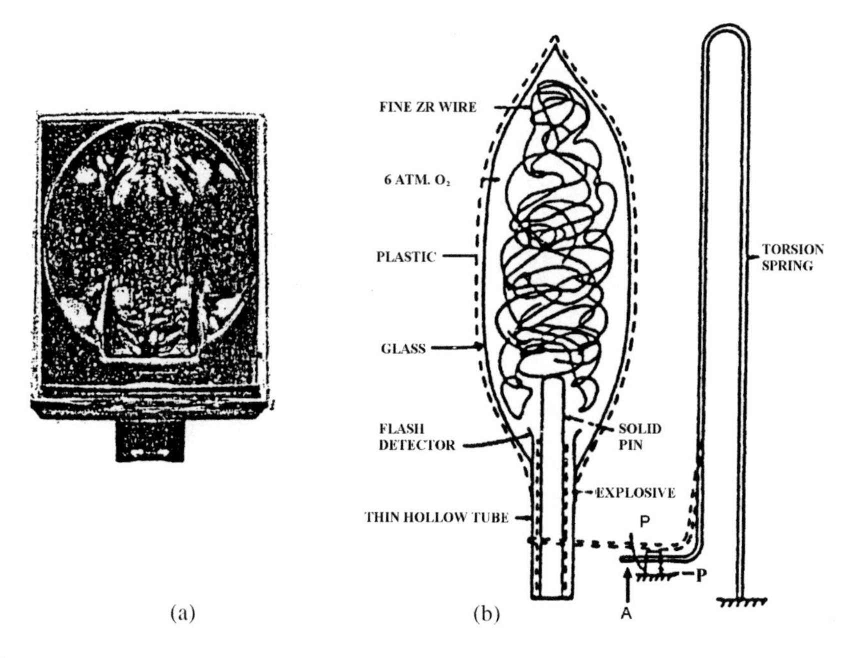

Figure 12.7. The Magic Cube *(a)* photo of entire cube and *(b)* schematic of one cube element.

The important fact this demonstrates is that an explosive need not involve the uncontrolled release of huge amounts of energy but, when properly controlled, may provide a very precise action. This has already been applied in the medical field. Dr. Watanabe of Kyoto, Japan has successfully developed a non-intrusive procedure for removing large kidney stones using a micro-explosive technique. The bladder is filled with water and a small catheter, containing a minute amount of explosive material, is inserted into the bladder without incision and when detonated, the large stone is broken into pieces small enough to pass through the normal body opening.

These applications suggest that micro-explosive action could be used in a process called precision crack-off. In 1912, Professor Bridgman (a Nobel Laureate) reported a phenomenon he termed "pinching off." He found that when a metal rod was subjected to high fluid pressure, but was unconstrained axially [Fig. 12.8 (*a*)], it failed by severe necking similar to the failure of a very ductile metal such as lead in uniaxial tension [Fig. 12.8 (*b*)]. A very brittle material such as glass behaves in a similar way, but fractures without necking [Fig. 12.8 (*c*)]. Professor Sato used this phenomenon for a useful purpose—to cleave brittle materials. By making a very small scratch or indentation in the surface of a brittle rod, and then subjecting the scratch to high balanced biaxial pressure [Fig. 12.9 (*a*)], the Bridgman situation is obtained. At a critical fluid pressure which intensifies the tensile stress at the tip of the imposed defect, a sharp crack initiates and once initiated, runs spontaneously in a straight line across the specimen producing two very smooth surfaces without the usual kerf loss associated with a cutoff operation. This is particularly attractive where high surface integrity is required and where the cost of material associated with kerf loss is high.

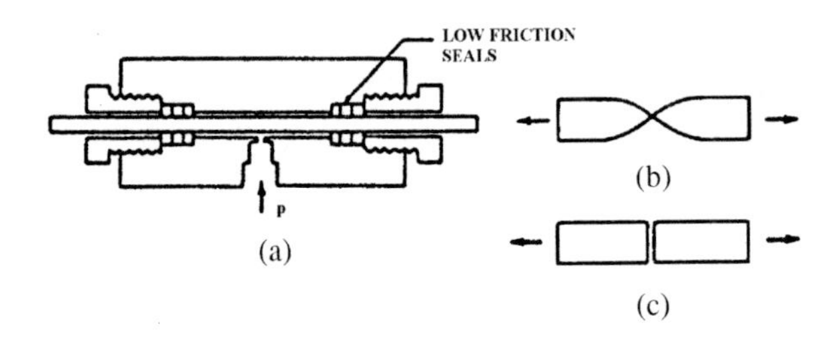

Figure 12.8. The Bridgman pinch-off phenomenon (*a*) Bridgman apparatus, (*b*) pinch-off of ductile material, and (*c*) pinch-off of brittle material.

By placing a small amount of explosive in the indentation, the equivalent of the high fluid pressure of Fig. 12.9 (*a*) will be obtained upon detonation [Fig. 12.9 (*b*)]. By applying pressure with extreme rapidity, ductile materials that would normally "pinch-off" should crack off due to the tendency towards brittleness with increased strain rate. This would be useful as an alternative to friction sawing in cropping operations in the steel mill, or for severing the individual pieces from the central stalk in precision casting operations [Fig. 12.9 (*c*)]. In the latter case, small notches could be cast into the stems near the points of attachment to the stalk, filled with explosive material, and detonated simultaneously in a safety cabinet. This would eliminate a very dangerous, costly, and difficult friction band sawing operation as presently practiced.

This example illustrates how one application of a concept may lead to other applications. Once controlled micro-explosive action has been demonstrated, it is a relatively straightforward task to find other uses. While it is not clear what prompted Dr. Watanabe to employ micro-explosive action in a surgical procedure, the idea could have come from an understanding of the principle involved in the magic cube had he been familiar with this device, and had he taken the trouble to learn how it functioned.

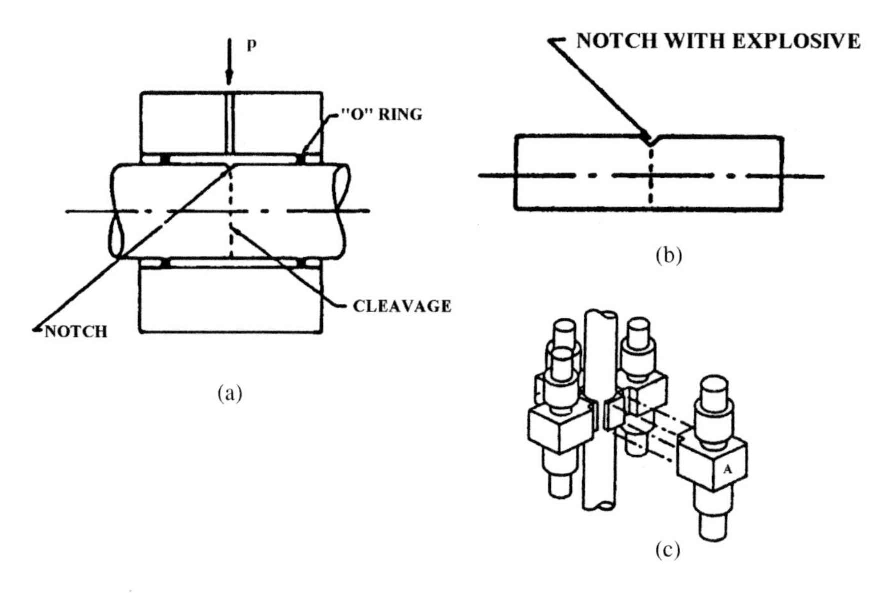

Figure 12.9. The precision crack-off process for brittle materials (*a*) Sato apparatus, (*b*) explosive method, and (*c*) application of (*b*) to removal of parts from precision-casting tree; *A* represents the part removed by precision crack-off.

Still another application of the micro-explosive concept is in wide use today in activating airbags in automobiles. In the event of a forward crash, the impact detonates a small explosive charge that releases the gas that fills the bag almost instantaneously.

3.0 DESIGN EXAMPLES

3.1 Wind Power Generation

Various types of wind-driven turbines have been used for power generation. The efficiency of a rotor such as that shown in Fig. 12.6 (*e*) is equal to the ratio of the energy developed to the energy content of the wind flowing across the projected area of the rotor (i.e., area ℓd where ℓ is the rotor height and d is the swept diameter). That is:

Eq. (12.3) $e = P/(0.5\rho \ell d V^3)$

The numerator of Eq. (12.3) is the power delivered to the shaft per unit time while the denominator is the product of the specific energy of the wind ($0.5\rho V^2$) and the volume rate of air flowing past the rotor ($\ell d V$). The efficiency (e) is, of course, nondimensional.

A dimensional analysis for the power (P) developed by any rotor of the S-type follows. Before dimensional analysis:

Eq. (12.4) $P = \psi_1(\rho, \ell, d, V, N)$

where ρ is the mass density of the fluid. The viscosity of the fluid is not included since it is assumed this is a high Reynolds Number device. After dimensional analysis:

Eq. (12.5) $P/(\rho \ell d V^3) = \psi_2(Nd/V, \ell/d)$

Combining Eqs. (12.3) and (12.5):

Eq. (12.6) $e = 2\,\psi_2(Nd/V, \ell/d)$

The quantity $(\pi dN)/V$ is the ratio of the tip speed to wind speed. Experimentally, it is found that maximum efficiency is obtained for a Savonius rotor when the gap (g) is about $d/3$ and ℓ/d is about 3. For a Savonius rotor of these proportions, the efficiency vs speed ratio curve shown in Fig.12.10 (a) is obtained. The overall efficiency is seen to be a maximum (about 33%) when the shaft is loaded so that the speed ratio $[(\pi dN)/V]$ is about unity. These results may be used to predict the behavior regardless of the fluid medium involved (air, water, etc.).

Also shown in Fig. 12.10 are efficiency vs speed ratio results for other types of wind devices. These fall into two categories—low-speed devices $[(\pi dN)/V = 1]$ and high-speed devices $[(\pi dN)/V > 5]$. The multi-blade windmill is the usual horizontal-axis type found on farms across the USA, which have about 16 blades. This is seen to have a peak efficiency of about 27% at a speed ratio $[(\pi dN)/V]$ of about 0.9. The "Dutch" type four blade windmill is a horizontal-axis machine having a peak efficiency of about 20% at a speed ratio of about 2.2 [Fig. 12.10 (c)]. These machines have canvas sails lashed to the rotor blades. The high-speed two blade propeller is a horizontal-axis unit having a typical airfoil shaped cross section [Fig. 12.10 (*d*)]. This has a much higher peak efficiency (~44%) but to achieve this, it must be operated at a speed ratio of about six which requires careful attention to bearings, rotor balance, and other mechanical details.

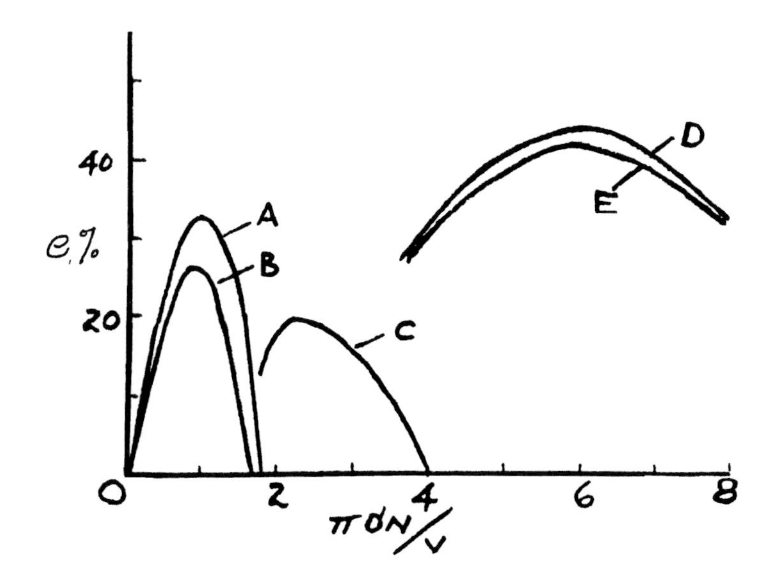

Figure 12.10. Efficiency (*e*) vs speed ratio [$(\pi dN)/V$] curves for several types of wind machines. (*a*) Savonius rotor (*A*), (*b*) multiblade U. S. type horizontal-axis windmill (*B*), (*c*) Dutch-type four blade windmill (*C*), (*d*) high-speed two blade propeller (*D*), and (*e*) vertical axis three blade high-speed wind turbine (*E*).

The vertical spindle, three blade machine [Fig.12.11 and curve (E) in Fig. 12.10] is a high-speed wind turbine having blades in the shape of airfoil sections like an aircraft propeller. However, instead of having vertical blades, they are bowed to form a catenary thus eliminating the bending moments that straight vertical blades would experience. The peak efficiency of the three blade machines is about 42% at a speed ratio of about six. This means that great care of construction is required for these machines, as in the case of conventional two blade horizontal-axis propellers. A 14 ft diameter machine, tested in a wind tunnel, had an output of 0.65 hp at 130 rpm with a wind speed of 18.2 fps.

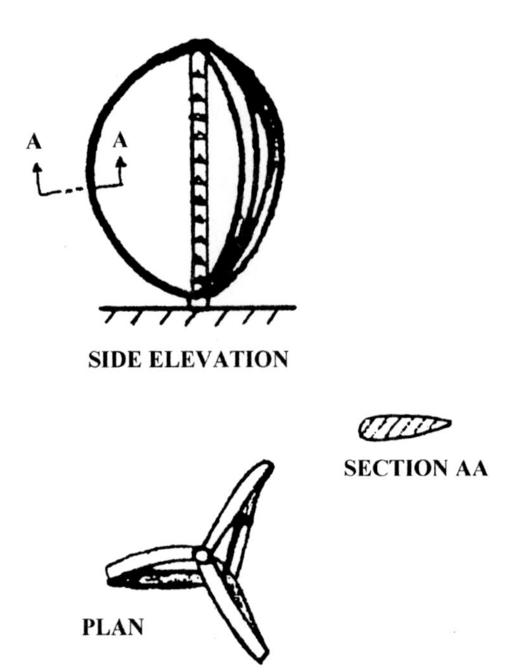

Figure 12.11. Three blade high-speed vertical-axis wind machine.

Other vertical spindle high-speed wind turbines are also under study. One of these uses a two blade bowed vertical-axis rotor with semi-rigid blades (called the sailwing design). The main advantage of this rotor construction is light weight (300 lbs, for a 25 ft diameter rotor). It is expected that a rotor of this type can produce 5 kw in a 12 mph wind. One of the disadvantages of the vertical high-speed wind machines is that they have low starting torque, hence they are not self-starting. Wind machines of the type discussed here could have a major impact on power generation in areas where there are constant average wind velocities of about 15 mph, such as in the Caribbean.

The low-speed machines, such as the Savonius rotor, can be used in water as well as in air. The Savonius rotor has been successfully used as an instrument to measure currents in the sea and is being considered as a low-head turbine to generate electrical power in regions where there is a high tidal rise (such as Passamaquoddy Bay, Maine, 18 ft; Irish Sea, 22 ft; West Coast of India, 23 ft; East Coast of Argentina, 23 ft; a river in France, 26 ft; Severn River in England, 32 ft; and the Bay of Fundy in Canada, 40 ft).

3.2 The Mechanical Fuse

Safety, in particular auto safety, has been receiving wide attention. Concern is not only with the welfare of the driver, but with the reduction in property damage associated with front and rear end collisions. The latter aspect was brought into sharp focus by the offer of an insurance company to reduce insurance rates 20% on cars capable of withstanding a front or rear end crash at a speed of 5 mph without damage.

An acceptable solution should provide a simple, inexpensive yet predictable and reliable means of absorbing energy associated with the crash, without subjecting the frame or bumper of the car to too high a force (P) or passengers to too high an acceleration (a). As in Fig. 12.12, the solution can be represented by a "black box" inserted between the bumper and car frame. The mechanism in the black box must absorb the kinetic energy of the vehicle by allowing the bumper to move a distance (L), while the resisting force is limited to a value (P).

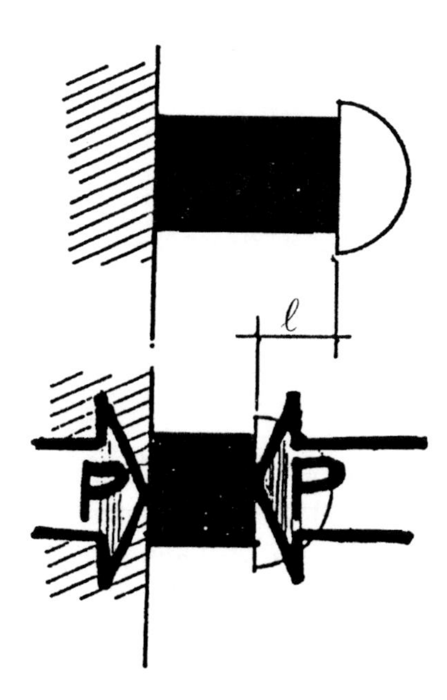

Figure 12.12. Energy-absorbing black box located between bumper and frame of car.

Equating the kinetic energy of the vehicle of weight (W) and velocity (V) to the work done by force (P) acting through a distance (ℓ):

Eq. (12.7) $(WV^2)/2g = P\ell$

where g is the acceleration due to gravity. From Newton's second law:

Eq. (12.8) $P = (W/g)a$

Combining these equations:

Eq. (12.9) $\ell = V^2/2a$

If acceleration (a) is limited to $3g$ (the highest safe acceleration for a passenger without a seat belt), then for $V = 5$ mph:

$$P = 3W$$

$$\ell = V^2/6g = 3.33 \text{ in.}$$

Thus, for a 4,000 lb car, the force on the frame and bumper would be 12,000 lb and a minimum of 3.33 in. of travel would be required between bumper and car frame. These values appear to be reasonable and may be taken as the design requirements.

The least expensive way of protecting electrical circuits is by means of a fuse consisting of a wire, which burns through when the current exceeds a certain safe limit. By analogy, the black box might be developed as an inexpensive expendable item that allows motion when the resisting force on the bumper reaches a value (P), and in so moving, absorbs the unwanted energy. This type of energy-absorbing device may be thought of as a "mechanical fuse." The energy absorbed should be irreversibly converted into heat. This rules out springs which return the energy stored to the car after the crash, causing a reverse acceleration which subjects passengers to whiplash.

When a metal is plastically deformed, practically all of the energy involved ends up as heat. While there are many ways of plastically deforming materials, such as bending, denting, shearing, drawing, extruding, or cutting, not all are equally efficient relative to energy absorption. The cutting process represents a very efficient use of metal for energy absorption. It is also easy to predict the forces and energy involved. For mild steel, about 300,000 in. lb are required to produce a cubic inch of chips, and this value is essentially independent of the speed of the operation.

Use of the cutting process to absorb energy is not initially attractive, since it normally involves a tool, guideways, power transmission, and a complex system of tool control. However, the energy-absorbing bumper design of Fig. 12.13 is a simple means of absorbing the unwanted energy. Before impact, the nut at the left is tightened against the spring and the assembly is capable of withstanding normal towing and bumper jack loads. Upon impact, the surface of the rod is cut after a critical force (P) develops. Equating the energy absorbed to the cutting energy gives:

Eq. (12.10) $P\ell = \ell(\pi d t)u$

where d is the diameter of the tool, t is the radial depth of cut, and u is the specific cutting energy (energy per unit volume cut). Then, solving for t:

Eq. (12.11) $t = P/(\pi d u)$

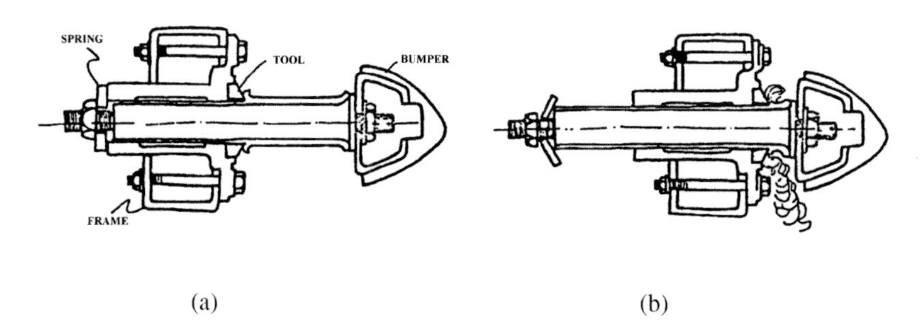

Figure 12.13. Metal cutting energy absorbing bumper unit (*a*) before impact and (*b*) after impact.

For the 4,000 lb car considered, the required force is 12,000 lb. When cutting mild steel in this way, u will be about 0.3×10^6 in.lb/in.^{3}. Thus, for a tool of 1 in. diameter and for two points of attachment on the bumper, the wall thickness (t) required is:

$$t = (12{,}000/2)/[\pi(1)(0.3 \times 10^6)] = 0.0064 \text{ in.}$$

Figure 12.13 (*b*) shows the device after the tool has cut 3.33 in. of steel and the car has been brought to rest. The rods are axially unrestrained and will now rattle in the guide bearings, thus warning the operator that the "mechanical fuse" should be replaced. To restore the bumper to its original condition, it is only necessary to remove two nuts, pull the entire bumper and guide rods out of the guide bearings, and replace the inexpensive expendable steel tubes. This could be done at any service station using color-coded tubes of different diameters for cars of different weight. The design is simple, inexpensive, easily predictable, reliable, and does not cause whiplash since over 95% of the energy ends up as heat in the chips. The tool is a simple carbon-steel washer, and the unit operates equally well over a wide range of velocities.

The only disadvantage of this design is that *all* of the tube is not converted into chips. An attempt was made to eliminate the extra material in the tube and have the tool slide directly on the steel guide rods with a small clearance [Fig. 12.14 (*a*)]. However, instead of cutting chips as shown in Fig. 12.13 (*b*), the tube buckled, accordion fashion [Fig. 12.15 (*b*)]. Progressive development of the accordion pleats is illustrated in Fig. 12.16. This result was completely unexpected.

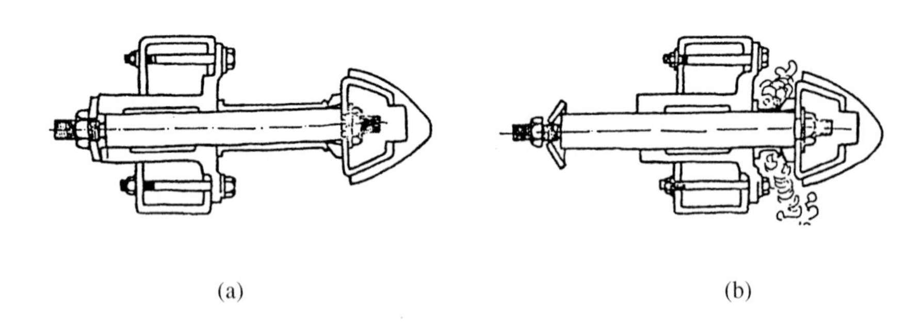

Figure 12.14. Expected action of metal cutting unit when entire tube is cut (*a*) before impact and (*b*) after impact.

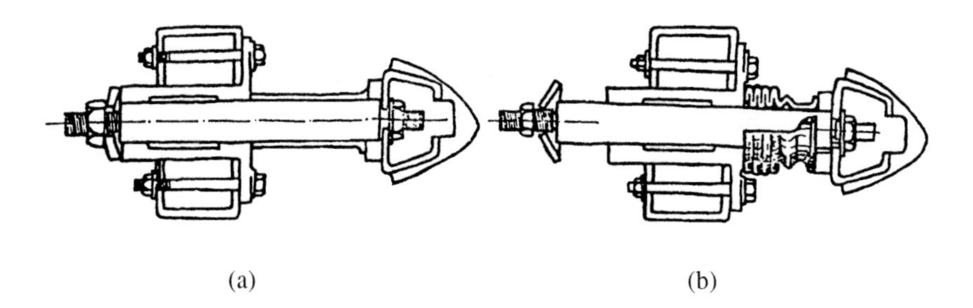

Figure 12.15. Actual performance of unit shown in Fig. 12.14 (*a*) before impact and (*b*) after impact.

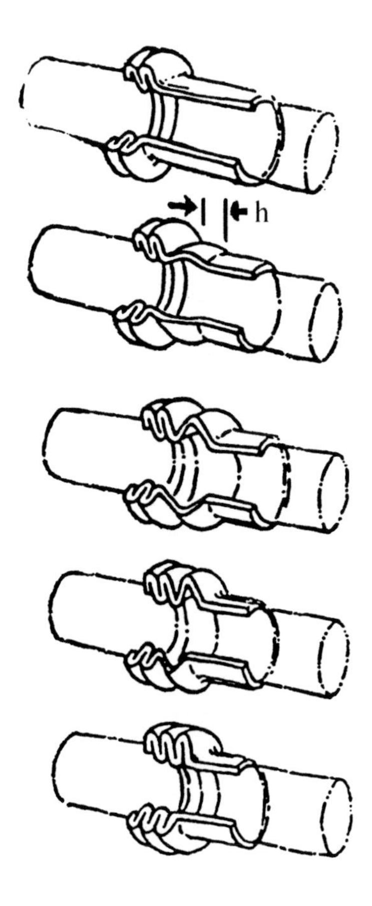

Figure 12.16. Progressive development of accordion pleats.

Since this method of absorbing energy is also of interest, it was decided to study the buckling action both experimentally and analytically. Experiments showed that perfectly uniform pleats were formed over a very wide range of thickness-to-diameter ratios as long as a central supporting mandrel was employed. In the absence of the mandrel, the action was unpredictable and the buckling was very complex. Dynamic tests run on welded, thin-wall, electrical steel conduit were very satisfactory. The 20,000 in. lb per unit required to stop a 4,000 lb car from 5 mph with a motion of 3.3 in. may be achieved using a 4 in. length of standard 1 in. thin-wall steel conduit (1.054 in. ID) having a wall thickness of 0.040 in.

This pleating energy absorption device was found to be well behaved, reproducible, and easily calibrated in simple drop tests. It appears to have many additional applications where unwanted energy is to be absorbed. The stability against gross buckling of the tube provided by the mandrel enables it to be used with long as well as short strokes. It would, therefore, provide a useful, inexpensive safety device for elevators. The folding tube concept is also applicable to collapsible steering wheel design, as well as for overload protection for press brakes, shears, and planers. It could be used to provide a soft-landing capability for spacecraft and as a safety backup for aircraft and helicopter landing gear. A multiple replaceable tube arrangement should be useful in the design of grinding wheel guards, shock-limiting shipping containers, and airdrop cargo containers. To absorb energy in a major crash where structural damage cannot be avoided, major portions of the car frame itself could incorporate the tube on mandrel concept.

It is not uncommon in engineering practice for progress to be blocked by arbitrary government regulations, construction codes, or trade union restrictions. This appears to be the case for automotive energy absorption devices. The U. S. Government ruled out the fuse concept that has been so successful in the electrical industry. Instead, they initially required automotive energy absorption devices to be self-setting and capable of 16 successive impacts without replacement or readjustment of any parts. This needlessly increases initial car cost, and substantially increases the cost of repair when the severity of impact exceeds the capacity of the energy-absorbing device. Later when automotive companies complained of the cost and added weight associated with the complex energy absorption devices, the government reduced the requirement to 3 crashes from 1.5 mph instead of 16 crashes from 5 mph.

Another example of overkill in design is the requirement that energy-absorption devices for use in elevators be capable of being tested and then returned to service without the replacement of any parts. It is not difficult to imagine how much the introduction of the electric light would have been hampered if the government had insisted that every user provide an automatically self-setting circuit breaker instead of a fuse.

It is, of course, important that the mechanical fuse is replaced after use, but this may be taken care of by periodic vehicle inspection and by having the bumper rattle after an impact. The problem of enforcement appears to be no more difficult than in the case of the electrical fuse. If we are to continue to improve the quality of life, it is important to eliminate waste associated with unnecessary design requirements and over restrictive construction codes.

3.3 Highway Crash Barrier

One of the methods used to gently decelerate an automobile about to strike a bridge abutment or other rigid obstacle head on, involves an array of 55 gal drums as shown in Fig. 12.17. Here, there are 29 steel drums. A car striking this assembly head on is brought to rest over a considerable distance, and the operator can escape injury due to excessive deceleration. It is desired to model the performance of the crash barrier shown in Fig. 12.17 on a scale of 1:25 for a car weighing 3,360 lbs, and traveling 52.5 mph.

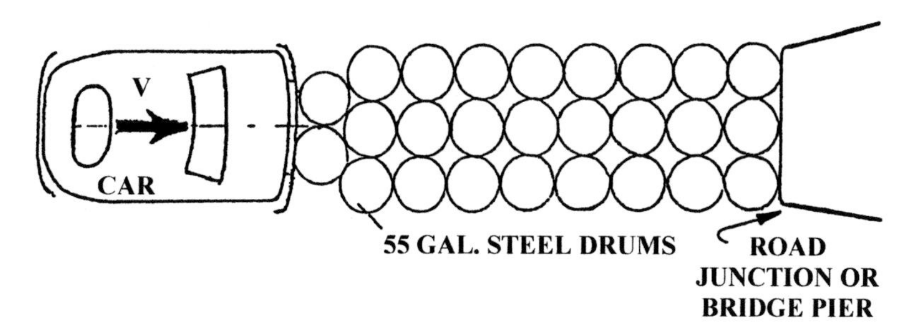

Figure 12.17. Crash barrier of steel drums.

The deceleration of the car (a) as it comes to rest after a head on impact with an array of drums (Fig. 12.17) will be a function of the following variables:

- Drum diameter, d
- Weight of car, W
- Acceleration due to gravity, g
- Velocity of car at time of impact, V
- Energy required to crush a single drum, E

Thus,

Eq. (12.12) $a = \psi_1(d, W, g, V, E)$

After performing a dimensional analysis the following result is obtained where d, V, and E are the dimensionally independent variables:

Eq. (12.13) $(ad)/V^2 = \psi_2\,[(Wd)/E, (gd)/V^2]$

Since three fundamental dimensions (F, L, and T) are involved in this dimensional analysis we may arbitrarily select three scale factors. It is convenient to select the following (where p stands for prototype and m for model):

$$d_p/d_m = 25$$

$$g_p/g_m = 1$$

$$E_p/E_m = (25)^3 = 15{,}625$$

In order to apply the principle of dimensional similitude:

Eq. (12.14) $[(Wd)/E]_p = [(Wd)/E]_m$

or

$$W_p/W_m = (E_p/E_m)/(d_m/d_p) = (25)^3/25 = 625$$

Therefore, the weight of the model car should be 3,360/625 = 5.38 lbs.

Similarly:

Eq. (12.15) $[(gd)/V^2]_p = [(gd)/V^2]_m$

or

$$V_m/V_p = [(gd)_m/(gd)_p]^{0.5} = (1/25)^{0.5} = 1/5$$

Therefore, the initial velocity of the model car should be 52.5/5 = 10.5 mph. When the values of W_m and V_m are determined as indicated above, then it follows from the principle of dimensional similitude that:

Eq. (12.16) $[(ad)/V^2]_p = [(ad)/V^2]_m$

or

$$a_p/a_m = [(V^2)_p/(V^2)_m](d_m/d_p) = 25(1/25) = 1$$

Thus, the deceleration of the model and prototype will be the same if $W_p/W_m = 625$, and $V_p/V_m = 5$.

The scale model test car was made from a weighted wooden block fitted with slot car wheels. The car was given its initial velocity of 10.5 mph by a compressed air launcher operated at 150 psi air pressure. A high-speed motion picture camera was mounted above the model test set up to record the action of the drums and to provide the data required to obtain the displacement time record for the model car.

It is not necessary to model the drums exactly, but only to provide the correct diameter and height (1/25 of the values for the actual drums), the correct weight [$1/(25)^3$] the value for an actual drum], and the correct crushing energy. The actual drums are made from 16 gage steel (0.0598 in. thick) and the model drums are but a few thousands of an inch thick. A convenient material to use is ordinary household aluminum foil, which is 0.003 in. thick. Model drums of the correct diameter may be produced by wrapping foil on a one in. diameter pin and applying cement to the seam. The heights of these tubes should be 1.4 in. Next, an actual steel drum is loaded diametrically in a test machine and the load-displacement curve (*B*) shown in Fig. 12.18 (*a*) determined. Stiffening strips are then cemented across the ends of the model drums as shown in Fig. 12.18 (*b*) and the load deflection curve for the model drum is determined and compared with that of the prototype drum, taking size (1/25) and force [(1/(25)^3] scale effects into account.

The width and thickness of the end stiffening strips are adjusted until the model and prototype load-displacement curves have the same appearance as shown in Fig. 12.18 (*a*). The weight of the model drum is then adjusted to correspond to that of the prototype drum by cementing axial strips of metal at the points where the drums touch. When the end strips on the model drums are proper, the ratio of energy to crush a prototype drum to that required to crush a model drum will be (25)^3. In addition, the load-deflection curves will have approximately the same shape as shown in Fig. 12.18.

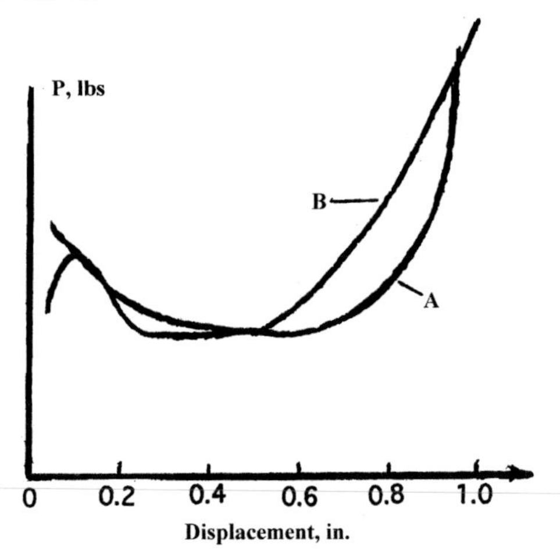

A - Model Drums
0.003 in. thick, soft Al cyl., 1 in. dia x 1.4 in. long
0.005 in. thick x 0.1 in. wide Al ends

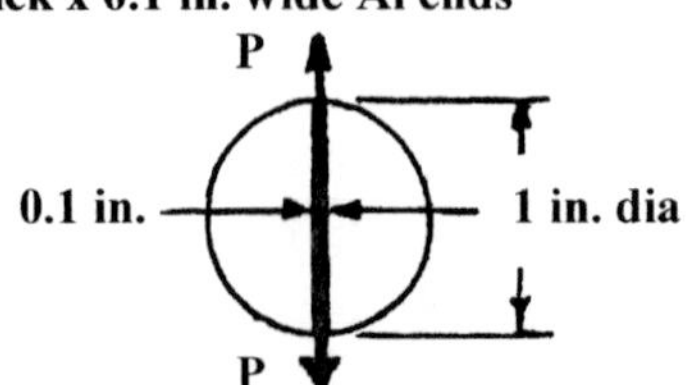

B - PROTOTYPE DRUMS scaled to model dimensions
16 gage steel, dia, 23.4 in. x 34.8 in. high
weight, 60 lbs, peak load = 7600 lbs (P)

Figure 12.18. Comparison of load-displacement curves for model and prototype drums. The displacements are modeled 25:1 and the loads (25)^3:1.

A full scale test involving actual 55 gal drums arranged in the pattern shown in Fig. 12.17 was performed at the Texas Transportation Institute. The initial car speed was 52.5 mph and the weight of the car was 3,360 pounds. The displacement-time curve for the model test is shown in Fig. 12.19 together with that for the prototype test adjusted to take care of differences in scale. The two curves are seen to be practically identical. At the same time, the motions of the model drums resembled those of the prototype drums in all respects, including a wave that propagated forward from drum to drum and the sidewise motion of the drums.

Figure 12.19 also shows the first derivative of the displacement-time curve for the model (the velocity time curve) as well as the second derivative (the acceleration curve). The maximum model deceleration is seen to correspond to about 8.5 g, and this will correspond to that for the prototype as well. This is a very reasonable deceleration considering the speed of the moving vehicle (52.5 mph). The maximum deceleration at which injury of a passenger without a seat belt is unlikely is 3 g, while the corresponding value with a seat belt is 12 g. Thus, the 29 drum configuration considered here is capable of stopping a car from a speed of 52.5 mph without danger to the operator, provided a seat belt is being worn. Of course, a similar study could be performed for polymer drums which are more widely used today.

3.4 High Speed Grinding Wheel Designs

Grinding is a very effective method of removing unwanted material and producing parts of desired geometry. The tool that is used is typically a composite disc with central hole which contains small very hard abrasive particles (typically Al_2O_3, SiC, cubic boron nitride = CBN or diamond = D) held together by a glass-like (vitreous), resin, or metal-bonding material. Figure 12.20 shows a grinding wheel with the point of maximum tensile stress due to centrifugal force at (A). It is desirable from the standpoint of the finish produced and overall cost that the operating speed of the wheel be as high as possible. However, grinding wheels are inherently brittle and brittle materials are much weaker in tension than in compression (up to a 6 to 1 ratio). Since the centrifugal stress at (A) in Fig. 12.20 is tensile and varies as the square of the surface speed of the wheel (V), grinding wheel speeds are limited by brittle fracture.

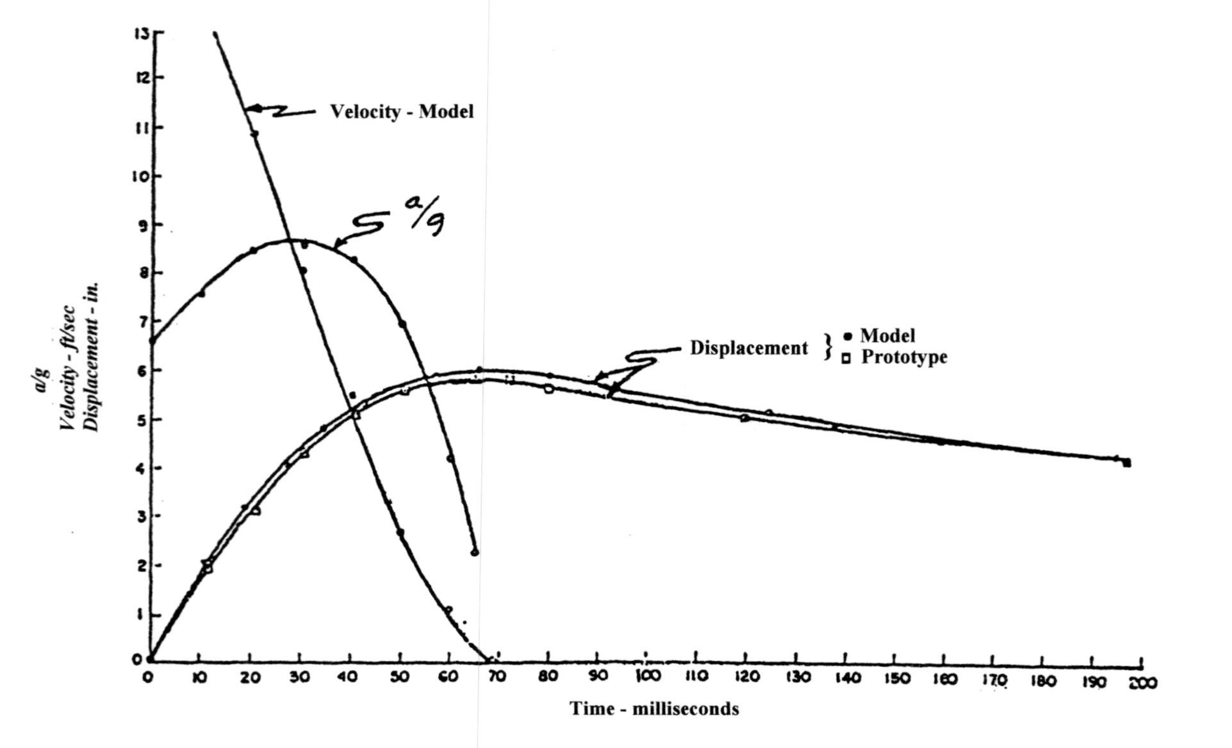

Figure 12.19. Results of scale model impact test compared with results of the actual test scaled down to the model dimensions of distance and time. Also shown is the velocity of the model in fps and the ratio of acceleration to gravity a/g for the model.

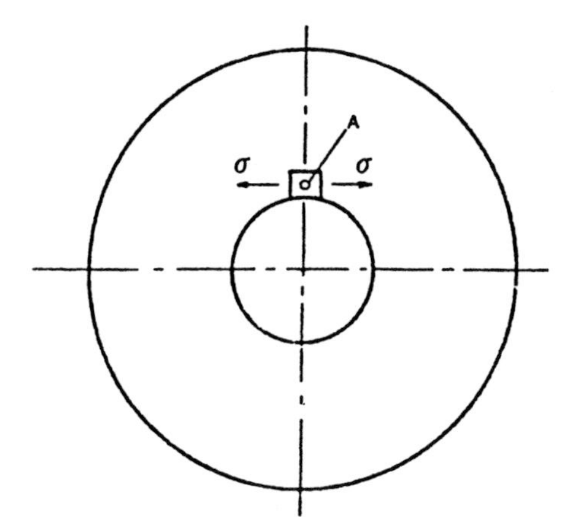

Figure 12.20. Grinding wheel showing point of maximum tensile stress (σ) at (A).

Resin bonded wheels are capable of higher speeds (V up to about 16,000 fpm) than vitrified wheels (V up to about 85,000 fpm). This is because even though a resin bonding material has lower strength, it has greater ductility and an ability to relieve points of local stress concentration by plastic flow.

The main reason a high wheel speed is important is because of a chip storage problem. In operation, a grinding wheel makes intermittent contact with the work as chips are generated. These chips have to be stored in the limited space between active grits as long as there is wheel-work contact. The stored chips are released when contact ceases and they become sparks. For a given material removal rate, the chip storage space between active grits increases in direct proportion to wheel speed (V). Thus, the rate of removal before chip-wheel interference occurs (leading to reduced wheel life) increases directly with wheel speed (V).

Chip storage space may be increased by use of a coarser grit size and less bonding material. However, both of these lower the tensile strength of the wheel and thus offer only a very limited solution to the chip storage problem. In order that safety not be sacrificed in order to inciease productivity, it is a Federal requirement that the maximum surface speed of a grinding wheel be limited to 2/3 the speed at which the wheel disintegrates due to brittle fracture. This provides a factor of safety based on fracture stress of $(1.5)^2 = 2.25$.

A radical departure in the design of a grinding wheel is to employ segments instead of a continuous disc and to substitute compressive retaining stresses for tensile ones. Figure 12.21 shows one way in which this may be done. To prevent the segments from moving out radially, they are provided with tapered faces. As the segments tend to move radially outward, heavy compressive retaining stresses develop (*F* in Fig. 12.21). Since grinding wheels are about 6 times as strong in compression as in tension, the design of Fig. 12.21 enables very much higher wheel speeds and greater productivity. The segmental wheel design offers safety features beyond the substitution of compressive stress for tensile stress. It is much easier to produce and ship small grinding elements without flaws than it is for larger ones. This is particularly true for large vitrified wheels that are apt to develop high thermal stresses on cooling.

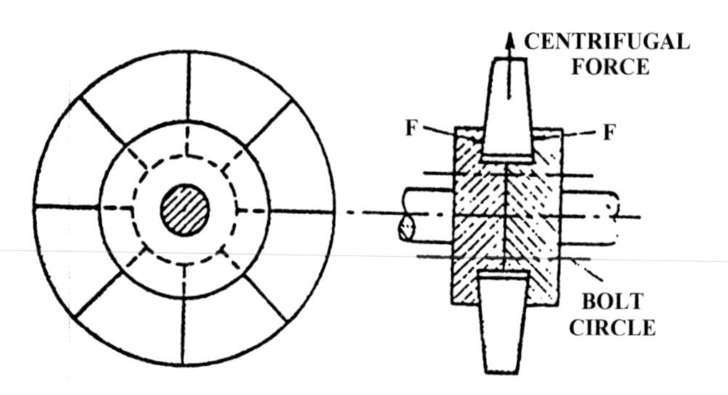

Figure 12.21. Segmental grinding wheel design.

If all of the increase in wheel speed made possible by substitution of compressive for tensile retaining stresses cannot be realized due to machine tool limitations, some of the remainder may be utilized by use of coarser grits and less binder material both of which enable an increase in removal rate. Figure 12.22 shows an alternative design based on the segmental principle. In this case, straight-sided wheel segments are used with tapered steel wedges. This design removes the need for clamping bolts required by the design of Fig. 12.21. It also provides the means for effective distribution of grinding fluid without the use of a high pressure pump. Fluid at atmospheric pressure introduced into compartments (*A*) will flow through passages (*B*) and (*C*) and be forced out between the surfaces of the segments at very high pressure resulting from centrifugal action.

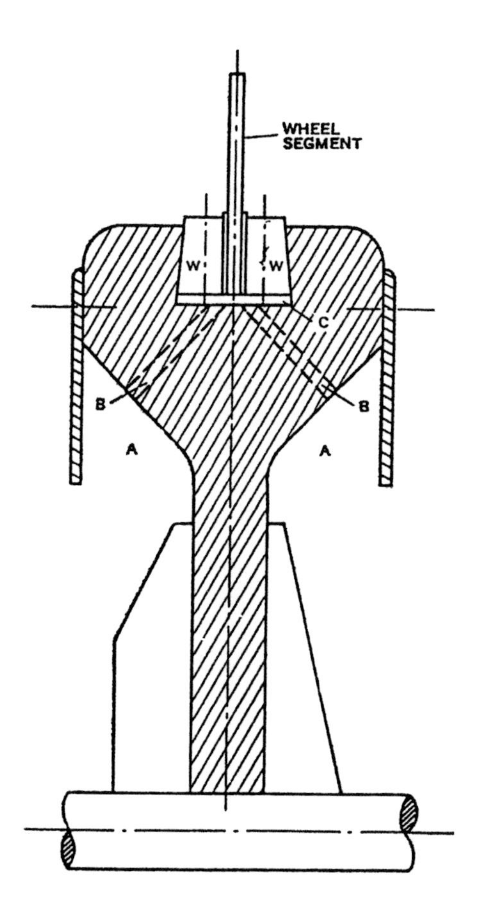

Figure 12.22. Section view of segmental grinding wheel with steel clamping wedges (*W*) and fluid distribution system *A*-*B*-*C*.

Another interesting possibility of the segmental wheel concept is to have alternate segments that have different wear rates (different binder, material, or content). After break-in, this will provide a very high frequency intermittent grinding action. The time when little grinding is taking place provides time for heat to flow with extreme rapidity into the work, thus giving rise to a lower surface temperature of the work.

3.5 The Speed Wand

The speed wand (Fig. 12.23) is an inexpensive device for measuring the speed of a motor boat as it moves through the water. The clear plastic tube is held vertically at the stern of a boat with the curved open end of the tube about three inches (h_3) below the surface of the water and pointing forward as shown in Fig. 12.23. Cap (A) should be tight so that air cannot escape out the top of the tube. The kinetic energy of the water moving with a velocity (V) causes fluid to rise a distance (h_2) up the tube. As the fluid rises in the tube, the air above is compressed. A steel ball check valve prevents water from running from the tube when it is taken from the water to note the level of the water in the tube, which is calibrated in miles per hour. After a test, water is removed from the tube by loosening cap (A) and allowing the compressed air above the liquid column to escape. The device should be held in the water long enough so that an equilibrium height (h_2) is obtained (this takes only a few seconds).

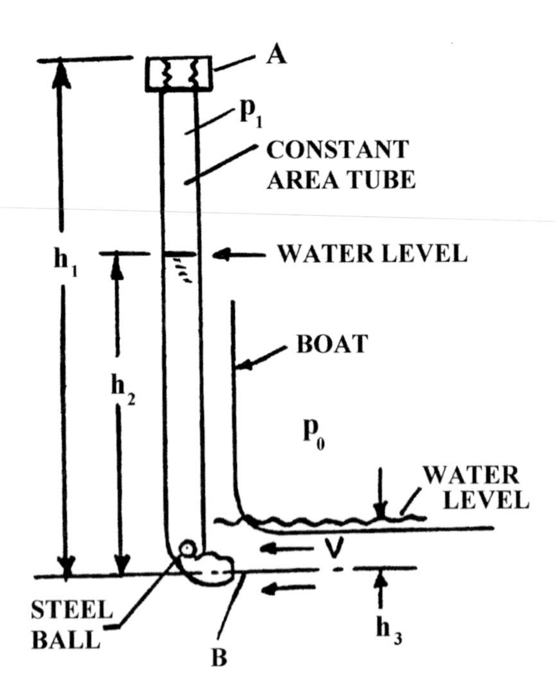

Figure 12.23. The speed wand.

The ideal gas law (Ch. 11) may be used to predict the pressure (p_1) of the air trapped above the liquid in the tube. Since the temperature of the air in the tube before and after a test will be the same, the ideal gas law states that:

$$p_0(\text{vol})_0 = p_1\,(\text{vol})_1$$

or, since the area of the tube is constant:

Eq. (12.17) $\quad p_0(h_1) = p_1\,(h_1 - h_2)$

The Bernoulli equation [Eq. (5.15)] may be applied to point (B) before and after equilibrium has been established:

Eq. (12.18) $\quad p_0/\rho + V^2/2 + gh_3 = p_1/g + 0 + gh_2$

From Eqs. (12.17) and (12.18):

Eq. (12.19)

$$[(h_2/h_1)/(1 - h_2/h_1)][p_0/(h_1\gamma)] + h_2/h_1 - h_3/h_1 = V^2/(2gh_1)$$

For the speed wand, the following constants pertain:

$h_1 = 16$ in.

$h_3 = 3$ in.

$p_0 = 14.7$ psi (atmospheric pressure)

$\gamma = 0.0361$ lb/in^3

$g = 32.2$ ft/sec^2.

After these constants are substituted into Eq. 12.19, a quadratic equation is obtained relating h_2 and V. For a speed of 20 mph (29.33 fps), h_2 is found to be 4.51 inches. When the calculation is repeated for $h_3 = 2$ in. instead of 3 in., h_2 for $V = 20$ mph $= 4.46$ inches (a difference of only 0.05 inch). Similarly, if $h_3 = 4$ in. instead of 3 in., h_2 for $V = 20$ mph $= 4.57$

(a difference of only 0.06 inch). From this it is evident that this instrument is not very sensitive to the depth at which it is placed below the surface of the water (h_3).

Of course, the markings on the tube should be based on a calibration rather than on calculations. However, the calculated values were actually found to be in very good agreement with experimental values, indicating that the foregoing analysis is adequate for initial design of this instrument.

4.0 CHEMICAL ENGINEERING

4.1 Introduction

Chemical engineering is a relatively recent activity (last 100 years). It involves producing a wide variety of chemical products at an ever increasing volume and at reduced cost. Until the beginning of the 20th century, Germany was at the forefront of chemical production largely based on coal. This position has been taken over by the United States largely due to the development of chemical engineering with its emphasis on continuous flow type processing rather than batch processing, and the development of new processes based on the abundant supply of raw materials, principally petroleum. Capital investment in a modern chemical plant is unusually high. This makes it difficult to introduce new technology even though the new technology may have a great potential for the future. An additional factor making it difficult to introduce new technology is that the old technology has proceeded down a learning curve that involves many cost-reducing small scale improvements while the new technology enters high up on its learning curve.

Benefits of large scale production are unusually high in the case of continuous flow chemical plants. When the size of a plant doubles, the unit cost of product approaches 0.5. This poses a difficult economic problem. While it is tempting to overbuild a new plant in anticipation of increased product demand in the future, this may prove disastrous if increased demand falls short of expectations due to unexpected competition, or emergence of a better product or process. Two design problems follow. The first involves the possibility of replacing a batch type process by a continuous flow type, while the second involves improvement of a typical unit process.

4.2 Mechanical Activation

Reactions involving metals and organic compounds have been used in chemical synthesis for a long time. For example, zinc is reacted with organic (carbonaceous) compounds (Frankland, 1849). Grignard (1871-1935) was awarded the Nobel prize in 1912 for his contributions to chemistry. Grignard reactions are among the most versatile available to the organic chemist. Practically any compound may be produced by a Grignard reaction. However, use of these reactions is limited because they are difficult to initiate and control, and are extremely dangerous, and labor intensive. Danger is associated with the highly combustible and explosive ethyl ether used as a catalyst/solvent. High cost is associated with the fact Grignard reactions are normally carried out as a batch rather than a continuous flow process. A challenging design problem is to devise a Grignard process that is controllable, safe, and economical. In the paragraphs below, the conventional Grignard operation is described followed by a description of an improved technique called Mechanical Activation.

The Grignard process involves two steps. First, an organic halide (organic hydrocarbon containing chlorine, bromine or iodine) is reacted with magnesium. This is normally done by placing magnesium chips in a vessel and adding the halide and ethyl ether. A condenser is then attached, and the mixture is refluxed (boiled/condensed) in the absence of air and moisture as shown in Fig. 12.24 (*a*), until reaction begins. There is usually an induction period during which no reaction occurs. However, once reaction begins, the source of heat must be quickly removed and an ice pack quickly applied to reduce the rate of reaction to a safe level. This is because these reactions are exothermic (release heat) and autocatalytic (the rate of reaction varies exponentially with the amount of product produced). If the reaction, once started, is not brought under control, the hazardous ether vapor may explode with serious consequences.

The following equation illustrates the first step in a Grignard synthesis.

Eq. (12.20)

$$\begin{array}{ccccccccccc} & & H & & H & & H & & H & & H \\ & & | & & | & & | & & | & & | \\ H & - & C & - & C & - & C & - & C & - & C - Br + Mg \\ & & | & & | & & | & & | & & | \\ & & H & & H & & H & & H & & H \end{array} \xrightarrow[\text{Heat}]{\text{Ether}} \begin{array}{ccccccccccc} & & H & & H & & H & & H & & H \\ & & | & & | & & | & & | & & | \\ H & - & C & - & C & - & C & - & C & - & C - Mg - Br \\ & & | & & | & & | & & | & & | \\ & & H & & H & & H & & H & & H \end{array}$$

amylbromide   amylmagnesiumbromide

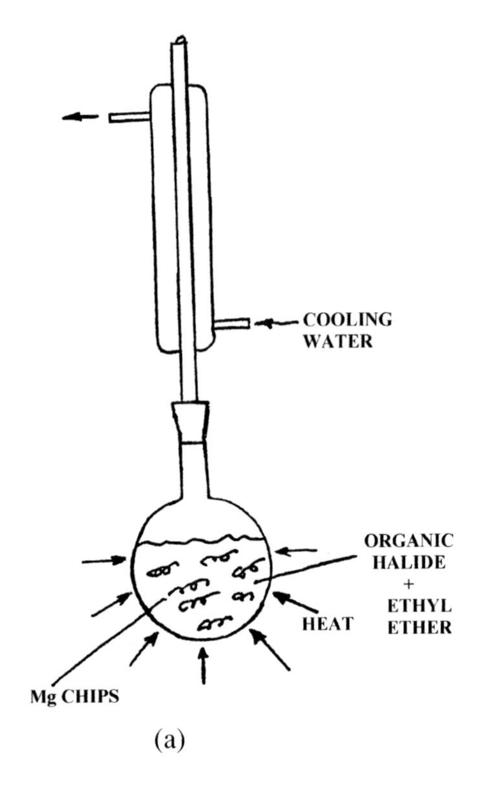

(a)

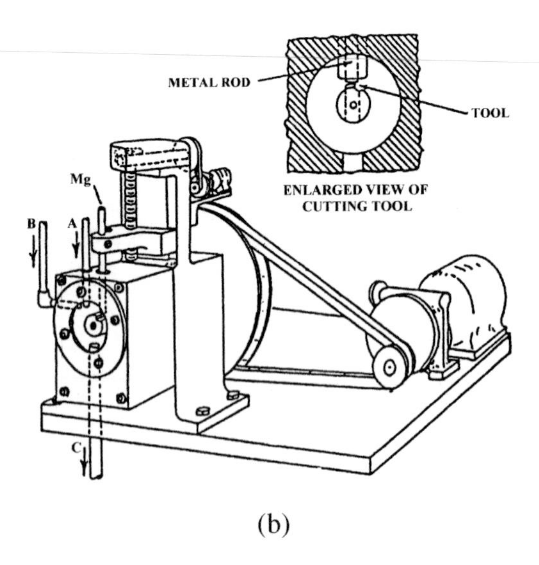

(b)

Figure 12.24. Apparatus for performing a Grignard reaction *(a)* conventional batch type arrangement and *(b)* mechanical activation apparatus.

The second step involves reacting the Grignard reagent with one of a wide variety of compounds to obtain the desired final product. For example, if formaldehyde and then water is reacted with the Grignard reagent in Eq. (12.20), the final product will be the normal primary alcohol hexanol:

Eq. (12.21)

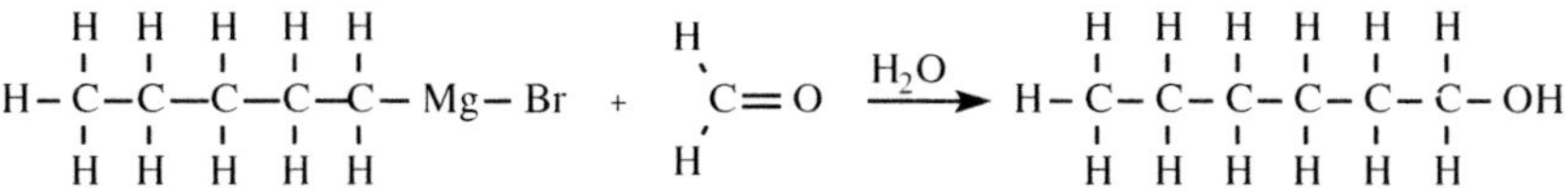

$$\begin{array}{c} \\ \text{H}- \\ \\ \end{array}\begin{array}{c} \text{H} \\ | \\ \text{C} \\ | \\ \text{H} \end{array}\!-\!\begin{array}{c} \text{H} \\ | \\ \text{C} \\ | \\ \text{H} \end{array}\!-\!\begin{array}{c} \text{H} \\ | \\ \text{C} \\ | \\ \text{H} \end{array}\!-\!\begin{array}{c} \text{H} \\ | \\ \text{C} \\ | \\ \text{H} \end{array}\!-\!\begin{array}{c} \text{H} \\ | \\ \text{C} \\ | \\ \text{H} \end{array}\!-\text{Mg}-\text{Br} \;+\; \begin{array}{c} \text{H} \\ \diagdown \\ \\ \diagup \\ \text{H} \end{array}\!\text{C}=\text{O} \xrightarrow{H_2O} \text{H}-\begin{array}{c} \text{H} \\ | \\ \text{C} \\ | \\ \text{H} \end{array}\!-\!\begin{array}{c} \text{H} \\ | \\ \text{C} \\ | \\ \text{H} \end{array}\!-\!\begin{array}{c} \text{H} \\ | \\ \text{C} \\ | \\ \text{H} \end{array}\!-\!\begin{array}{c} \text{H} \\ | \\ \text{C} \\ | \\ \text{H} \end{array}\!-\!\begin{array}{c} \text{H} \\ | \\ \text{C} \\ | \\ \text{H} \end{array}\!-\!\begin{array}{c} \text{H} \\ | \\ \text{C} \\ | \\ \text{H} \end{array}\!-\text{OH}$$

amylmagnesiumbromide formaldehyde hexanol

Equation (12.20) may be generalized as follows:

Eq. (12.22) $$R - X + Mg \xrightarrow[\text{heat}]{\text{ethyl ether}} R - Mg - X$$

where R represents any aromatic or aliphatic hydrocarbon and X is a halide (Cl, Br, or I). Similarly, the second reaction may involve one of a wide variety of secondary items. For example, water alone will lead to a hydrocarbon, $CO_2 + H_2O$ will lead to an organic acid, etc.

An improved procedure for carrying out a Grignard reaction is illustrated in Fig. 12.24 (*b*). Here, a magnesium rod is converted into small chips under the surface of the liquid reactants (halide and ethyl ether) as shown in the insert in Fig.12.24 (*b*). Liquids (A) and (B) are introduced continuously into the sealed reaction chamber along with dry nitrogen to exclude air and water vapor. First stage reaction products are extracted continuously at (C). Small chips (having an undeformed thickness of about 0.001 in. (25 μm) are generated under the surface of the liquid reactants, and begin to react immediately as they fall toward the outlet at (C).

There are three reasons why reaction begins immediately:

- High pressure—pressures up to the hardness of the metal cut are present on the surface of the chip as it is generated.

- High temperature—surface temperatures approaching the melting point of the metal being cut will be present at the chip-tool interface.
- Highly reactive surfaces are continuously generated.

When a new surface is generated, there is an excess of electrons in the new surface. For new surfaces generated in vacuum, these excess electrons are expelled to the atmosphere as the surface comes to equilibrium and are called exoelectrons. For new surfaces generated in air, an oxide will normally form. For new surfaces generated under the surface of a liquid, an appropriate reaction will occur induced by the presence of excess free electrons. Collectively, the three items mentioned above (high temperature, high pressure, and enhanced surface reactivity) are referred to as *Mechanical Activation.*

This method of organometallic synthesis is preferable to conventional synthesis for the following reasons:

- Safety—a very small quantity of material is reacting at any one time, and even this is contained in a steel reaction chamber.
- Control—if the rate of reaction becomes too great, the cutting speed need only be reduced by reducing the rpm of the cutter.
- Continuity—the conventional batch arrangement is replaced by a continuous one.

When properly set up, inputs at (A) and (B) are in equilibrium with the output at (C).

When the two steps of a Grignard synthesis are combined, this is called a Barbier reaction. Such a procedure, which is normally not possible in a conventional batch procedure is frequently made possible by mechanical activation. This constitutes an added advantage.

Processes such as mechanical activation emerge when the conventional activities of a chemist or chemical engineer are combined with those of a mechanical engineer. This example illustrates why it is advantageous for an engineer involved in the development of new technology to be familiar with fundamentals in several areas of engineering rather than specializing in a single field. Both types of engineers (generalists and specialists) are equally important. The one best suited to a particular individual depends upon his or her background, interest, and inherent capability.

4.3 Static Mixer

In chemical engineering, a number of unit processes are normally involved in the design of a system. These include distillation, filtration, recrystallization, mixing, etc. In industrial mixing, two or more materials are blended together to yield a homogeneous product. Mixers that employ propellers or turbines are quite common. Such dynamic mixers involve power driven elements that tend to make them expensive.

Motionless or static mixers are advantageous in that they have no moving parts, but mixing action is not always adequate. The simplest static mixer is a straight pipe 50–100 diameters in length. If operated at a Reynolds Number greater than a few thousand, flow will be turbulent and some mixing will result. More complicated static mixers involve a series of baffles that repeatedly divide the stream. This is equivalent to cutting a deck of cards n times where n is the number of baffles. While such devices perform well they are relatively expensive to manufacture.

Figure 12.25 shows a static mixer based on the Bernoulli equation [Eq. (5.15)] that is relatively simple and inexpensive to manufacture. Two fluids to be mixed enter Box (B) at (A_1) and (A_2) respectively. The box is divided into two compartments by an undulating member (C). A series of holes (H) located at peaks and valleys of (C) cause cross flow between the two compartments. At section (1) in Fig. 12.25 (*b*), the flow velocity will be high in the lower compartment but low in the upper compartment, causing a downward cross flow of fluid through the holes located at section (1) in accordance with the Bernoulli equation. At section (2), conditions are reversed and cross flow will be upward. The extent of mixing will depend upon the relative velocities in the two compartments (upper and lower area ratio), the ratio of transverse area for flow to total longitudinal flow area, and the number of undulations.

The central element (C) may be easily produced from a flat sheet by punching followed by a forming operation. The mean pressure will be approximately the same for the upper and lower compartments requiring support of member C only at its ends. A welded support at the entrance end and a spider (multiple radial wires) support at the exit should suffice.

Application of the Bernoulli and continuity equations will enable the ratio of transverse area to total longitudinal area and number of sections required for any degree of mixing to be determined.

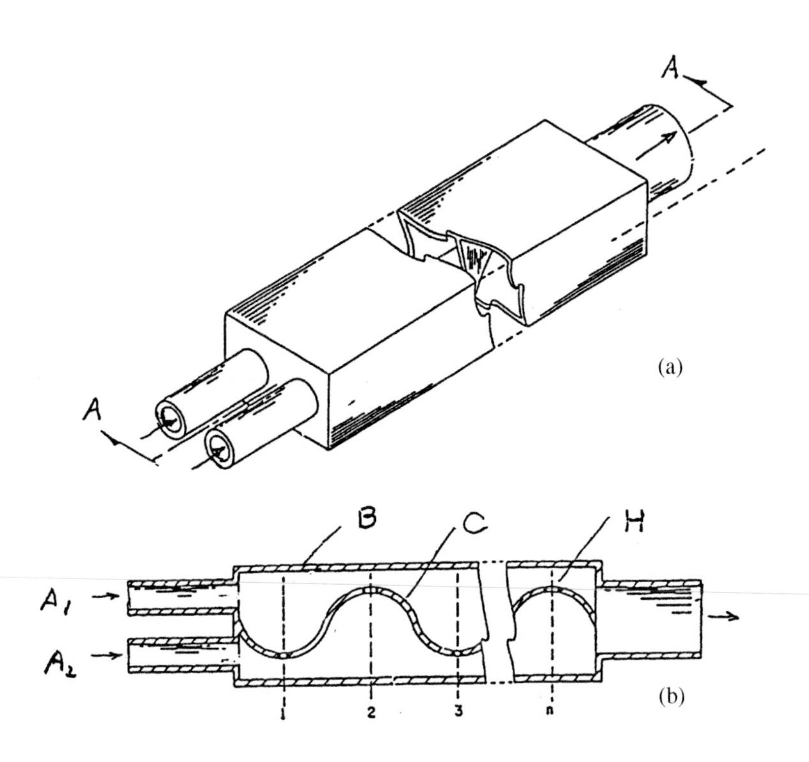

Figure 12.25. Bernoulli type static mixer *(a)* overall view of box type design and *(b)* section A-A showing central perforated baffle.

Figure 12.26 shows an alternative design for the Bernoulli type mixer of Fig. 12.25. This consists of an outer tube and center rod with an undulating tube with punched holes at peaks and valleys between the tube and rod. Other configurations of a Bernoulli type static mixer are possible.

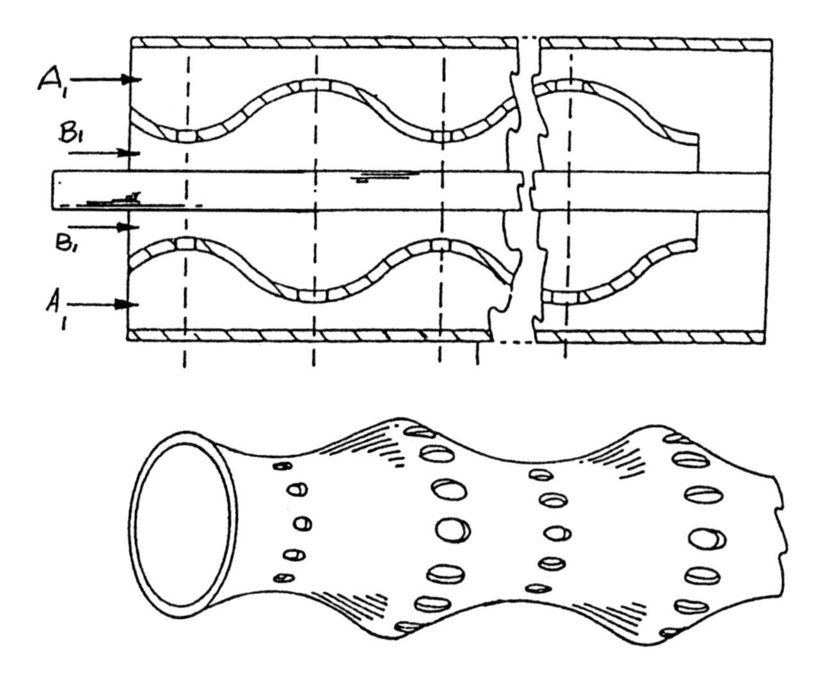

Figure 12.26. Tube type Bernoulli mixer (*a*) sectional view; and (*b*) detail of central perforated baffle.

PROBLEMS

12.1 Briefly describe a new application for the principles involved in one of the following areas discussed in this chapter:

- Falling paper
- Micro-explosive action
- Wind power generation
- The mechanical fuse
- Mechanical activation
- Static mixer

12.2 A weir (described in Problem 5.25) is being designed for use in sea water (5% denser than fresh water). Perform a dimensional analysis for the rate of flow and explain how the calibration for the same weir designed for fresh water should be altered for use with sea water.

12.3 A 3×5 index card glides through the air at a speed of 1,000 ipm at an angle of 30°.

a) Estimate the rpm of the card.

b) Estimate the lift-to-drag ratio.

12.4 The kite shown in Fig. 12.6 (*a*) weighs 1.5 oz, rotates at 100 rpm in a 10 mph horizontal wind, the tensile force on the kite string is 5 oz, and makes an angle of 45° to the horizontal.

a) Find the lift and drag forces on the kite.

b) Estimate the diameter of the rotor.

12.5 A Savonius rotor 10 ft in diameter and 30 ft high operates at a speed of 93 rpm in a wind of 10 mph. Using the efficiency curve of Fig. 12.10 (*A*) and γ_{air} from Table 5.1, estimate the horsepower generated. If the wind velocity increases to 30 mph, estimate the horsepower.

12.6 Design a water turbine using a Savonius rotor for a stream of water five ft deep, 10 ft wide, and flowing with a mean stream velocity of:

a) 5 mph.

b) 10 mph.

Estimate the power output for your design for these two fluid velocities.

12.7 As described in *Popular Science* (Oct. 1974, p. 4), a Savonius rotor has been built from oil drums (Fig. P12.7). Ordinary 55 gallon drums measure 23 in. diameter by 35 in. high. Estimate the power generated in the rotor shown in a 10 mph wind.

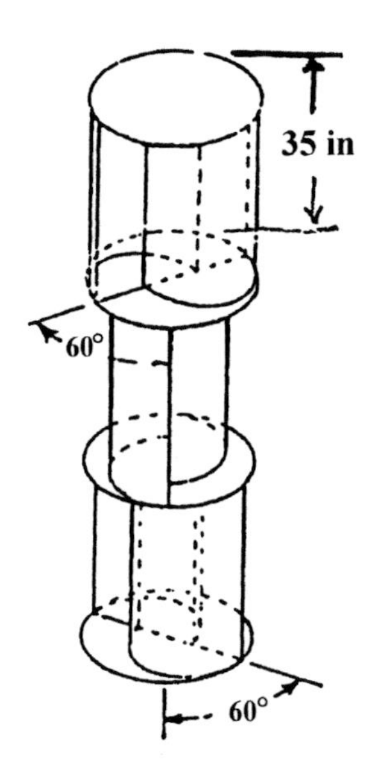

Figure P12.7.

12.8 A multiblade U. S. type horizontal-axis wind mill operating at the optimum value of tip speed ($\pi DN = V$) has an efficiency of about 27% [Fig. 12.10 (*B*)]. If the wind velocity is 30 mph and the diameter of the rotor is 3 ft, find the output horsepower of the rotor and the optimum rpm.

12.9 For a rotor as in Fig. 12.11 having a diameter of 14 ft, the efficiency vs speed ratio is given in Fig. 12.10 (*E*).

a) Estimate the rpm for maximum efficiency at a wind speed of 10 mph.
b) Estimate the corresponding output horsepower.

12.10 Design an energy-absorbing bumper attachment (attached at two points) that will absorb the energy of a 2,500 lb small car operating at 10 mph if the allowable deceleration during energy absorption is five times the acceleration due to gravity.

12.11 A vehicle that weighs 2,000 lbs has a velocity of 10 mph when it strikes a rigid stone wall. If the bumper moves back a distance of 6 in. as the car is brought to rest, estimate the mean deceleration of the vehicle in ft/sec^2. Repeat the problem if the initial velocity is 20 mph.

12.12 Two identical automobiles each weighing 1,000 lbs collide head-on when each has a velocity of 10 mph. Simulate this situation by use of two identical 1:10 model cars made from the same materials as the prototype. Perform a dimensional analysis for deformation (δ). Assuming the same properties for model and prototype, what should the velocity ratio and weight ratio for model and prototype be in order that the change in dimensions after collision of model and prototype scale in a ratio of 1:10?

12.13 The deflection (δ) of a cylindrical spring per turn when subjected to a tensile force (F) is considered to be a function of the following variables:

$$\delta = \psi_1(F, D, d, G)$$

where: D = diameter of the coil

d = diameter of the wire

G = shear modulus of the wire

a) Perform a dimensional analysis.

b) Simplify this result by use of an obvious fact.

c) Using a straightened paper clip wire, form two circular rings one having a diameter (D) and the other twice as large (dia = $2D$). Load the rings with the same tensile force (F) as shown in Fig. P12.13. Use this relationship to further simplify the relation for δ. A convenient way of applying the same force (F) is to hold one end of a wire and load the other with a soft rubber band. When the deflection of the rubber band is the same in the two cases the forces will be the same. Care must be taken not to load the rings into the plastic region.

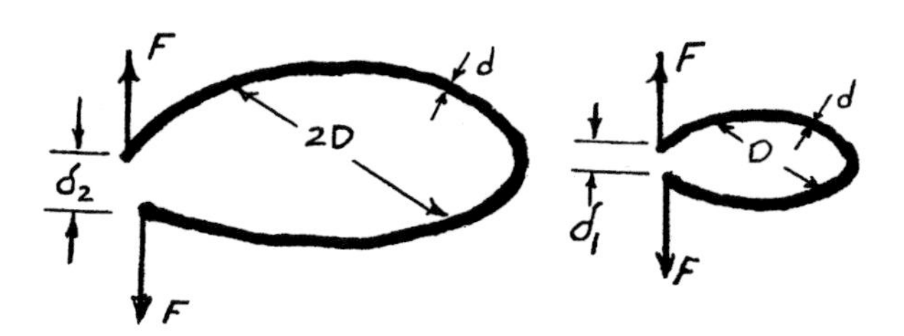

Figure P12.13.

12.14 A gondola lift at Sulfur Mountain in the Canadian Rockies near Banff has the following specifications (Fig. P12.14):

- Elevation of Lower Terminal, 5,194 ft above sea level
- Elevation of Upper Terminal, 7,486 ft above sea level
- Horizontal distance traveled, 4,498 ft
- Velocity of gondola, 10 fps
- Maximum passengers per hour, 450 in each direction
- Electric motor drive at Upper Terminal, 180 hp
- Track cable: 1-11/32 in. diameter, 46,587 lbs
- Hauling cable: 1-3/32 in. diameter, 21,091 lbs
- Overall mechanical efficiency, 75%

a) If the capacity of a gondola is 20 people, what is the total number of gondolas required?

b) If the weight of a single gondola is 2,000 pounds, estimate the horsepower required for the extreme case where all gondolas are ascending full and those descending are empty.

c) Is the horsepower available adequate?

d) What is the average incline in percent?

Figure P12.14.

12.15 During the period when the Suez Canal was closed, it became necessary for the British to quickly devise a means for increasing petroleum transport capacity. This was done at Cambridge University under the leadership of Prof. W. R. Hawthorne (now Sir William). The solution was to use a large nylon reinforced rubber "sausage" called a Dracone that was towed behind a small 50 hp launch. In one test, a Dracone measuring 100 ft in length by 4.75 ft in diameter filled with 35 tons of kerosene was towed at a speed of 8 mph (7 knots) with a steady 1,000 lb pull on the nylon tow rope. After being pumped out, the empty 2,300 lb Dracone was rolled up into a package measuring only 9 × 6.5 ft.

a) Determine the horsepower required to tow the Dracone under the conditions described above.

b) Estimate the drag coefficient (C_D) for this Dracone.

c) Estimate the drag force on the Dracone at a speed of 10 mph.

d) Estimate the horsepower dissipated at the Dracone at a speed of 10 mph.

12.16 A thin circular ring is a useful dynamometer element. When loaded radially and elastically as shown in Fig. P12.16, $\delta_1 = \delta_2$. It is reasonable to expect that $\delta = \psi_1(W, E, r, I)$, where:

W = applied tensile or compressive load

E = Young's modulus of the ring material

r = radius of ring

I = moment of inertia of rectangular section of ring about the neutral axis (i.e., $I_N = 1/12\ bt^3$)

b = width of ring

t = thickness of ring

a) Perform a dimensional analysis.

b) Simplify a) by noting that $\delta \sim W$ and $\delta \sim b$.

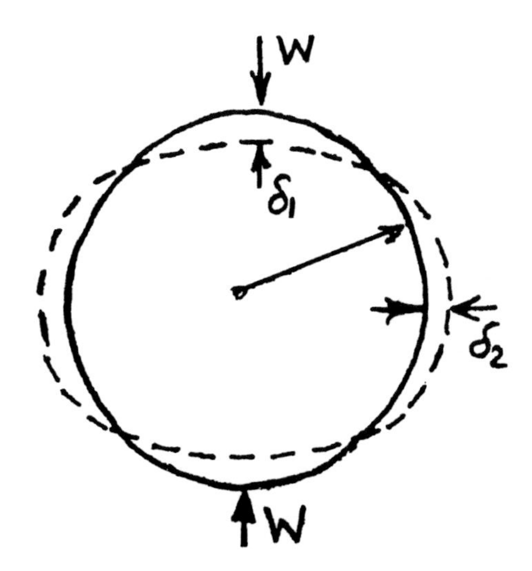

Figure P12.16.

12.17 In general, the volume rate of flow (Q) [$L^3 T^{-1}$] of a fluid through a straight pipe will depend upon the following variables:

d = diameter of pipe [L]

ρ = mass density of fluid [$FL^{-4}T^2$]

μ = viscosity of fluid [FTL^{-2}]

G = pressure difference between two points a distance (ℓ) apart = pressure gradient = $\Delta p/\ell$, [FL^{-3}]

That is: $Q = \psi_1(d, \rho, \mu, G)$

a) Perform a dimensional analysis with d, r, and μ as the dimensionally independent group.

b) Simplify this equation for flow at a very low Reynolds Number (R).

c) Simplify equation (a) for flow at a very high Reynolds Number (R).

13

Engineering Economics

1.0 INTRODUCTION

The broad field of economics may be divided into macro and micro economics. Macroeconomics involves problems associated with nations such as trade, trade deficits, monetary policy, national productivity, growth of the economy, inflation, budget deficits, national debt, unemployment, tariffs, etc. Microeconomics involves problems of firms and of individuals. Engineering economics is a special branch of microeconomics largely involved with the analysis of engineering alternatives and their performance.

The two most important macroeconomic theories are those espoused by Adam Smith (1723–1790) and J. M. Keynes (1883–1946):

- Smith suggested (*Wealth of Nations*, 1776) that a free economy driven by market forces and free of government intervention is best.
- Keynes suggested (*General Theory of Employment, Interest, and Money*, 1936) that government intervention is important, particularly in times of economic stagnation and inflation. His views had a strong influence on actions taken by President Roosevelt in formulating "The New Deal" of the 1930s.

Debate continues today, fueled by those two points of view, on the basic question of whether the nation is better served by a big or small federal government?

The cost of designing, producing, or using an engineering structure or device is an extremely important engineering variable and should be quantitatively treated just as carefully as strength, rigidity, output, or efficiency. There is no more compelling reason for basing the solutions to engineering economic problems on hunch than there is for adopting such an approach in making any other type of engineering decision.

Many engineering economics problems involve the choice, based upon cost, between two or more alternative solutions. When the alternatives are not exactly identical, it is necessary to assign a money value to differences which exist. When these differences are intangible, this may call for considerable judgment. When considering alternative solutions involving different costs, it is important that none of the possibilities be overlooked. Accounting is not to be confused with engineering economics. While accounting is concerned with the apportionment of costs and what money spent was actually used to achieve an end, engineering economics is concerned with the cost of alternative solutions to a problem, mainly projected into the future.

It is important to recognize that economic considerations may lead to a design or structure that is less perfect than could be achieved if costs were not considered. For example, it may be advisable to specify a piece of energy conversion equipment that operates at an efficiency that is less than the maximum attainable value, provided in so doing the actual unit cost of power is less over the life of the machine. The cost of achieving the last increment of improvement frequently costs more than the average unit cost and, in such cases, it usually is not justifiable to include this last increment of output when designing the device. Despite these observations, many textbooks on hydraulic machinery and similar equipment discuss the problem of choosing the type of machine best suited to a given set of operating conditions (head, flow rate, speed, etc.) from the standpoint of efficiency without regard for the relative costs involved in the several designs. This infers that the engineer's job is to design or to select a machine that operates at maximum efficiency, regardless of cost. While this may be the case under special circumstances (for example, where the weight of the device is of major importance), the more usual situation corresponds to the case where the best solution is a minimum overall unit cost without regard for efficiency or other secondary variables except as they influence unit cost.

Frequently, a design variable that effects unit cost can be changed continuously over a range of values. In such cases, an optimum value of the design variable may exist for which the unit cost is a minimum. It is usually advantageous to operate as close to this minimum as convenient.

The main variable of engineering economics is cost (C, or unit cost, ¢) measured in units of dollars. In dimensional analysis problems in economics, the dollar is a fundamental quantity having the dimension [$]. Money is a tangible device for paying costs. Unlike the inch or the pound, the dollar is a unit that changes in absolute value with time for two reasons:

1. Money has a self-multiplying character. A dollar invested today is worth more than a dollar ten years hence. The growth in the amount invested (called the principal, P) is called *interest*.
2. The goods or services that a dollar will buy changes with time due to fluctuations in supply and demand and technological development. This shift in absolute purchasing power with time tends to be in a direction corresponding to a reduction in the absolute value of the dollar, and is called inflation.

The changing value of money with time makes it necessary that all comparisons of cost be referred to a fixed point in time. This is usually taken as the beginning or the end of the period of time under consideration.

2.0 INTEREST

Interest is money charged for the use or potential use of borrowed money. While interest is of two types—simple (fixed yearly charge for the use of money over a period of years, where interest is charged on principal only) and compound (where interest is charged on previous interest in addition to the principal). The latter is the only realistic quantity, and will be the only one considered here.

If the amount borrowed (the principal) is designated (P) and the interest due at the end of the period (usually one year) is (i percent), then, the amount due at the end of the first period for the use of P during this period will be $P(1+i)$. The total amount due after two periods will be $P(1+i)^2$, while the total single payment (S) due for the use of P for n periods will be:

Eq. (13.1) $$S = P(1+i)^n$$

The quantity (S) is the equivalent value (n) periods hence, corresponding to P right now.

Instead of repaying (P) with compound interest by a single payment (S) after n periods of use, another equivalent system of repayment is frequently used in which the same payment (R) is made at the end of each of the n periods. The uniform payment (R) made at the end of the first period will be worth $R(1+i)^{n-1}$ at the end of the nth period. The second payment will be worth $R(1+i)^{n-2}$ at the end of the nth period, etc. The sum of the n payments each equal to R will have the following total value (S) at the end of the nth period:

Eq. (13.2) $$S = R + R(1+i) + R(1+i)^2 + \ldots R(1+i)^{n-1}$$

Eq. (13.3) $$(1+i)S = R[(1+i) + (1+i)^2 + \ldots (1+i)^n]$$

Subtracting Eq. (13.2) from Eq. (13.3) and solving for R:

Eq. (13.4) $$R = (Si)/[(1+i)^{n-1}]$$

From Eq. (13.1):

Eq. (13.5) $$R = \{[i(1+i)^n]/[(1+i)^{n-1}]\}P = F'P$$

where $F' = \psi(i, n)$ is called the *capital recovery factor*.

The quantity (i) appearing in the above equations may be the interest charged by a bank for the loan of money used to purchase a house (usually from 4–8% or more for this purpose). This value includes the expected rate of inflation. In such a case, (i) is called the interest rate. It is customary to repay the loan with compound interest by a series of n uniform payments each equal to (R) as determined from Eq. (13.5), where n may vary from 5–30 years or more.

In engineering economics problems, (i) has a somewhat different character and is called the minimum attractive return. In addition to the actual bank value of the money, (i) must include the following in such instances:

- Cost of securing and administering a loan if necessary
- Added risk associated with a business's use of the funds
- An amount that reflects the fact that funds for investment are not without limit, and that the proposal under consideration must compete for the funds available with other proposed projects
- Inflation

These factors cause the value of (i) in engineering economics problems to be significantly higher than in ordinary banking situations (i = 10 to 20%).

3.0 CAPITAL

Capital is money that must be obtained to pursue an engineering enterprise, and it is of two types—equity and debt. Equity is money usually obtained by issuing stock that makes the one providing the funds a part owner in the enterprise. The provider expects to obtain a return in the form of increased value of the stock or in dividends. However, neither of these is guaranteed. Debt is money that is borrowed with an obligation to repay the principal plus a fee for the use of the borrowed funds (interest). Equity involves low risk to the company, but high risk to the provider while debt involves heavy risk to the company, but low risk to the provider.

New projects usually require new capital. To encourage growth, it is important that equity funds be available. Availability of such funds depends upon the savings habits, the tax system, and the stock pricing policies obtainable in a given country. The following conditions in Japan create a greater opportunity for entrepreneurial development than in the USA.

- Greater per capita savings
- Essentially no capital gains tax on growth of equity
- A higher average price/earnings ratio for stock offerings

The first two of these are self-evident. The last requires explanation. The average price earnings/ratio of stocks in the USA is about 14 while in Japan, it is close to 40. This forces CEOs in the USA to favor projects having a much shorter development period. This practice rules out innovative solutions that require more development, but which eventually may prove more productive. Another important factor that depletes equity funds for industrial development is the national debt. A large national debt siphons off large amounts of money (interest) in the form of federal bonds required to process the debt. This depletes the equity funds available for growth of the economy.

4.0 INFLATION

Inflation is usually a few percent (2–5%) per year. However, in the late 1960s and 1970s, there was double-digit inflation (up to 20% per year). High inflation has an important negative effect on engineering development. High inflation not only increases the interest rate for a loan, but also requires a larger loan since the value of money decreases over the period of the loan.

Inflation is frequently measured by the Consumer Price Index (CPI) published periodically by the Labor Department. The CPI indicates the increase in cost of a set of products over a fixed period of time. The commercial interest rate charged by banks is regulated by the U. S. Federal Reserve as a means of controlling inflation. A decrease in this controlled interest rate moves investments toward stocks, while an increase moves investments toward bonds.

5.0 DEPRECIATION

Depreciation over the life of a machine is the difference between initial cost and the final value (i.e., $C_i - C_f$). The yearly depreciation (d) is sometimes calculated by a straight line method:

Eq. (13.6) $$d = (C_i - C_f)/y$$

where y is the period involved in years.

A sinking fund method is also sometimes used:

Eq. (13.7) $d = (C_1 - C_F)/F'$

where F' is the capital recovery factor in Eq. (13.5).

6.0 SIMPLE COMPARISONS

If two different machines (α and β) are under consideration, the most attractive one to purchase may be decided on the basis of minimum annual cost. The following example will illustrate the principles involved.

The two machines have different values of initial cost and different values of annual cost (yearly operating and maintenance costs) as indicated in Table 13.1. If each machine has an expected life of 8 years with zero salvage value and the expected rate of return (i) is 10%, then the annual cost associated with the investment (R) will be as indicated in Table 13.1 as computed from Eq. (13.5). The total annual cost (investment plus operating cost) is seen to be least for machine β, which is the most attractive one to purchase provided all items of importance have been considered.

Table 13.1. Comparison of Total Annual Cost of Two Machines (α and β) with the Same Expected Life, Expected Rate of Return on the Investment, and with Zero Salvage Value

Quantity	α	β
Initial cost (P)	$12,000	$9,000
Mean annual operating cost	2,500	3,000
$R = 0.188\,P$ [Eq. (13.5)]	2,250	1,690
Total annual cost	4,750	4,690

If machine α has a salvage value of $2,000 while machine β still has no salvage value, the calculation of R for machine α must be changed. The capital recovery factor (0.188) should be applied to only the part of the initial investment that will not be recovered ($12,000 - 2,000 = $10,000) while only the interest need be paid each year on the $2,000 that is recovered at the end in salvage. Thus, the item $2,250 that appears in the α column should be replaced by [(0.188)(10,000) + 0.1(2,000) = 2,080] and the total becomes $4,430. Machine α now represents a more attractive purchase than machine β.

The foregoing calculation can be extended to cases where the expected life is different for each case, or the minimum expected return (i) differs (different risk or variation in i with amount of capital required). Comparisons may also involve more than two alternatives. However, these are merely extensions of the simple cases already considered.

Often the best alternative is the one requiring the lowest capital investment. It is also often best to build the largest facility the present market will absorb, and even somewhat larger if there is strong evidence for growth. However, such estimates are risky and should take the possibility of the emergence of new technology into account.

7.0 KELVIN'S LAW

In many engineering problems, the total cost of an activity (C) depends upon more than one function of a design variable (x) (which may be area, speed, resistance, etc.). If one function causes an increase in cost when x is varied while another causes a decrease, it is possible for an extreme to exist which may represent an optimum value of x (designated x^*), if the extreme happens to be a minimum.

Lord Kelvin considered the optimum cross-sectional area (A) for a copper wire carrying a direct current of magnitude (I). The criterion of selection was minimum yearly total cost of an installation including the loss of power due to resistive loss in the wire and the value of capital invested in copper.

When a current (I) flows through a wire of resistance (R), the voltage drop from one end of the wire to the other (R) is from Ch. 10:

Eq. (13.8) $V = IR$

and the power loss in watts will be:

Eq. (13.9) $VI = I^2R$

The resistance of a long wire (R) may be expressed in terms of its specific resistance (ρ), its length (ℓ), and area (A) as follows:

Eq. (13.10) $R = (\rho\ell)/A$

The value of the power loss in one year will be:

$$I^2RhC_2 = (I^2\rho\ell hC_2)/A$$

where h is the number of hours current (I) flows per year and C_2 is the value of power per W/hr.

The value of the money invested in copper per year will be ($F'C_1\gamma\ell A$) where C_1 is the cost of copper per pound, F' is the capital recovery factor (computed from Eq. (13.5) for reasonable values of n and i such as 25 yrs and 10% respectively), γ is the specific weight of copper (weight per unit volume), and ℓ and A are the length and area of the conductor.

The total yearly cost will be:

Eq. (13.11) $$C = \underbrace{(F'C_1\gamma\ell)A}_{\text{I}} + \underbrace{(I^2\rho\ell hC_2)/A}_{\text{II}} + d = \underbrace{aA}_{\text{I}} + \underbrace{b/A}_{\text{II}} + d$$

where d includes inspection and other costs that are not a function of the area of the wire (A).

When A is made large, the cost of the lost power [cost II in Eq. (13.11)] will be low, but the cost of the copper [cost I in Eq. (13.11)] will be high and vice versa. There should be an optimum value of A for which C will be a minimum. This may be found by equating $\partial C/\partial A$ to zero (see App. A, if needed):

Eq. (13.12) $\partial C/\partial A = a - b/A^2 = 0$

The optimum value of A (i.e., A^*) will be:

Eq. (13.13) $$A^* = (b/a)^{0.5} = [(I^2 \rho h C_2)/(F' C_1 \gamma)]^{0.5}$$

It should be noted that the length of the conductor (ℓ) that appears in Eq. (13.11) has cancelled out in going from A to A^*

The corresponding value of the yearly cost of investment in copper and power loss (C^*) may be found by substituting A^* into Eq. (13.11) and ignoring d which is independent of A:

Eq. (13.14) $$C^* = a(b/a)^{0.5} + b/(b/a)^{0.5} = \underset{\text{I}}{(ab)^{0.5}} + \underset{\text{II}}{(ab)^{0.5}}$$

That is, total cost is a minimum when copper cost and cost due to power loss are equal.

Kelvin's law in electrical engineering states that the economic size of a conductor for transmitting current is that for which the annual investment charges for copper just equal the annual cost of the power lost. The similarity between this law and the concept of impedance matching discussed in Ch. 10, Sec. 12, should be noted.

Example. The optimum diameter of a transmission line is to be determined for the following conditions:

Current carried (I) = 300 amps

Specific resistance of copper per inch (ρ) = 0.678×10^{-6} ohm

Use per year (h) = (360)(24) = 8,640 hr

Value of power per Wh = $\$10^{-5}$

Cost of copper conductor per lb (C_1) = $1.00

Expected return on investment in copper (i) = 10%

Life of copper conductor (n) = 25 yrs

Specific weight of copper (γ) = 0.32 lb/in^3

The value of the capital return factor (F') is found from Eq. (13.5) to be 0.11, and hence the value of copper per inch length $= F' C_1 \gamma$. Substituting into Eq. (13.13):

$$A^* = [(300)^2(0.678 \times 10^{-6})(8{,}640)(10^{-5})/(0.11)(1)(0.32)]^{0.5} = 0.387 \text{ in}^2.$$

It should be noted that both numerator and denominator are per inch length. For a circular conductor, this corresponds to a diameter of 0.702 inch.

If all conditions remain unchanged except the use per year (h) which is reduced to 1,000 hr (5 hr/day for 200 days), the optimum diameter is reduced to 0.409 inch.

8.0 DIMENSIONAL ANALYSIS

The fact that the dollar ($) is just as respectable a variable as force, length, viscosity, etc., may be demonstrated by performing a dimensional analysis for the optimum area (A^*) in Kelvin's conductor problem. The quantities of importance in this problem are listed in Table 13.2 together with their dimensions, in terms of the fundamental set L, F, T, Q, and $.

Table 13.2. Dimensions Involved in Kelvin's Conductor Problem

Quantity	Symbol	Dimensions
Cost optimum conductor area	A^*	$[L^2]$
Current	I	$[QT^{-1}]$
Specific resistance	ρ	$[FL^2Q^{-2}T]$
Hours of use per year	h	$[T]$
Power cost per watt hour	C_2	$[\$F^{-1}L^{-1}]$
F' × copper cost per lb	$F'C_1$	$[\$F^{-1}]$
Specific weight of copper	γ	$[FL^{-3}]$

Before dimensional analysis for the cost optimum area (A^*):

Eq. (13.15) $$A^* = \psi_1 (I, \rho, F'C_1, C_2, h, \gamma)$$

The length of the conductor need not be included since both the cost associated with power loss and the investment in copper vary linearly with conductor length (ℓ). The total cost associated with the investment in copper is $(F'C_1)$, hence F' and C_1 are important only as the product $(F'C_1)$. The current flowing in a conductor is measured by the number of units of charge (Q) passing a given point per unit time. Electrical problems require the addition of the fundamental unit (Q) to F, L, and T. The dimensions of current (I) are thus $[QT^{-1}]$. Quantities involving cost require an additional fundamental dimension ($).

The units for specific resistance (ρ) require a word of explanation. From Eq. (13.7):

Eq. (13.16) $\quad I^2R = VI = \text{Work per unit time}$

This work per unit time can be expressed in mechanical or electrical units since equivalence exists between these two forms of energy. Thus,

Eq. (13.17)

$$R = \text{work per unit time}/(I)^2 = [FLT^{-1}/(QT^{-1})^2 = [FLQ^{-2}T]$$

or

Eq. (13.18) $\quad \rho = RA/\ell = [FL^2Q^{-2}]$

Five fundamental quantities appear in Table 13.2 (F, L, T, Q, $) hence, we should expect to find a maximum of five dimensionally independent quantities that cannot be combined to form a nondimensional group. Five dimensionally independent quantities do, in fact, exist $(I, h, F'C_1, C_2, \gamma)$ and thus there are two π quantities. Dimensional analysis yields the following equation:

Eq. (13.19) $\quad A^*/[(FC_1)/C_2]^2 = \psi[(\rho I^2 h)/\gamma(FC_1/C_2)^5]$

This is as far as we can proceed by dimensional analysis.

PROBLEMS

13.1 What uniform annual year-end payment for five years is necessary to repay a present value of $5,000 with interest at 6% compounded annually?

13.2 If you received $1,000 now with the understanding that you will repay the $1,000 and all interest due five years from now, how much should you pay if the agreed rate of interest is 4% compounded every six months?

13.3 If instead of making one payment in Problem 13.2, you make an equal payment every six months until 10 payments are made, what is the amount you should pay every six months?

13.4 If you receive $2,000 today that is to be repaid in six years with all interest due (8% compounded every 6 months) making 12 equal payments, one every 6 months, how much would each of these payments be?

13.5 Two machines are under consideration (A and B):

	Machine A	**Machine B**
Initial cost	$10,000	$8,000
Salvage value	$2,000	$1,000
Life, years	10	7
Return on investment	12%	10%
Annual operating cost	$500	$800

Which is the most attractive based on total annual cost?

13.6 The following information is available for two alternative machines for a given service. The minimum attractive return on the investment is 10%.

	Machine A	Machine B
First cost	$4,000	$10,000
Life	4 years	10 years
Salvage value	none	$2,000
Annual operating cost	$900	$700

a) Find the true equivalent uniform annual cost for each.

b) Which would you choose?

13.7 You purchase something now for $3,000 that has a salvage value of $1,500 in two years. Your money is worth 7% per year to you since that is the return you could receive if you purchased a bond instead. How much is it costing you to own the device per year considering loss of principal and loss of interest on both the lost principal and the amount recovered in salvage?

13.8 The following information is available for two alternative machines for a given service. The minimum attractive return on the investment is 10% for machine A and 15% for B.

	Machine A	Machine B
First cost	$3,000	$10,000
Life, years	4	10
Salvage value	none	$2,000
Annual operating cost	$1,400	$700

Compare the true equivalent uniform annual cost for these two machines and indicate which one you would choose.

13.9 A manufacturer has machine (A) that he had planned to retire in two years. His uniform yearly disbursements for this machine are $1,200. The present value of this machine installed is $2,000, while the net salvage value right now after moving out the old machine is $1,500. The salvage value in two years will be $700. A new machine (B) has just appeared on the market that runs faster and occupies less space. This machine costs $12,000 installed and the estimated uniform yearly disbursements for the new machine are $400. Its useful life is estimated at 10 years and the salvage value at $4,000. The minimum attractive return on the investment is 15%. Should the manufacturer buy the new machine or retain the old machine? The operating and maintenance cost per year for A is $2,000, that for B is $1,500.

13.10 You now own an automobile that costs $500 per year for gas, oil, insurance, and repairs. You can sell it immediately for $1,000. A different car can be purchased for $4,000, and, if kept for 4 years, will have a salvage value of $1,000. It will cost $300 per year to operate. If your money is worth 6% per year, should you buy the new car now?

13.11 You now own an automobile that you paid $5,000 for three years ago. You can sell this car right now on the open market for $2,000. You can purchase a new car that interests you for $6,000. It is estimated the sales value of this new car will be $4,000 after two years. You drive about 10,000 miles per year and the current mileage on your present car is 30,000 miles. If you keep your current car for two more years you will have to purchase tires, a battery, reline the brakes, and do other maintenance work, all of which will cost you $800. This may be figured as a $400 expenditure at the end of each year. It is estimated that your present car will be worth $1,000 in two years. Operating costs for the two cars are estimated to be the same except for the maintenance charges enumerated above. The greater reliability and prestige value of the new car is estimated to be worth $500 to you for each of the next two years. This may be figured as a $500 award at the end of each year if the new car is purchased. The money you have invested is considered to be worth 8% per year. Determine the cost per year for each plan.

13.12 You now own a large automobile that gets 10 miles per gallon of gasoline, and gasoline costs $1.40/gallon. You can sell this car on the used market for $6,000, or keep the car for two more years when it is estimated it will be worth $4,000. A new subcompact car that would serve your needs would cost $12,000. You estimate that under your conditions of use, this car would last five years and be worth $2,000 on the used market after five years. This car gets 35 miles per gallon. You drive 20,000 miles per year. Maintenance will cost $400 per year on your present car (you just purchased new tires and had an overhaul six months ago), but will average only $200 per year on the new car. Should you keep your present car or buy the new one if inexpensive transportation is your objective? Your money is worth 7% compounded yearly.

13.13 A DC rectifier with an output of 300 amps is used to start jet engines. A two-wire cable, 100 ft long is needed to connect the rectifier to the starter motor on the engine. The cost of the rectifier, which has a useful life of one year, depends upon the voltage capacity, and, in the vicinity of the 30 volts DC required, increases in cost at the rate of $20 per volt. The cost of the finished cable is $2 per pound of copper, and the cable also has a useful life of one year.

a) Determine the diameter of the copper conductor in order that the total cost (rectifier plus cable) is a minimum. The specific weight of copper is 0.32 lb/in^3. and the specific resistance is 0.678×10^{-6} ohm-in. The value of power is 10^{-4} $/Wh = C_2, and the expected return (i) = 15%. It is estimated the unit will be in use 800 hrs/yr.

b) What is the answer if 200 ft of cable are needed?

13.14 The cost of a large concrete pipe to conduct irrigation water is $\$25\ d^2$ per foot of length. Pumps used to move water through the 10 miles of pipe at the desired flow rate will cost $\$1{,}600 \times 10^6/d^3$ (total cost). In both cases d is the diameter of pipe in feet. Maintenance for the buried pipeline will be negligible while that for the pumps will be 5% of the pump cost per year. If the pipeline is considered to have a 50 year life with no salvage value and the pumps have a 10 year life with 10% of the initial cost as salvage value:

a) Calculate the best pipe size if the expected return on the investment is 10%. Maintenance for the pumps = 5% of the initial pump cost per year.

b) What is the ratio of initial pipe cost to initial pump plus pipe cost?

13.15 A pipeline is being designed to carry liquid over a long distance. If large pipe is used, the initial capital investment will be too great, whereas if small pipe is used, the annual fuel cost to pump the fluid against the greater pipe friction will be excessive. There is an optimum pipe size for minimum cost and this is known to lie between 6 and 12 inches Reference to the pipe manufacturer's catalog shows that the initial cost of the pipe to cover the required distance varies with diameter (d, in.) as follows:

$$\text{Total cost of pipe} = \$1{,}500\ d^{0.9}$$

The initial cost of the pumping stations required to pump the fluid against pipe friction will also depend on pipe diameter in inches as follows:

$$\text{Total cost of pumping stations} = \$15 \times 10^6/d^4$$

The pipeline and pumping stations are considered to have a life of 30 years with no salvage value and the expected return on the investment is 10% per year. The maintenance cost for the pipeline per year is 5% of its initial cost, while the maintenance cost of the pumping stations and pumps is considered to be 10% of their initial cost per year. The cost of fuel per year is estimated to be $\$(2.5 \times 10^6)/d^4$.

a) If pipes come in even inch sizes (6, 8, 10 in., etc.), determine the best pipe diameter for this situation.

b) Estimate the percentage of the total yearly cost that should be associated with the pipe (pipe cost and pipe maintenance) when the cost optimum pipe diameter is used.

13.16 Banks now advertise continuous compounding of interest. How much more money will you make in a year on $1,000 with continuous compounding than with semiannual compounding at a 6% annual rate?

13.17 It was claimed (*Popular Science*, p. 26, Aug. 1976) that a 20 ft diameter, three blade, high-speed rotor designed by the National Research Council of Canada would develop 23,000 kWh over one year on Block Island, where the mean wind velocity is 18 mph, and the cost of electricity was 0.145 $/kWh. If the initial cost of the unit was $10,000, estimate the break-even time in years. Assume the maintenance cost per year to be $1,000, and the salvage value to be zero.

14

Engineering Statistics

1.0 INTRODUCTION

There are a number of important applications of statistical theory in engineering. These include:

- Analysis of data
- Production sampling and quality control in manufacturing
- Fitting a continuous function to experimental data
- Planning of experiments

In this chapter, these topics will be introduced in order to give a flavor for this activity. Statistics is a broadly ranging, highly mathematical endeavor, and many of the details are beyond the scope of this book. The objective here is merely to introduce the subject and present a few applications that illustrate its importance to engineering. Statistics is a core subject in industrial engineering and this subject is usually associated with such departments.

Engineering frequently involves experimental observations and the analysis of data collected. When measurements are made in the process of experimentation, two types of errors may be involved:

- Systematic errors
- Random errors

A *systematic error* is an error that is consistently present in an instrument while a *random error* is associated with a system that is out of control or one that is due to human error. An example of a systematic error is one associated with the zero setting of a scale (ohmmeter, postage scale, etc.). An example of random error is one associated with the difficulty of precisely interpolating between markings on a scale or an instrument. In general, some random errors are positive while others are negative.

In metrology (measurement technology), a measurement is said to be precise when associated with a small random error, but accurate when associated with a small systematic error.

2.0 STATISTICAL DISTRIBUTIONS

When a given measurement (length, temperature, velocity, etc.) is made several times, a number of different values may be obtained due to random errors of measurement. In general, the observed magnitudes will tend to cluster about a central value and be less numerous with displacement from this central value. When the frequency of occurrence (y) of a given observed value is plotted vs the value of the variable of interest (x), a bell shaped curve will often be obtained particularly when the number of determinations (N) is large. In statistical parlance, such curves are called distributions.

The central value of a distribution may be expressed in several ways. The arithmetic mean ($\bar{x}$) is the most widely used measure of central tendency. This is the sum of all of the individual values divided by the number of values (N) and is expressed as follows:

Eq. (14.1) $$\bar{x} = \Sigma x_i / N$$

where: x_i = individual values as i goes from 1 to N
Σ indicates the sum of all N values to be taken.

Sometimes some of the N values of x involved in calculating the mean are more important than others. This is taken care of by determining a weighted mean value ($\bar{x}_W$). Equation (14.1) is then written as follows:

Eq. (14.2) $$\bar{x}_W = \Sigma W_i x_i / \Sigma W_i$$

where W_i = the weighting factor for each item. When all items are equally weighted:

$$\Sigma W_i x_i = \Sigma x_i \text{ and } \Sigma W_i = N$$

The median is another way of expressing the central value of a distribution. This is the measurement for which equal numbers of measurements lie above and below the value. The mode is still another method of measuring central tendency. This is the most frequently observed value. For a distribution that is perfectly symmetrical, the arithmetic mean, the median, and the mode are all equal. However, this will not be the case for distributions that are not symmetrical.

The extent of a distribution is another characteristic that is of importance. This may be expressed as the spread which is the difference between the maximum and minimum values ($x_{max} - x_{min}$). The extent may also be expressed as the mean deviation ($\bar{d}$) of the N values from the mean. That is:

Eq. (14.3) $$\bar{d} = \Sigma |x_i - \bar{x}|/N$$

Since some deviations will be plus and some minus, the values of ($x_i - \bar{x}$) to be used in Eq. (14.3) are those without regard for sign. These are called *absolute values* and the mathematical way of indicating this is to place the quantity between vertical bars as in Eq. (14.3).

A more useful way of measuring dispersion and also one that gets around the sign problem is to determine the square root of the mean of the squares of the individual dispersions. This is called the *standard deviation* (σ) and is expressed mathematically as follows:

Eq. (14.4) $$\sigma = [\Sigma(x_i - \bar{x})^2/N]^{0.5}$$

The standard deviation is sometimes called the root-mean square dispersion. The dimensions of σ will be the same as for x. In general, the arithmetic mean ($\bar{x}$) and the standard deviation (σ) of a distribution are two parameters that are sufficient to characterize the distribution.

3.0 THE NORMAL DISTRIBUTION

The normal distribution of the frequency of occurrence of a group of readings is a bell-shaped curve that is symmetrical about the mean. Figure 14.1 is an example where y is the frequency of occurrence of a measurement x_i.

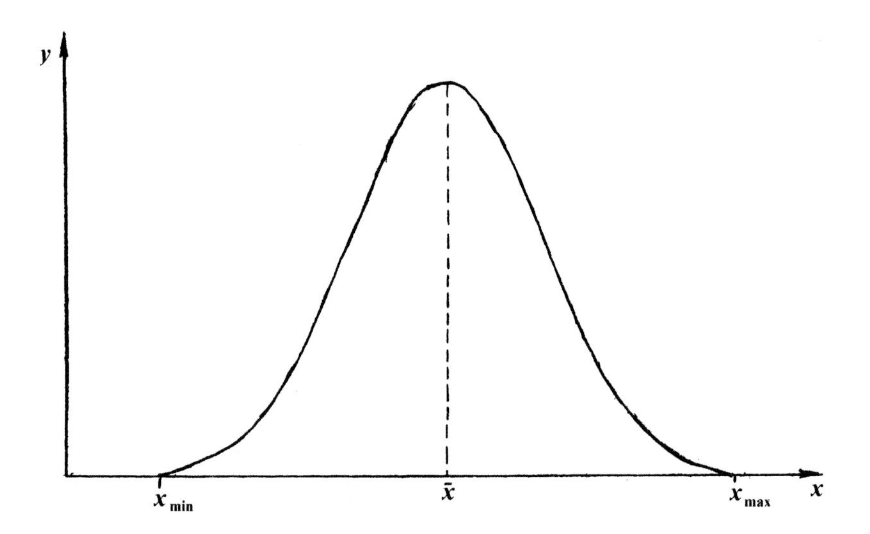

Figure 14.1. Representative normal distribution curve showing frequency of observed values (y) vs the value of some variables (x). The mean value of x for this symmetrical distribution (x) corresponds to the maximum value of y.

DeMoivre was the first to show (in 1793) that the following equation closely represents experimental bell-shaped normal curves:

Eq. (14.5) $$y = \left[\frac{1}{\sigma(2\pi)^{0.5}}\right] e^{-\left[\frac{(x-\bar{x})}{2\sigma^2}\right]}$$

where e is the base of Napierian logarithms = 2.7183.

The mathematics associated with the normal distribution was independently developed by Gauss and La Place at about the same time that

DeMoivre did his work. The normal distribution is sometimes called the *Gaussian distribution* and sometimes the *Law of Errors*.

Equation (14.5) may be simplified and made more useful by letting $t = (x - \bar{x})/\sigma$, $\bar{x} = 0$ and $\sigma = 1$. Equation (14.5) then becomes:

Eq. (14.6) $$y' = \left[\frac{1}{(2\pi)^{0.5}}\right] e^{-\left(\frac{t^2}{2}\right)}$$

and the value of $x = t\sigma$.

Figure 14.2 shows a plot of y' vs $t\sigma$. This is known as the standard normal distribution corresponding to $\bar{x} = 0$ and $\sigma = 1.0$. Tables are available that give values of y' for different values of t and a few of these values are given in Table 14.1. It is thus seen that although the standard normal distribution curve extends from $-\infty$ to $+\infty$, essentially all experimental points should lie within $t = \pm 3\sigma$ (i.e., <1% should lie beyond $\pm 3\sigma$).

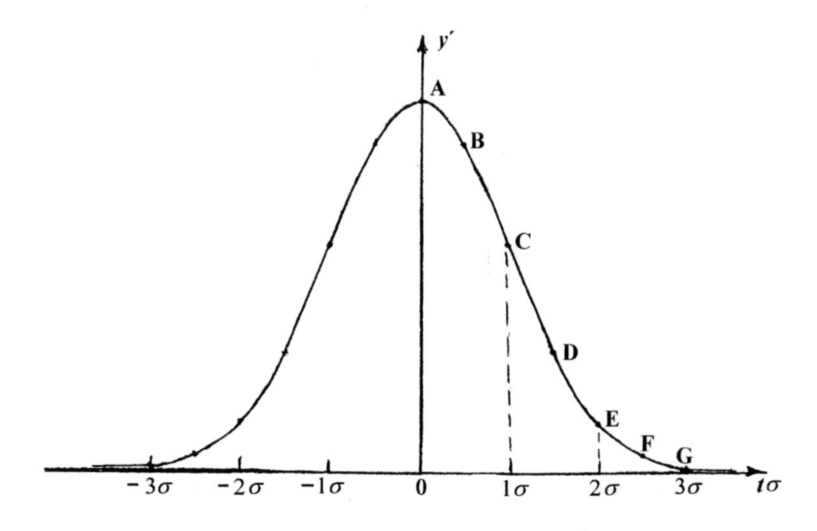

Figure 14.2. The standard normal distribution curve.

The derivation of Eq. (14.5) and (14.6) and Table 14.1 are beyond the scope of this presentation, but may be found in many texts concerned with statistics. The results presented here are merely to illustrate the nature of the topic and its application.

Table 14.1. Values of Standard Normal Distribution of Frequency of Occurrence y'

t	0	0.5	1.0	1.5	2.0	2.5	3.0
Point in Fig. 14.2	(A)	(B)	(C)	(D)	(E)	(F)	(G)
y'	0.399	0.352	0.242	0.130	0.054	0.018	0.004

The values of y' are for one tail of the distribution only.

Example: Normal Distribution. When a metal rod was measured ten times, the following values of length (x) and deviation (d) were obtained:

Test No.	x, in.	$d = x - \bar{x}$	d^2
1	3.56	0.055	0.0030
2	3.47	0.035	0.0012
3	3.51	0.005	0.0000
4	3.49	-0.015	0.0002
5	2.51	0.005	0.0000
6	3.48	-0.025	0.0006
7	3.50	-0.005	0.0000
8	3.53	0.025	0.0006
9	3.48	-0.025	0.0006
10	3.52	0.015	0.0002
	$\bar{x} = 3.505$ in.		$\bar{d^2} = 0.00064$
			$\sigma = (0.00064)^{0.5} = 0.026$ in

The values of frequencies (y) for this example at multiples of σ may be obtained by multiplying the values in Table 14.1 for y' by N/σ (i.e., by $10/0.0026 = 384/6$). The corresponding values of x are obtained from $x = \bar{x} \pm \sigma$. For example, the coordinates for point A in Fig. 14.3 will be:

$$x_A = \bar{x} + \sigma = 3.505 + 0.026 = 3.531 \text{ in.}$$

$$y_A = (0.242)(334.6) = 93.07$$

The frequency distribution for this example is given in Fig. 14.3 (*a*). Figure 14.3 (*b*) shows the individual points that were observed plotted along the x axis with their test numbers to the left. It appears that point one has an unusually high value of x relative to the other values. However, this is within the right hand tail of the distribution, and, therefore, should cause no alarm of being a rogue point.

4.0 PROBABILITY

Probability, which predicts the number of times a specific event will occur in N trials, plays an important role in statistics. It is usually introduced by considering results of flipping coins, removing balls from a bag, throwing dice, or drawing cards from a shuffled deck.

When an unbiased (unbent) coin is tossed, the probability of a head coming up is one-half. This is based on zero corresponding to a complete impossibility and one to complete certainty. The probability of obtaining a head or a tail will be ½ + ½. This is an example of the *addition law of probability* which states that the probability of an event occurring in one of several ways is the sum of the individual probabilities. This law is the basis of the old joke "heads I win, tails you lose."

When two coins are tossed, the probability of obtaining two heads will be (½)(½) = ¼. This is an example of the *multiplication law of probability* which states that the probability of several independent events occurring simultaneously is the product of their individual probabilities.

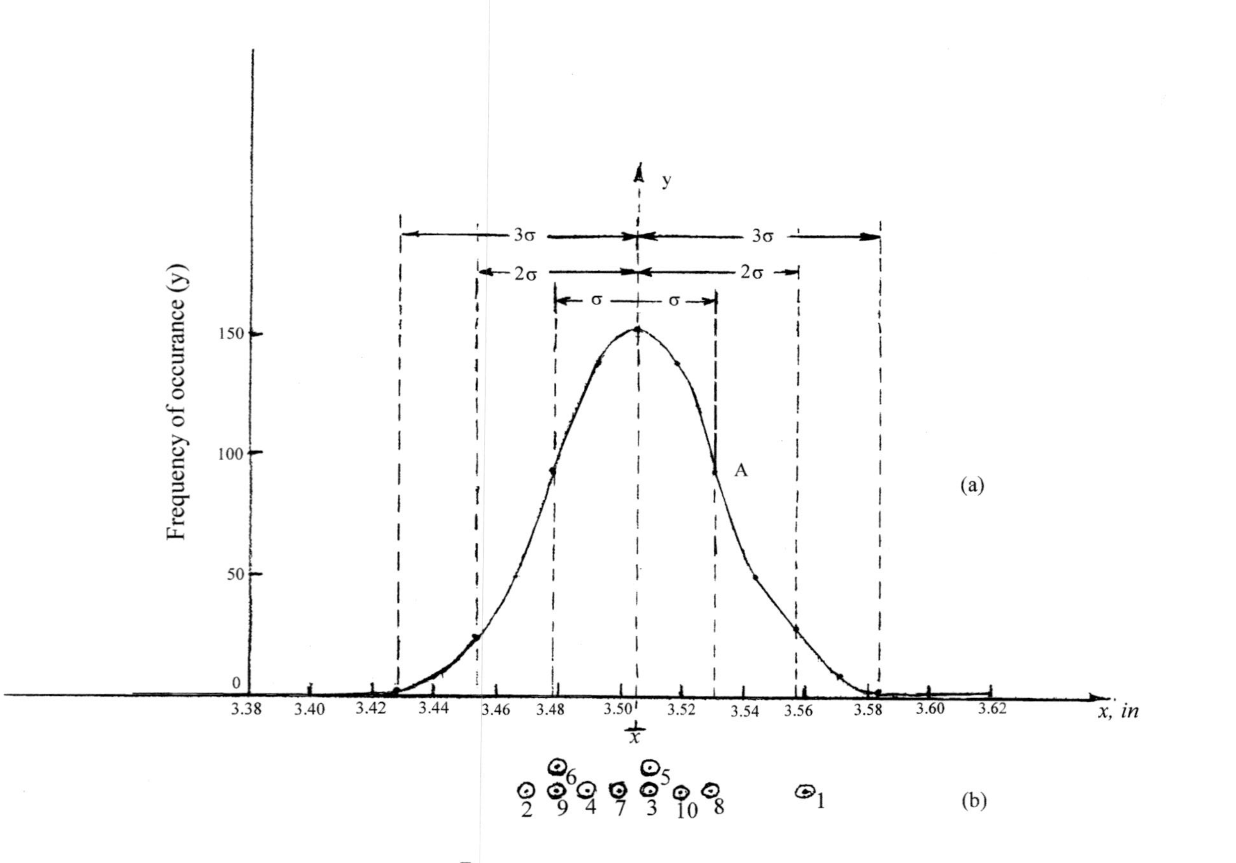

Figure 14.3. *(a)* Normal frequency density curve derived from $\bar{x}$ and σ obtained from experimental values. *(b)* Observed values of x plotted along the x axis.

Next, consider the various probabilities when two coins are tossed. The results will be as follows:

Result		Probability
Coin 1	Coin 2	
H	H	(½)(½) = ¼
H	T	(½)(½) = ¼
T	H	(½)(½) = ¼
T	T	(½)(½) = ¼
		Σ = 1.0

The probability of obtaining two heads or two tails = ¼ while the probability of obtaining one head and one tail = ½. This simple example illustrates both the addition and multiplication laws of probability.

Next consider the various probabilities when five coins are tossed simultaneously. Table 14.2 gives the results with their probabilities. The probability of obtaining three tails and two heads or three heads and two tails is 5/16 = 0.313. The laborious work of solving such problems can be greatly simplified by introducing mathematical conventions called permutations and combinations.

Table 14.2. Possible Results Obtained and Corresponding Probabilities when Five Coins are Tossed

Coin No.					
1	**2**	**3**	**4**	**5**	**Probability**
T	T	T	T	T	$(½)^5 = 1/32$
T	T	T	T	H	
T	T	T	H	T	
T	T	H	T	T	$5(½)^5 = 5/32$
T	H	T	T	T	
H	T	T	T	T	
T	T	T	H	H	
T	T	H	T	H	
T	T	H	H	T	
T	H	T	T	H	
T	H	H	T	T	
T	H	T	H	T	$10(½)^5 = 10/32$
H	H	T	T	T	
H	T	H	T	T	
H	T	T	H	T	
H	T	T	T	H	
H	H	H	T	T	
H	H	T	H	T	
H	H	T	T	H	
H	T	H	H	T	
H	T	T	H	H	
H	T	H	T	T	$10(½)^5 = 10/32$
T	T	H	H	H	
T	H	T	H	H	
T	H	H	T	H	
T	H	H	T	H	
T	H	H	H	T	
H	H	H	H	T	
H	H	H	T	H	
H	H	T	H	H	$5(½)^5 = 5/32$
H	T	H	H	H	
T	H	H	H	H	
H	H	H	H	H	$(1/2)^5 = 1/3125$
					$32/32 = 1.0$

5.0 PERMUTATIONS AND COMBINATIONS

Consider the number of ways that *a*, *b*, and *c* may be arranged two at a time if order is important (i.e., *ab* and *ba* are considered to be different). The answer is: *ab*, *ac*, *bc*, *ba*, *ca*, and *cb* (six ways). This corresponds to what is called a *permutation* of three ways taken two at a time, which is written:

$$3P2 = 3!/1! = 6$$

where a number followed by ! is called a *factorial* and $n! = (n)(n - 1)(n - 2)\ldots(1)$. In general, the number of different ways that r objects can be taken from n objects is:

Eq. (14.7) $$nPr = n!/(n - r)!$$

The number of ways of arranging *a*, *b*, and *c* three ways at a time if order is important is:

$$3P3 = 3!/(3 - 3)! = 6/0! = 6$$

When the factorial of zero (0!) is encountered, it is equal to one. The six permutations of *a*, *b*, and *c* taken 3 at a time will be *abc*, *acb*, *bca*, *bca*, *cab*, and *cba* (= 6).

When order is not important (i.e., *ab* and *ba* are considered to be the same), the number of arrangements is called a *combination*. The combination of n items r at a time will be:

Eq. (14.8) $$nCr = rPr/r! = n!/[(n - r)!\,(r)!]$$

where $r!$ in the denominator is the number of ways r items may be arranged r at a time. The number of combinations will always be less than the corresponding number of permutations by the factor $r!$ which adjusts for the change when order is considered important (permutation) to when order is unimportant (combination).

Thus, the number of combinations of three items taken 2 at a time is:

$$3C2 = 3!/[(1!)(2!)] = 3$$

The number of combinations of *A*, *B*, and *C* taken two at a time is *AB*, *AC*, and *BC*. The number of combinations of *A*, *B*, *C* taken three at a time will be:

$$3C3 = 3!/[(0!)(3!)] = 1 \text{ (i.e., } ABC\text{)}.$$

Consider the number of ways in which the letters in the word EG_IG_{II} may be arranged to from three letter words. The two Gs have been distinguished for reasons to be explained. The possibilities would appear to be: EG_IG_{II}, $EG_{II}G_I$, $G_{II}EG_I$, G_IEG_{II}, $G_IG_{II}E$, $G_{II}G_IE$. This corresponds to $3P3 = 6$. However, when the distinguishing marks are removed, it is evident that EG_IG_{II} and $EG_{II}G_I$ are the same as are $G_{II}EG_I$ and G_IEG_{II}, and $G_IG_{II}E$ and $G_{II}G_IE$. Thus, there are only three possibilities: EGG, GEG, and GGE. This corresponds to $3P3/2P2 = 3$.

In problems where order is important (permutations), but some of the *n* items involved are alike, this must be taken into account. If *n* items are to be arranged as permutations, but s_1, s_2, s_3, etc. of the *n* items are alike, then, in general, the number of possible permutations of the *n* items will be:

Eq. (14.9) $$\text{No. of permutations} = n!/[(s_1!)(s_2!)(s_3!)(\text{etc.})]$$

The number of ways in which three tails and two heads or three heads and two tails may be arranged in the foregoing problem where five coins were tossed simultaneously may be found by use of Eq. (14.9). This will give $P = 5!/(3!)(2!) = 10$. This was found to be the case by writing out all possible results. To further illustrate the use of Eq. (14.9), find the probability of obtaining three heads and three tails when six coins are tossed. This will give $[6!/3!^2](1/2)^6 = 0.31$.

6.0 NORMAL PROBABILITY DISTRIBUTION

The normal probability distribution may be obtained from the standard normal frequency distribution (Fig. 14.2). The probability that a reading will fall between $(t\sigma)_1$ and $(t\sigma)_2$ in Fig. 14.4 will be the area under the frequency curve between these two points. Table 14.3 gives the probability for a value to fall between $x = 0$ and $x = t\sigma$ on the standard normal distribution curve. This corresponds to the area under the curve between these two points.

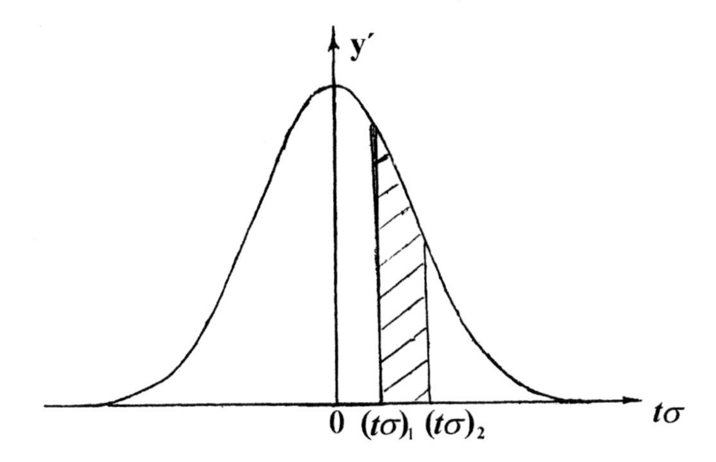

Figure 14.4. The standard normal distribution curve showing a cross-hatched area below the curve between $(t\sigma)_1$ and $(t\sigma)_2$. The area of this cross-hatched region is the probability (P) that a point will fall within this area.

Table 14.3. Values of Probability that a Point will Fall Between $x = t\sigma$ and $x = \infty$

t	0	0.5	1.0	1.5	2.0	2.5	3.0
P	0.500	0.309	0.159	0.067	0.023	0.006	0.001

Note: The values of P are for one tail of the distribution only.

For the foregoing example that was worked to illustrate the use of the standard normal distribution curve, the probability of a measurement falling between 3.531 (1σ) and 3.583 (3σ) is found to be 0.159 from Table 14.3. Similarly, the probability of a measurement falling above 3.583 would be 0.0001 (essentially zero). By symmetry, the probability of a point falling between 3.479 (-1σ) and 3.427 (-3σ) would be 0.159 while the probability of a point falling below 3.427 would be essentially zero (0.001).

The *Intelligence Quotient* (IQ) of a person is defined as (mental age/chronological age) times 100. To a very good approximation, the distribution of IQ in the population is found to be a normal one with $\bar{x} = 100$ and $\sigma = 13$. This is shown plotted in Fig. 14.5 where the cross-hatched regions correspond to the areas from $\bar{x}$ to $(\bar{x} + \sigma)$, from $(\bar{x} + \sigma)$ to $(\bar{x} + 2\sigma)$, and from $(\bar{x} + 2\sigma)$ to $(\bar{x} + 3\sigma)$. The areas of these three cross-hatched regions correspond to 34%, 14%, and 2% of the total population respectively. These values may be obtained from Table 14.3 as follows:

$$\begin{aligned}(0.500 - 0.159)100 &= 34\% \\ (0.159 - 0.023)100 &= 14\% \\ (0.023 - 0.001)100 &= \underline{\ 2\%} \\ & \quad\ 50\%\end{aligned}$$

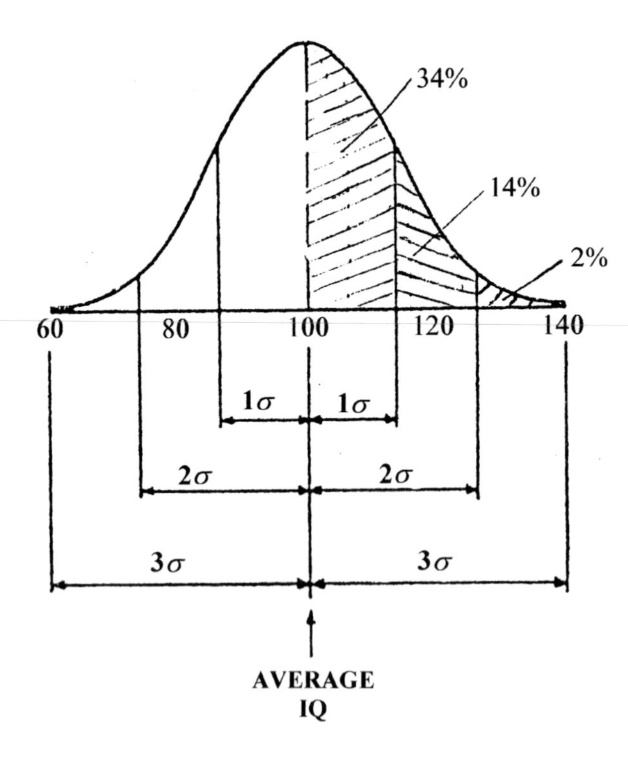

Figure 14.5. Normal distribution for Intelligence Quotient (IQ) which has a mean value of 100 and a standard deviation of 13. The distribution has a range extending from 60 to 140. The cross-hatched regions give the % of the population falling 1σ above and below the mean (34%) from 1σ to 2σ and from -1σ to -2σ (14%) and from 2σ to 3σ and -2σ to -3σ (2%). Approximately 0.1% will be above 140 and below 60 (see Table 14.3).

7.0 BINOMIAL DISTRIBUTION

The binomial distribution forms the basis for quality control in mass production manufacturing. Consider a large batch of parts having 5% that are defective and 95% that are satisfactory. The probability that the first part taken in a sample of two parts is satisfactory will be 0.95 (call this s). The probability that the first part is defective will be 0.005 (call this d). The probability that both parts will be satisfactory will be $s^2 = (0.95)^2$ while the probability that both are defective will be $d^2 = (0.05)^2$. There are two other possibilities: the first may be satisfactory while the second is defective, or the first may be defective and the second satisfactory. Each of these will have a probability of $(sd) = (0.95)(0.05)$. Thus, the probability of one satisfactory and one defective = $2sd$ = 2(0.95)(0.05). These results for a sample of two from a large batch (population) with 5% defective may be summarized as follows:

Result	Probability	
Both satisfactory	$s^2 = (0.095)^2 =$	0.9025
1 s and 1 d	$2sd = 2(0.095)(0.05) =$	0.0950
Both defective	$d^2 = (0.05)^2 =$	0.0025
		$\Sigma = 1.0$

Consider next, a sample of three from a population with 5% defective drawn in sequence 1, 2, 3. These results are summarized below:

Sequence			Result	Probability	
1	2	3			
s	s	s	$3s$	$s^3 = (0.95)^3 =$	0.857375
s	s	d			
s	d	s	$2s, 1d$	$3s^2d = 3(0.95)^2(0.05) =$	0.135375
d	s	s			
d	s	d	$1s, 2d$	$3sd^2 = 3(0.95)(0.05)^2 =$	0.007125
s	d	d			
d	d	d	$3d$	$d^3 = (0.05)^3 =$	0.000125
					$\Sigma = 1.0$

It should be noted that the coefficients of the s^2d and sd^2 terms are permutations of three items at a time with two that are the same:

$$3P2 = 3!/2! = 3$$

For a sample of four items under the same conditions:

Result	Probability	
4*s*	$s^4 = (0.95)^4 =$	0.814506
3*s*, 1*d*	$4!/3!\ s^3d = 4(0.95)^3(0.05) =$	0.171475
2*s*, 2*d*	$4!/2!\ s^2d^2 = 6(0.95)^2(0.05)^2 =$	0.013538
1*s*, 3*d*	$4!/3!\ sd^3 = 4(0.95)(0.05)^3 =$	0.000475
4*d*	$d^4 = (0.05)^4 =$	0.000000
		$\Sigma = 1$

The probabilities in each of the three cases considered above are:

Sample of 2: $s^2 + 2sd + d^2 = (s + d)^2$

Sample of 3: $s^3 + 3s^2d + 3sd^2 = (s + d)^3$

Sample of 4: $s^4 + 4s^3d + 6s^2d^2 + 4sd^3 + d^4 = (s + d)^4$

These are seen to correspond to the binomial expansion of $(s + d)^n$ with the exponent n equal to the number of items in the sample. The possibilities for a sample of 10 items taken from a population with 5% defective are as follows:

Number Defective	Probability		
0	$s^{10} = (0.95)^{10}$	=	0.598736
1	$10!/9!\ s^9d = 10(0.95)^9(0.05)$	=	0.315125
2	$10!/8!2!s^8d^2 = 45(0.95)^8(0.05)^2$	=	0.074635
3	$10!/7!3!s^7d^3 = 120(0.95)^7(0.05)^3$	=	0.010475
4	$10!/6!4!s^6d^4 = 210(0.95)^6(0.05)^4$	=	0.000965
5	$10!/5!5!s^5d^5 = 252(0.95)^5(0.05)^5$	=	0.000003
etc.	etc.		etc.
			$\Sigma = 1.0$

The results of the four cases considered (samples of 2, 3, 4, 10), are plotted in Fig. 14.6. This type of bar plot is called a *histogram*. In this case, the heights of the bars show the percentage of selections showing 0, 1, 2, etc. defects for samples of 2, 3, 4, and 10 when 5% of the total population is defective. The probability of finding just one defective item is found to increase with sample size, but only very slowly (approximately from 10% to 14% to 17% to 32% as the sample size goes from 2 to 3 to 4 to 10 respectively. A very large sample size would be required in order to predict the percentage of defects present in the entire population of parts.

8.0 CONTROL CHARTS

In order to monitor a production process with regard to some variable (length, surface finish, etc.), it is not practical to use large samples. However, it is found that small samples taken at fixed intervals may be used cumulatively to get around this difficulty. An inspector may take a relatively small group of measurements of a variable (x) periodically and record the mean for the group ($\bar{x}$). After about 20 such means have been obtained under stable manufacturing conditions, the mean of these means ($\bar{x}_\theta$) and their standard deviation (σ_θ) may be determined. A control chart is produced showing the target value (T) as a horizontal straight line plotted vs the number of parts (N) (Fig. 14.7). Other horizontal lines corresponding to $T \pm k\sigma_\theta$ are then added. As each new group value of $\bar{x}$ is obtained, it will be plotted on this figure. These will fall above and below T by $\pm k\sigma_\theta$ where k is small if the process is stable. However, if values of $\bar{x}$ fall above or below a predetermined multiple of σ_θ due to too! wear, etc., then action is needed. Usually, horizontal lines above and below $\bar{x}_\theta$ are labeled "warning" and "action" as in Fig. 14.7. The value of k corresponding to these designated lines will depend upon the limits set by the design engineer, stability of the process, or the significance of producing defects relative to cost, performance, and other considerations. The use of control charts is an important tool for controlling production in industrial engineering with many more details than the brief introduction presented here might indicate.

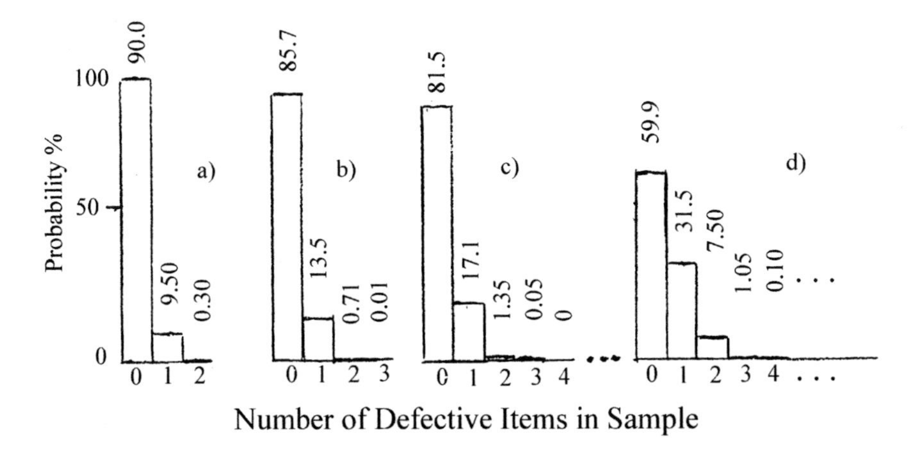

Figure 14.6. Histograms based on binomial distribution for probability of obtaining satisfactory items from a large number of items containing 5% that are defective when the sample size is (*a*) 2, (*b*) 3, (*c*) 4, and (*d*) 10.

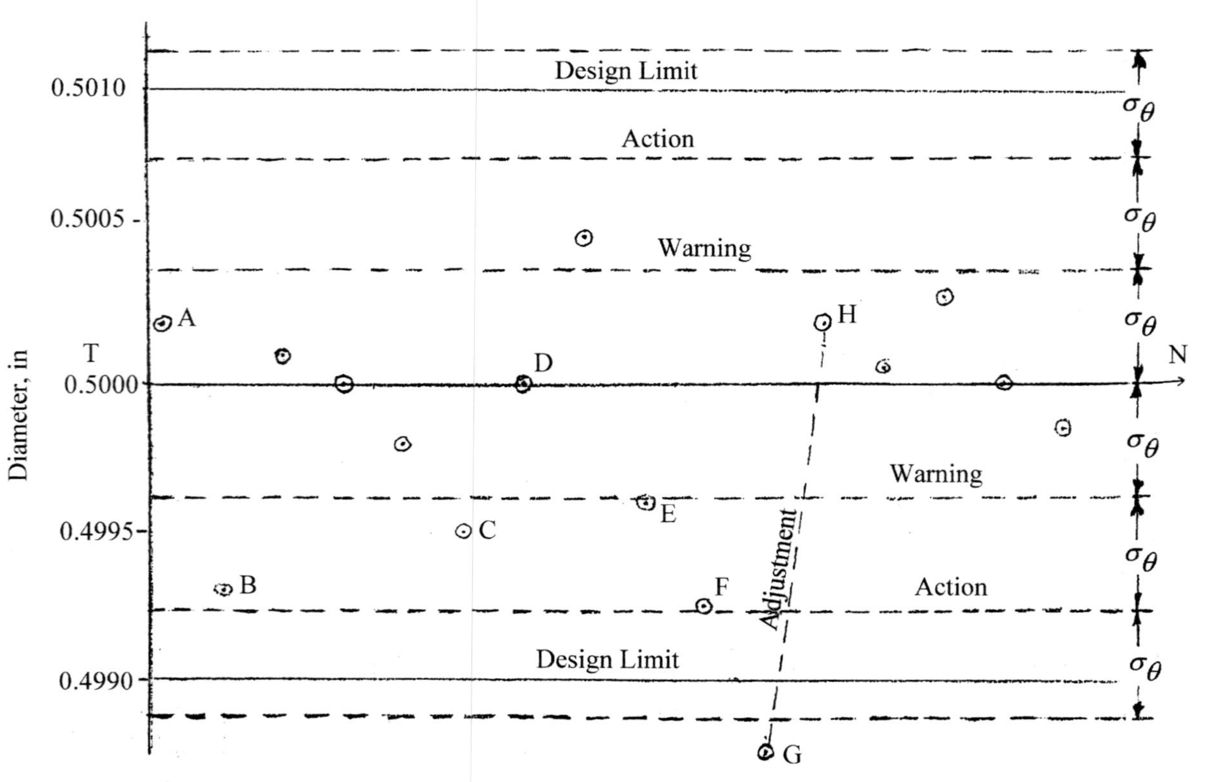

Figure 14.7. Control chart for monitoring a machining operation.

Example: Control Chart. The control chart (Fig. 14.7) shows the diameter of a part having a nominal value of 0.500 inch. The design engineer in conference with the manufacturing engineer has set the acceptable tolerance at ±0.0010 (±25 μm). The dimension on the drawing reads 0.500 ±0.001. After the machine is adjusted and producing parts, it is decided to measure five parts out of each 1,000 produced. In order to establish the control chart limits, 10 sets of five measurements were recorded. The averages of each sample of five parts were found to vary as indicated below:

Sample No.	Meas. Dia ($\bar{x}$)	$\bar{d} = \bar{x} - \bar{x}_\theta$	$\bar{d}^2 \times 10^8$
1	0.5002	0.00037	13.69
2	0.4993	0.00053	28.09
3	0.5001	0.00027	7.29
4	0.5000	0.00017	2.89
5	0.4998	0.00003	0.09
6	0.4995	0.00033	10.89
7	0.5000	0.00017	2.89
8	0.5005	0.00067	44.89
9	0.4996	0.00023	5.29
10	0.4993	0.00053	28.09
	$\Sigma = 4.9983$		$\Sigma = 144.1 \times 10^{-8}$
	$\bar{x}_\theta = 0.49983$		$\sigma_\theta = [(144.1 \times 10^{-8})/10)]^{0.5}$
			$= 0.00038$ inch

The horizontal lines on Fig. 14.7 were derived from $\bar{x}_\theta$ and σ_θ:

- The horizontal line (T) at 0.5000 is the nominal value desired
- The horizontal warning lines are at $T \pm\sigma_\theta$ (0.50038 and 0.49962)
- The action lines are at $T \pm 2\sigma_\theta$ (0.50076 and 0.49924)
- The design limits are at 0.5000 ±0.0010

After the control chart has been produced, the mean diameter for each succeeding sample of five parts is plotted. The mean for this first set is shown at A and the second at B. The mean value for each set of five samples is seen to fall on either side of the nominal 0.5000 value until point C is reached. This point falls below the warning line which suggests the next value should be given particular attention. However, the next point D falls in the satisfactory range suggesting that point C was just an unusually low value and the fact that action was not taken was the correct decision. However, point E falls just below the warning line. No action is taken, but the next point F is just above the action line. Action could have been taken at F, but it was decided to see what the next set of samples revealed. This point G fell well below the design limit and the machine was adjusted (tool, depth of cut, etc. changed) so that the next point H fell in the safe range and subsequent points again fluctuate in an acceptable range above and below the nominal 0.5000 value.

9.0 OTHER DISTRIBUTIONS

In addition to the normal and binomial distributions, there are others that are better suited to other purposes. In medical and biological research, the Poisson distribution is useful in expressing the distribution of organisms in a sample. For example, the hemocytometer enables the number of red corpuscles to be counted in each of a number of small volumes when observed under the microscope. For these particular applications, the Poisson distribution is useful. The Poisson distribution is also employed in nuclear physics involving the extent of radioactive decay in a given interval of time. It is also used in agricultural research.

The Weibull distribution is useful for studies involving brittle fracture and fatigue of metals. It is widely used in the testing and analysis of ball and roller bearings. In the application of Weibull statistics, specially ruled paper that results in plotted data falling on a straight line, is quite convenient.

10.0 CURVE FITTING

In many cases, experimental data points fall on either side of a straight line (Fig. 14.8). While it is often sufficient to estimate the line that best fits the data by eye, this can be done more precisely by a method of least squares. This involves determining the slope and intercept of the straight line that best fits the data analytically. The dispersion in the y direction (d_i) is determined for each point from the straight line being sought. The sum of the squares of these dispersions will be a minimum for the line that best fits the data. This procedure involves considerable calculation when the numbers of experimental points involved are large.

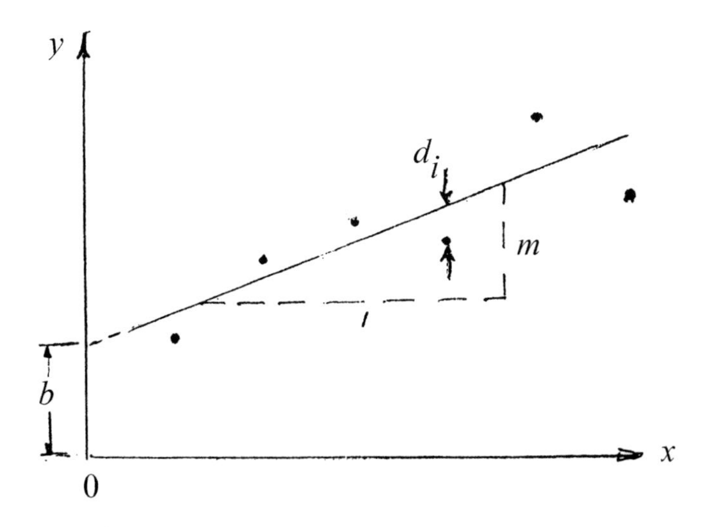

Figure 14.8. Experimental points with straight line $y = mx + b$ fitted by minimizing the sum of dispersions (d_i) from the point to the line.

Today, software for use on a personal computer reduces the task of least squares curve fitting to a simple operation. Before the appearance of the personal computer in the early 1980s, most statistical texts contained a chapter on short cuts in performing calculations. Fortunately special software for statistical calculations makes such short cuts unnecessary today.

In cases where experimental points indicate a curve, it is usually profitable to explore the use of specially ruled paper (log, semi-log, log-log, etc.) to determine whether the data gives a straight line when plotted on differently ruled papers.

11.0 FACTORIAL DESIGN OF EXPERIMENTS

When the influence of a large number of potentially important variables upon a single factor, such as surface finish in surface grinding is to be studied, the effect of changing one variable at a time is generally employed. This, however, is not the most efficient procedure for studying large systems. The method known as factorial design of experiments is an alternative procedure that is frequently useful in such cases. By this method, all of the variables to be studied are changed at once in a systematic way. In the simplest applications, each variable (factor) is assigned two values: a high one and a low one. The range spanned by these values should be as large as possible, consistent with an assumed linear increase or decrease of the main dependent variable (surface finish for example) over the range. Tests are performed using all combinations of the high and low values of all factors. The total number of tests will be 2^n where n is the number of factors investigated and 2 is the number of levels studied for each factor—hence the name 2^n factorial design.

In order to keep track of the individual tests, a special naming system is employed. Each factor is assigned a letter, such as a, b, or c in the case of a three factor experiment. When the high value of the factor is used, a letter appears in the symbol for this experiment, but if the low value is used, unity appears. The letters are simply multiplied together to obtain the symbol designating the experiment. For example, if the high values of variables a and b are used with the low value of c in a three factor experiment, this experiment would be referred to as (ab). The multiplication by unity for c is inferred. If, on the other hand, all variables had their respective low values, the symbol for this experiment would be (1).

Experiments should be done in a random order which may be determined by putting 3^n slips of paper, each one containing the symbol corresponding to a different experiment, in a hat and then removing the papers one at a time without prejudice. After all experiments are performed, the average may be computed as follows:

Eq. (14.10) $\text{Av} = (abc + ab + ac + bc + a + b + c + 1)/2^3$

Here we are assuming a three factor experiment for which there will be $2^3 = 8$ tests.

The direct effect of some variable (say variable a) may be found as follows:

Eq. (14.11)

$$\text{Direct effect of } a = \pm[abc + ab + ac + a - (bc + b + c + 1)]/8$$

Similarly, the direct effect of variables (b) and (c) will be:

Eq. (14.12)

$$\text{Direct effect of } b = \pm[abc + ab + bc + b - (ac + a + c + 1)]/8$$

Eq. (14.13)

$$\text{Direct effect of } c = \pm[abc + ac + bc + c - (ab + a + b + 1)]/8$$

The ± signs in this and subsequent equations refer to the variation due to the variable in question falling above and below the mean.

The relative effects of variables (a) and (b) may be obtained by comparing the magnitudes of their two direct effects. It should be noted that the effect of variable (a) will be more representative as determined in this away than in the classical method where one variable is changed at a time since variables (b) and (c) cover a range of values instead of corresponding to a fixed combination. Similarly, the relative effects of variables (a) and (c) and (b) and (c) may be obtained by comparing their corresponding direct effects.

Example: Factorial Experimental Design. Application of the factorial design technique is best illustrated by a very simple example. This will involve a 2^2 experiment where the influence of two of the many variables associated with the finish of a ground surface is involved.

Figure 14.9 is a schematic of a surface grinding operation. The grinding wheel of diameter (*D*) consists of hard abrasive particles of diameter (*g*) held together by a bonding material. Before use, the wheel is dressed by passing a diamond across the surface of the rotating wheel. This trues the wheel, sharpens the individual grits, and provides space between adjacent particles to accommodate chips. A cut is made by moving the wheel downward a small distance (*d*) and feeding the work horizontally against the wheel at a velocity (*v*). The variable (*d*) is called the wheel depth of cut, *v* is the work speed, and *V* is the wheel speed. In the case considered here, the axial width of the wheel is greater than the width of the work.

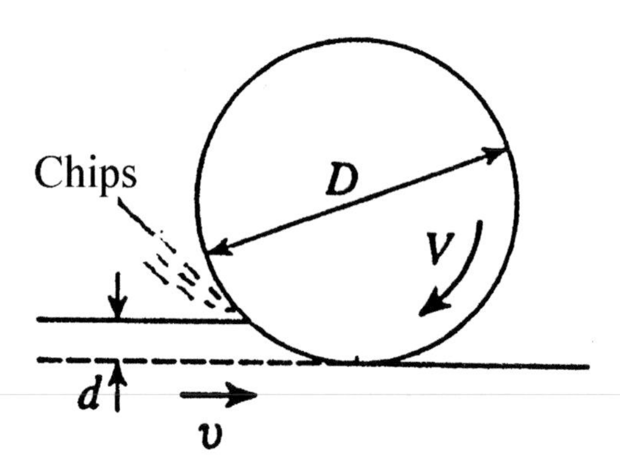

Figure 14.9. Schematic of surface grinding operation.

The roughness of the surface produced depends upon many variables including (*g*) and (*d*). The roughness of the finished surface is measured by an instrument that causes a diamond stylus to traverse the ground surface in the axial direction (i.e., perpendicular to the direction of *v*). The diamond moves up and down as it traverses peaks and valleys in the finished surface. The motion of the diamond is greatly amplified and converted to a number called the arithmetic average roughness of the surface (R_a in microinches). Figure 14.10 shows a greatly magnified ground surface. The centerline is positioned so that the areas of the peaks above the line equal the areas of the valleys below the line. R_a is the mean of the distances to peaks and valleys.

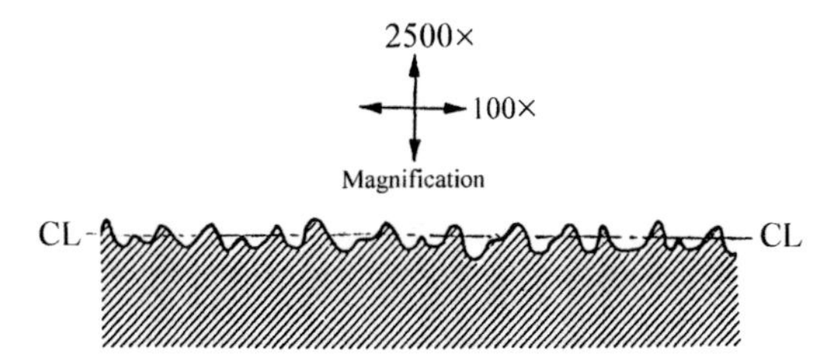

Figure 14.10. Typical surface finish trace showing the centerline from which the arithmetic average roughness value (R_a) is determined. The magnification is different in the horizontal and vertical directions since surface asperities are generally about 25 times as wide as they are high.

All other variables except (v) and (d) were held constant at representative values for this type of grinding:

Wheel diameter (D): 8 in. (203 mm)
Grit type: white Al_2O_3
Work material: hardened ball bearing steel
Wheel speed (V): 6,000 fpm (30.5 mps)
Work speed (v): 48 fpm (0.24 mps)
Grinding fluid: none

The high and low values of the two variables considered were:

Variable	Symbol	High value	Low value
Nominal grit dia.	g	#24 screen size	#60 screen size
		(0.030 in. = 75 μm)	(0.011 in. = 275 μm)
Wheel depth of cut	d	0.0004 in. (10 μm)	0.0002 in. (5 μm)

Downfeed readings were measured using an air gage accurate to 50 microinches (1.25 μm).

The ground surfaces were traced using a tracer instrument having a stylus of 0.0005 in. (12.5 μm) radius of curvature. All CLA (arithmetic average) values were obtained using a 0.03 in. (750 μm) tracing length (cut off). Surface finish traces were made in a direction transverse to the direction of grinding. Centerline average (R_a) values for the surfaces produced for the four combinations of grit diameter (g) and downfeed (d) were as follows:

Test Designation	R_a, μin. (μm)
gd	32.5 (0.813)
g	26.5 (0.663)
d	23.6 (0.590)
1	20.6 (0.515)

These R_a values are averages for 10 measurements in each case. The average value for all tests was 25.8 μin. (0.645 μm). The direct effect of grit size was:

$$\pm\tfrac{1}{2}''[(gd - d) + (g - 1)] = \pm\tfrac{1}{4}[(32.5 - 23.6) + (26.5 - 20.6)]$$
$$= \pm 3.7 \text{ μin. } (0.93 \text{ μm})$$

The order of signs in the answer indicates that the high value of grit size (#24 screen size) yields greater roughness than the low value (#60 screen size) as might be expected.

The direct effect of downfeed is:

$$\pm\tfrac{1}{2}''[(gd - g) + (d - 1)] = \pm\tfrac{1}{4}[(32.5 - 26.5) + (23.6 - 20.6)]$$
$$= \pm 2.25 \text{ μin. } (0.56 \text{ μm})$$

This indicates that the high value of downfeed (0.0004 in. = 10 μm) yields a greater roughness than a low value (0.0002 in. = 5 μm).

Comparing the direct effects of grit size and downfeed, it is seen that grit size is about 60% more important than downfeed in determining surface roughness over the range of the two variables considered.

This small experiment indicates that:

- Both grit size and downfeed have a significant influence on the R_a roughness produced in surface grinding
- Both grit diameter (g) and downfeed (d) should be small for fine finish, and grit size is more important than downfeed for the range of variables considered

In the foregoing example, a very small number of variables (n) has been considered. In a more realistic study of this sort, a larger number of variables would be involved (say 10 instead of two). In such a case, 2^{10} = 1,024 tests would be involved. For such cases, a fractional factorial design may be used to reduce the experimental effort required. However, the amount of computation would still be troublesome. In this connection, it should be mentioned that software programs are available that greatly reduce the computational effort involved. Also, in some instances, the assumption that the variation from the low to the high value is linear may not be a reasonable assumption. In such cases, intermediate experimental points may be employed leading to a 3^n or even 4^n factorial design.

PROBLEMS

14.1 In Ch. 3, the period (p) of a simple pendulum was found to be:

$$p = C\,\ell/g$$

where $C = 2\pi$. The acceleration due to gravity (g) may be found as follows:

$$g = 4\pi^2\ell/p^2$$

If ℓ and p may be measured with the following accuracy:

$$\ell = \pm 0.01 \text{ m}$$

$$p = 2 \pm 0.02 \text{ sec}$$

a) What is the maximum error in g when so determined?

b) What is the minimum value?

c) Is the uncertainty of ℓ or P most important?

14.2 Given the following values for a variable x:

1.03, 1.15, 1.07, 1.20, 1.01, 1.18, 1.05, 1.18, 1.04, 1.17

Find:

a) The arithmetic mean ($\bar{x}$).

b) The median.

c) The mode.

14.3 For the data of Problem 14.2, find:

a) The spread or range.

b) The mean deviation ($\bar{d}$).

c) The standard deviation.

14.4 The diameter of a pin is measured six times and the following values are obtained:

0.243, 0.241, 0.239, 0.2426, 0.2433, 0.2450

The first three values are measured with an instrument considered to be one-third as accurate as that used for the last three, find:

a) The average value for each instrument.

b) The weighted average value for all six values.

14.5 Four cards are drawn from a shuffled deck of 52 cards.

a) What is the probability that they will be four aces (without replacement)?

b) Repeat if each card is replaced before the next one is drawn.

14.6 If two cards are drawn from a shuffled deck of 52 cards, what is the probability that both will be spades?

14.7 A bag contains five black balls and 95 white balls.

a) What is the probability that the first ball drawn will be black?

b) What is the probability that n balls drawn successively will be black?

c) What is the probability that n balls drawn successively will be white?

In b) and c), the balls are replaced and shaken up after each draw.

14.8 What is the probability that two proper dice will come up with two ones when tossed?

14.9 For the letters A, B, C, and D, how many combinations are there, taken two at a time (like AB, CD, etc. where AB and BA are considered the same)?

14.10 Find the number of permutations for Problem 14.8.

14.11 How many seven letter words may be formed from the word *country*?

14.12 How many 10 letter words may be formed from the letters in the word *statistics*?

14.13 a) If the arithmetic average for the data of Problem 14.1 is 1.108 and the standard deviation is 0.0704, plot the normal distribution for the frequency of occurrence (y) vs the magnitude of the measurement (x) in the manner of Fig. 14.3 (*a*).

b) Plot the observed values of x beneath the x axis as in Fig. 14.3 (*b*).

14.14 The value of the mean ($\bar{x}$) and standard deviation (σ) of a set of data is found to be 25 and 5 units respectively.

a) Within what limits of x should essentially all of the values lie?

b) Within what limits of x should approximately 2/3 of the values of x lie?

14.15 A large batch of parts has 5% defective. If samples of four are drawn at random:

a) What is the probability of 0 being defective, one defective, two defective, three defective, and four defective?

b) Draw a histogram of these results.

14.16 A very large population of parts has 1% defective. If a sample of 12 items selected and tested on a go, no-go basis:

a) What is the probability that none of the 12 will be defective?

b) That one will be defective?

c) That two will be defective?

d) That three will be defective?

14.17 A 2^n factorial experimental design is used to study the relative influence on surface roughness of a change of two variables: wheel speed

Variable	Symbol	High value	Low value
Wheel speed	V	6,000 fpm	4,500 fpm
Work speed	v	83 fpm	13 fpm

(V) and work speed (v), when all other variables are maintained the same at average conditions. The range of values employed is as follows:

Test, fpm	Designation	CLA, μin.
V v		
6,000, 83	Vv	30.1
6,000, 13	V	19.3
4,500, 83	v	35.3
4,500, 13	1	22.3

The average values of centerline average (CLA) surface roughness obtained for the four tests of this case were as follows:

a) Find the direct effect for the changes in V and v.

b) Which variable has the greatest influence over the ranges considered?

c) For good finish (low roughness), should V and v be high or low?

15

Computers In Engineering

1.0 INTRODUCTION

A *computer* is any electronic device that solves problems by processing data according to instructions from a program.

Today, computers influence all aspects of life:

- Information transfer (including library activities)
- Communications (written, visual, and oral)
- Travel (airline, rail, sea, and hotel schedules and reservations)
- Commerce (banking, investment, marketing, and advertising)
- Data collection and analysis
- Design activities of all sorts
- Computations of all sorts
- All aspects of manufacturing (design, planning, production)
- Linguistics and the arts
- Entertainment and leisure
- Etc.

Computers are having important effects on engineering, some of which will be discussed in this chapter. However, before covering some of these, a number of computer related developments that brought about the present state of technology will be considered together with a brief review of some of the more important details of computer technology. The introduction of computers has greatly influenced the English language by adding many new words and extending the meaning of many others. This has been accompanied by the introduction of a large number of acronyms. An attempt will be made to bridge this obstacle to understanding by defining many of these new terms and interpreting some of the more important acronyms.

2.0 HISTORICAL BACKGROUND

The *abacus* (Fig. 15.1) is a calculating device developed in the ancient kingdom of Babylonia in SW Asia and widely used in early Greek and Roman times. It involves moving beads along spindles mounted in a rectangular frame. The frame is divided into two regions by a stationary member (AB in Fig. 15.1). Each spindle in upper region C contains two beads while each spindle in lower region D contains five beads. Each bead in region C has a value of five units while each bead in region D has a value of one unit. The spindles have different values of scale, the first spindle covers numbers from 1 to 10, the second spindle tens, the third hundreds, etc., as shown in Fig. 15.1. Beads are activated by moving them up and down to a position close to transverse member AB. When none of the beads is close to AB this corresponds to zero. The configuration of beads in Fig. 15.1 corresponds to the number 9,639. Numbers may be added to or subtracted from this by moving the appropriate number of beads toward or away from AB. For example, if the beads on the first two spindles were moved away from AB this would correspond to a subtraction of 39. The origin of the abacus is unknown but the fact that it is based on a system of ten suggests that it might be an extension of counting on one's fingers, each bead in region C corresponding to all of the fingers on one hand with each bead in region D corresponding to one finger. The abacus is still used in parts of the Asia-Pacific region and those skilled in its use can operate at a speed comparable to someone using a modern electronic hand held calculator.

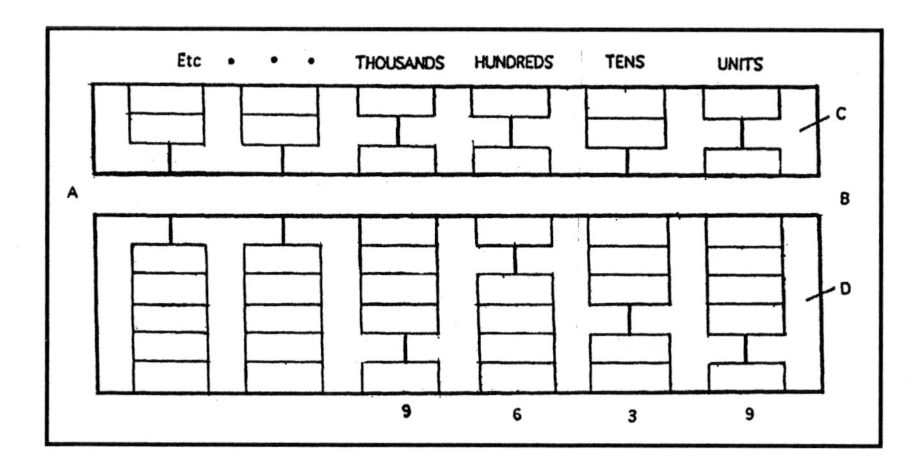

Figure 15.1. Abacus set to the number 9,639.

In 1642 B. Pascal (1623–1662), a French mathematician and philosopher, invented the first automatic calculator consisting of wheels and gears. The first wheel had 10 notches corresponding to numbers 0 to 9. The second wheel had 10 notches each worth 10. The third wheel had notches each worth 100, etc. When the first wheel had moved 10 notches a gear was engaged which moved the second wheel one notch as the first wheel returned to zero. This device was capable of only addition and subtraction and is apparently a mechanical abacus. In the 1670s G. von Leibnitz improved Pascal's machine to include multiplication and division.

Another device widely used in the 20th century for making engineering calculations, first developed by E. Gunter in 1620, is the slide rule. In its simplest form, it consists of two scales—one stationary and one movable—each with numbers spaced in proportion to logarithms to the base 10 of the numbers (called a log scale). Figure 15.2 shows the sliding scale C positioned to yield the product of 2 × 3. The answer (6) is read from the stationary D scale opposite 3 on the sliding member. A transparent cursor is usually used to give improved transfer from one scale to the other. The above multiplication is, of course, based on the fact that:

Eq. (15.1) $\log 2 + \log 3 = \log (2 \times 3)$

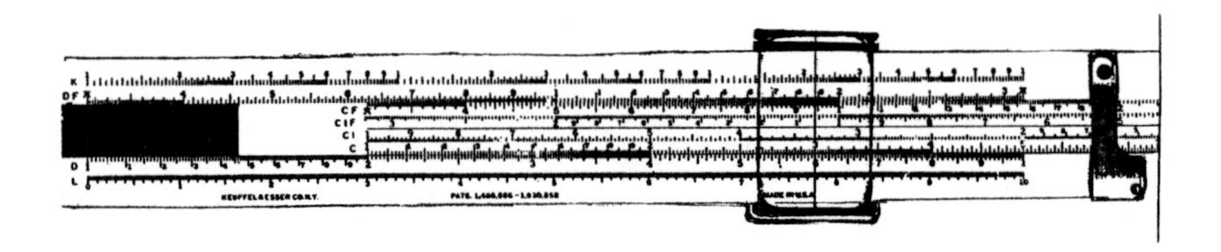

Figure 15.2. Engineering slide rule set to give the product of 2 × 3 = 6.

Division is accomplished by the inverse operation. Engineering slide rules also provide means for raising a number (n) to a power (p). This employs scales based on the fact that:

Eq. (15.2) $$n^p = p \log n = \log p + \log \log n$$

This requires a log scale used in conjunction with log log scales. Engineering slide rules also include means for approximating the sines and tangents of angles in degrees. The most widely used engineering slide rules are approximately 10 inches in length, and when carefully used give results accurate to three significant figures. When more accurate results are required larger slide rules good to four significant figures or log tables (accurate to five or more significant figures) as well as sine and tangent tables are used.

Throughout World War II calculating machines that operated with complex systems of racks and pinions were used to perform engineering calculations. Addition and multiplication involved rotational input in one direction while subtraction and division required rotation in the opposite direction. These rotations were achieved by DC motors in large (Marchand and Monroe) machines, but by hand in smaller machines such as the Facit. Facit calculators manufactured in Sweden were widely used for currency exchange calculations in Europe well into the 1960s. A hand rotated Facit (Fig. 15.3) about the size of a small typewriter weighing 15 pounds was capable of addition, subtraction, multiplication, division and, with patience, extraction of square roots. However, all of these mechanical calculating machines gave results to a much larger number of significant figures than slide rules.

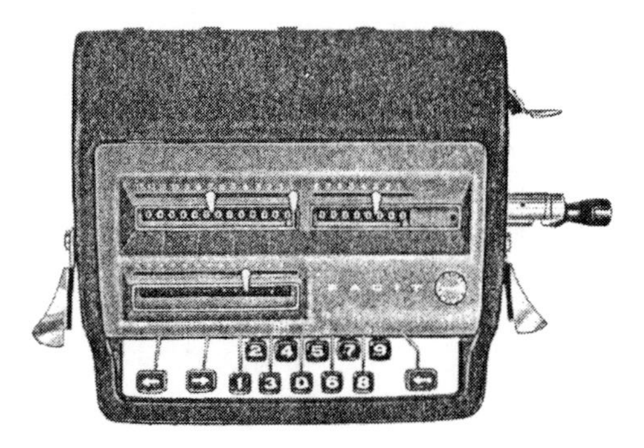

Figure 15.3. Facit calculating machine capable of addition, subtraction, multiplication, division, and extracting a square root.

C. Babbage, while serving as Lucas Professor of Mathematics at Cambridge University in the UK, developed plans for a digital "analytical engine" in the 1830s. Augusta Ada King, daughter of poet Lord Byron, was a mathematician associated with Babbage. She created a program for Babbage's proposed computer, which is said to be the world's first computer program. She was married in 1833 to William King, who later became the Earl of Lovelace, making her a countess. Hence, she is best known as the Countess of Lovelace. Babbage's machine was not built at the time due to lack of funds.

An analog device is one that uses data presented as a continuously changing variable. A digital device is one that operates with data presented as discrete blocks of information. A thermometer is an analog device since the column changes continuously with temperature. Pressure gages and strain gages are also analog devices as are oscillograph traces, voltmeters, ammeters, and micrometers. The abacus is a digital device since each bead represents a discrete number. It is a digital system with a base of 10 (a decimal base). Practically all modern computers are digital, but involve a base of 2 (1 for on, 0 for off). Inputs to such computers will usually be analog and must be converted to digital before being processed. This is accomplished by an analog to digital converter (an ADC).

Punched card input systems were first used by J. M. Jacquard in 1805 to control weaving looms. Such systems were used in many early computer systems into the 1950s.

Morse, an artist, invented the telegraph in 1844. This employs a digital binary device, based on a number system of 2, that sends messages in terms of combinations of long and short electrical pulses obtained by tapping a key. Each letter of the alphabet corresponds to a combination of dots and dashes. For example, the well-known SOS distress signal is shown in Fig. 15.4. In 1856 Western Union was founded based on Morse's invention. This was the first monopoly business in America.

Figure 15.4. Morse code for distress signal S O S.

The telephone, invented by Alexander G. Bell in 1876, was an analog device and was another very important communications breakthrough. The speaker's voice vibrates a diaphragm producing a variable resistance in a capsule containing compressed carbon particles in contact with the diaphragm. The change in electrical resistance is proportional to the acoustical pressure on the diaphragm produced by the speaker's voice. The changing resistance causes a variable current in a DC circuit that is transmitted over the phone line to the receiver. At the receiver the variable current causes an electromagnet to vibrate a diaphragm that reproduces an acoustical signal identical to that of the input voice signal.

H. Hollerith was the inventor of a tabulating machine that was used in the U.S. Census of 1890. This machine electronically recorded statistics by reading and sorting punched cards. In 1896 Hollerith organized the Tabulating Machine Company in New York which eventually grew into the International Business Machine Company (IBM).

There were not many computer-related developments before World War II. However, in the 1930s V. Bush, an electrical engineer, developed the first analog computer that would solve differential equations with up to 18 independent variables. This was a forerunner of the electronic computers developed in the late 1940s after World War II.

H. H. Aiken invented the Harvard Mark I automatic calculating machine that was completed in 1944. This was 31 feet long, 8 feet wide and weighed 35 tons. It solved problems using a punched paper tape with coded instructions, and was used to solve ballistic problems for the U.S. Navy. An improved model (Mark II) was completed in 1947.

A. M. Turing, an English mathematician, made important contributions to the theory of computers that were developed after World War II. In 1945 he developed a large electronic digital computer at the National Physical Laboratory in the UK. It was called the Automatic Computing Engine (ACE). Later, at the University of Manchester, he helped develop the Manchester Automatic Digital Computing Machine (MADCM) which had the largest memory capacity at the time (1948). Turing also made important contributions to the area that has since become known as Artificial Intelligence (AI).

J. P. Eckert was co-inventor with J. W. Mauchly of the first general purpose electronic digital calculator/computer that was the prototype of most computers used today. In 1946, under government contract, they completed a machine known as the Electronic Numerical Integrator and Calculator (ENIAC). It had a huge number of vacuum tubes. While it was 1,000 times as fast as the Mark I, it was very difficult to program. In 1949 Eckert and Mauchly produced the Binary Automatic Computer (BINAC) which stored information on magnetic tape rather than on punched cards. In 1951 Eckert and Mauchly produced the Universal Automatic Computer (UNIVAC I) that was used to predict Eisenhower's win 45 minutes after the election poles closed.

J. Forrester, an electrical engineer, invented the random access magnetic core memory (RAM) now used in most digital computers. In 1945 he founded the Digital Computer Laboratory at MIT and participated in the development of Whirlwind I, a general purpose digital computer. Forrester worked on the application of computers to management problems which led to computer simulation where real world problems represented by interconnected equations are solved by a computer.

In the late 1940s J. von Neumann introduced the idea of using internally stored programs in a computer.

In 1948 Norbert Weiner, a mathematical physicist at MIT, published a book concerned with the relationship between humans and machines. This was the origin of the concept of cybernetics and eventually terms like cyberspace, etc.

3.0 THE BIRTH OF NUMERICAL CONTROL (NC)

In the late 1940s, J. D. Parsons of Traverse City, MI obtained a contract with the U.S. Air Force to conduct a feasibility study of a pulsed servomechanism activated by punched cards or tape that would input data to control lead screws on a machine tool to produce complex shapes. This was to be done by approximating the motion of the tool by very small straight line motions. In June of 1949, Parsons consulted with the Servomechanism Laboratory at MIT to develop the controller for this machine. Under a subcontract, MIT conducted a complete systems study that revealed that a punched card unit was impracticably slow to control the lead screws and suggested use of a digitally controlled approach instead. At this point, Parsons' Air Force contract was up for review and the Air Force decided to continue the development with a contract directly to MIT. In 1950 W. Pease was placed in charge of the project with J. McDonough to assist Pease and with Parsons acting as a consultant. By early 1952 a controller had been designed and built, and a machine tool procured to test the design. The machine was a Cincinnati Milling Machine Company Hydrotel Milling machine from the Air Force Surplus Storage Facility in Georgia. The controller was capable of approximating any three dimensional shapes by straight line elements as small as 0.0005 inch (12.5 μm) in length. The controller was very large and involved a total of 792 vacuum tubes that emitted a large amount of heat. Tube life was a problem, and, in order to employ tubes with a sufficiently long life, they were used off line for many hours to remove weak ones.

In March 1951, it was decided that a simple name was needed for the concept of digital control of machine tools, and a contest to obtain a name was conducted in the Servomechanisms Laboratory. McDonough was the winner with "Numerical Control" (NC).

In the summer of 1950, a professor from the Mechanical Engineering Department, with two members of MIT's Industrial Liaison Office, visited several aircraft companies in California to obtain drawings of typical parts the companies would like to see produced by the new system.

At this time, the Air Force decided it was time to introduce the public to numerical control. This was done in three one-day sessions, which included lectures and demonstrations as follows:

1. Sept. 15, 1952 for machine tool builders
2. Sept. 16, 1952 for those interested in control technology
3. Sept. 17, 1952 for aircraft companies and those producing parts

Over 200 representatives from 130 organizations attended these meetings. The programs went off without difficulty, but reactions were mixed. Machine tool builders were very pessimistic, but aircraft builders (Boeing, etc.) were very optimistic. The adverse questions had to do with reliability and economics.

These demonstrations brought an end to phase 1 of the MIT/Air Force project which took only 40 months at a cost of $360,000 billed by MIT to the Air Force. However, this was only a proof of concept. Within a few years, relays and punched cards were replaced by magnetic storage and vacuum tubes by transistors, which were invented at the Bell Labs by J. Bardeen, W. H. Brattain, and W. B. Shockley in 1947.

Parsons, who had initiated the project, left it in 1951. He was rather bitter concerning the fact the Air Force chose to continue pursuing his idea at MIT because he had no practical way of carrying it out. Parsons deserves credit for having recognized the need for NC, of course, as does MIT for demonstrating a practical way of carrying it out.

The key feature of the numerical control project was the use of a single command to produce a 3D straight cut of any length at constant speed. Initially, mathematically designed cuts were programmed on the MIT Whirlwind I computer. Otherwise, an electromechanical desk calculator was used. By 1952 it had become clear the next problem to be solved was improved programming. The objectives of the second phase of the project (1953–1959) were:

- To further inform industry of the potential of NC
- To conduct an economic study of ten representative complex parts
- To develop an improved programming procedure

The main finding in the economic study was that whereas production cost was less, the high programming cost resulted in the product cost being greater. This intensified work on the third objective. The result was a programming procedure that used simple English commands. The new programming procedure was called the Automatic Programmed Tool (APT). This important development, completed in 1959, was headed by D. T. Ross, a mathematics major at MIT. The total cost of the program from the beginning to 1959 was about $\$10^6$. In 1959 the name of the laboratory at MIT was changed from the Servomechanisms Laboratory to the Electronic Systems Laboratory since, by that time, a much wider range of work was being conducted than servomechanisms indicated.

3.1 Japanese Contributions

Important contributions to NC and robotics have been made by Dr. Eng. S. Inaba, who is now President and CEO of Fanuc Ltd. He graduated from The University of Tokyo's Ordanance Department (now the Precision Engineering Department) in 1946, and joined Fuji Tsushinki Seizo KK, which became Fanuc Ltd. in 1972. In late 1952, he learned of the development of NC at MIT, was impressed, and requested permission from the company to develop a NC turret punch press. Several problems were encountered over a number of years. Originally an electrohydraulic pulse motor, based on previous experience with warship gun turrets, was used. This eventually proved to be too noisy and to require far too much energy. This was solved in 1974 by negotiating a license to use an electro-servomotor that had been developed in the USA. A deadline was set to have a NC machine tool exhibited at the Osaka Machine Tool Show in September 1974. This was successful and the Japanese NC endeavor began to grow rapidly.

By 1988 the Japanese market share for NC machine tools, as well as for robots, was 50% worldwide. The Japanese entrance into the NC field had an important influence on similar activity in the USA by providing important competition. In addition to NC machines and control motors, Fanuc is also a leader in the area of robotics, which it entered in 1978.

In his autobiography (*Management of a Company in the Advanced Information Age* by S. Inaba, U. Tokyo Press, 1987), Dr. Inaba stresses the following principles of management:

- Need for engineers to be creative and to seek new solutions rather than being controlled by past achievements and methods
- Importance of patience in developing a new product
- Importance of retraining workers displaced by automation to perform jobs that only humans can do
- Importance of producing products related to a company's strengths
- Importance of a global outlook and a horizontal strategy (seeking partners) rather than a vertical strategy (acting alone)

Examples of Fanuc's application of the last principle are the 50–50 joint ventures of Fanuc and General Motors relative to robots, the 50–50 venture between Fanuc and General Electric relative to servomotors, and overall factory automation. The recent (1999) merger leading to Daimler - Chrysler is an indication that Dr. Inaba's horizontal strategy is being globally adopted.

With Dr. Inaba, we have a well-trained engineer with remarkable management skills. This is an unusual combination. In most cases, there are strengths in one of these areas only. One of the objectives of this book is to make those in management involved with engineers more conversant with engineering and the relation between engineering and science.

4.0 CALCULATORS

In 1965 Sharp introduced an all transistor calculator in Japan. This was capable of addition, subtraction, multiplication, and division, was the size of a small typewriter, weighed 55 pounds, and cost $2,500. This was state-of-the-art for an electric calculator in 1965. In the same year, P. Haggerty, President of Texas Instruments (TI), decided that what he called a pocket calculator would represent a good application for Integrated Circuit (IC) technology that had been invented in 1958. J. Kilby, co-inventor of the IC, was assigned this project. Kilby decided to divide the project into parts that would be worked on simultaneously to decrease the time required and to have design inputs from different points of view (a procedure known as concurrent engineering). J. Merriman was placed in charge of developing the circuits and other microelectronic details. J. Van Tassel in charge of designing the keyboard, housing and other similar details, while J. Kilby coordinated the project. A rechargeable Ag-Zn battery was used. By the end of 1966, the first prototype housed in an aluminum box 4¼ × 6½ × 1¾ inches (108 × 85 × 44.5 mm), weighing 4½ oz (1.25 N) was produced. The output involved a paper printout operating at twelve characters per second.

The project took only two years. The U.S. patent, filed in 1967, was granted in 1974, and the first calculators were sold in Japan (the Pocketronic) in the late 1970s by Canon for $395, under a license from TI. In 1972 TI produced its four function "Data Math" (TI 2500) that used a light emitting diode (LED) output display. This was initially priced at $150. After 1972 many companies began producing hand-held calculators, and by 1980, the price of a four-function unit had dropped to about $10.

Figure 15.5. Current Hewlet-Packard 32S RPN Scientific Calculator.

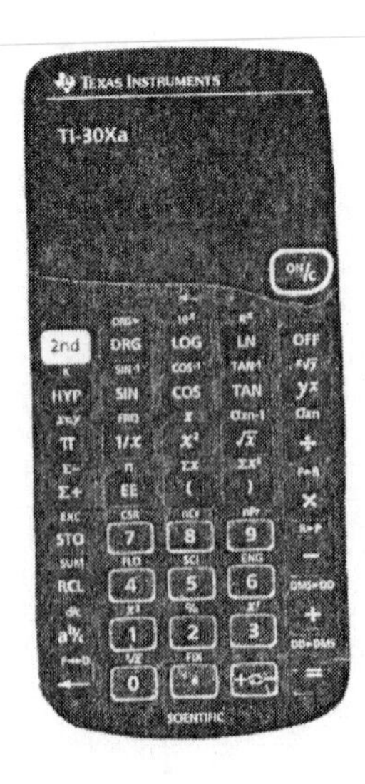

Figure 15.6. Current Texas Instruments TI 30X Scientific Calculator.

In 1972 Hewlett-Packard (HP) introduced its "Scientific Calculator" (HP 32S, Fig. 15.5). This greatly extended the range to include logarithms, trigonometric functions, numbers raised to powers, and memory units, all to over 10 significant figures with a price of $400. Soon after the HP 32S appeared, TI produced a similar scientific calculator (Fig. 15.6).

The architecture of the HP and TI calculators was quite different. For example, to multiply 2 × 3 on the TI unit the following items were keyed in and the underlined answer appeared in the register.

Eq. (15.3) $2 \times 3 = \underline{6}$

The HP calculator used a different sequence called Reverse Polish Notation (RPN). Jan Lukasiewicz (1878–1956), a Polish mathematican, developed a parenthesis-free system of entering numbers that places the operator before the numbers to be manipulated. This is known as Polish Notation (PN). Conventional algebraic notation places operators between the numbers, as in Eq. (15.3). Hewlett-Packard (HP) calculators employ a memory stack and an entry procedure where the operator follows the number (the reverse of PN). Hence, the term Reverse Polish Notation (RPN). This provides a parenthesis-free procedure that requires fewer keystrokes, in complex algebraic applications, than the conventional algebraic approach. Equation (15.3) would be carried out on the HP 32S as:

Eq. (15.4) $2 \text{ enter } 3 \times \underline{6}$

Early in the 1970s, hand-held calculators appeared that could solve many engineering problems using programmed magnetic strips that could be fed into the calculator to carry out a series of complex calculations after introducing a few input variables that defined the parameters of the problem. Figure 15.7 shows the HP 35 that was introduced by Hewlett-Packard in 1974. A simple example would be to find the maximum elastic stress and strain in a rectangular cantilever beam subjected to a concentrated load. After feeding the magnetic strip into the calculator and keying in pertinent parameters, pressing a particular key would give the maximum stress while a second key would give the maximum strain. These memory strips could be purchased programmed to solve a series of problems in a certain branch of engineering (mechanical, civil, electrical, chemical, etc.), or they could be programmed by the user to meet a particular need. Texas Instruments also produced a programmable calculator called the Programmable Slide Rule (Fig. 15.8).

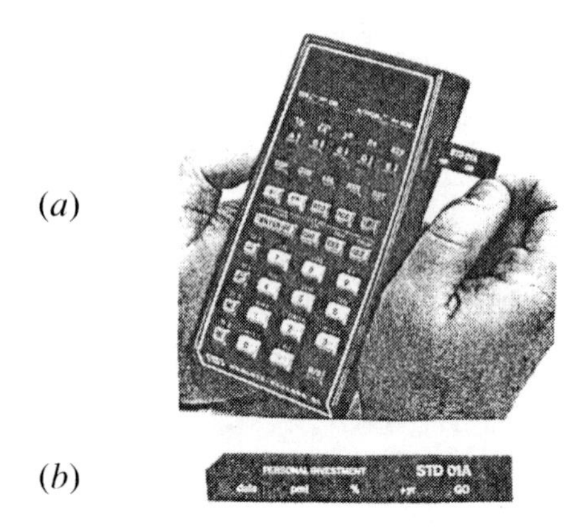

Figure 15.7. *(a)* Hewlett-Packard HP 35 Programmable Pocket Calculator, April 1974, and *(b)* magnetic program card for HP 35.

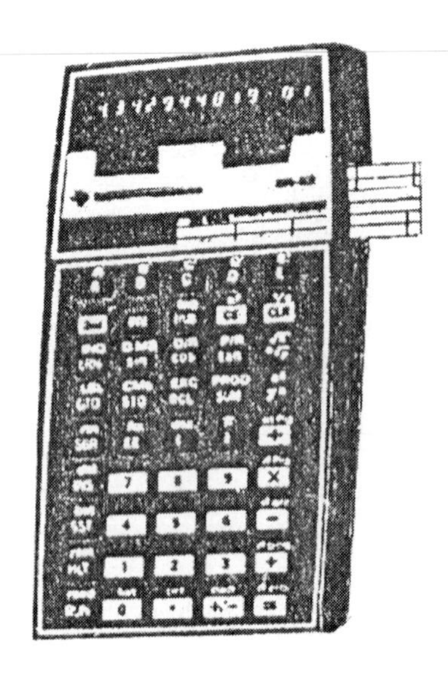

Figure 15.8. Texas Instruments SR 52 Programmable Slide Rule Calculator, 1975.

The appearance of these scientific calculators in the 1970s rendered the slide rule obsolete. However, the significance of programmable calculators was limited due to the appearance of personal computers (PCs) about 1980.

5.0 THE BIRTH OF COMPUTER-AIDED DESIGN

The third and last phase of the Air Force/MIT project (1960–1970) was concerned with quite a different area—use of the computer in design. This involved several members of the Mechanical Engineering Department as well as those in the Electronic Systems Laboratory and led to what is now called Computer-Aided Design (CAD). This included the use of the computer monitor to produce sketches during the creative phase of design and finished drawings. In addition, the computer was used for stress analysis of all sorts, standard parts selection, storage of catalog information related to standard parts, and the design of electrical circuits and networks. A detailed account of the entire Air Force/MIT project is presented in *Numerical Control: Making a New Technology* by J. F. Reintjes, 1991, Oxford University Press.

6.0 THE PERSONAL COMPUTER (PC)

Analog computers programmed to solve equations related to continuously varying systems were used mainly in the 1940s. By the mid-1940s, it became clear that digital computers held much more promise for the future.

Until the early 1980s, large computers were built that served several hard-wired workstations at a company or university. These were called main frame computers. The largest, most powerful, and expensive of these were called supercomputers.

An extremely important development was the independent discoveries of the solid state integrated circuit concept in 1958 by J. Kilby of Texas Instruments Corp. and R. Noyce of Fairchild Semiconductor Corp. This made it possible to produce extremely small and powerful digital data processing units that are the centerpiece of a computer (the processor). For this extremely important development, they shared the prestigious C. D. Draper prize awarded by the National Academy of Engineering. In 2001, Kilby was awarded the Nobel Prize, but this was not shared with Noyce since the Nobel Prize is not awarded posthumously.

Integrated circuit devices are called solid state when semiconductors (Ch. 10) are used. The integrated circuit concept ushered in a whole new branch of electrical engineering—microelectronics. An integrated circuit uses film deposition techniques applied to a small nonconducting single crystal of silicon dioxide (SiO_2) called a chip. By infusion of small amounts of phosphorus or boron, p- or n-type semiconductors are produced in the form of transistors (Ch. 10). The infusion process is called doping and the regions that are converted to semiconductor status or remain unchanged are controlled by an elaborate series of masking operations using a process known as photolithography.

Computers may be characterized as follows:

- Super computers - fastest, most powerful, most expensive
- Main frame computers - large with many workstations
- Microcomputers - medium size with a few workstations
- Stand alone personal computers (PCs)
- Dedicated computers built into a variety of instruments
- Portable computers:

Lap top	Notebook
Palmtop	Hand held

By the mid-1970s, integrated circuit technology was beginning to be used in the production of personal computers (PCs). There were those in the computer industry who opposed this trend believing that PCs could never challenge large main frame computers. There were others who saw the potential of PCs. Of course, as integrated circuit technology has progressed, the PC has become well established and main frame computers have faded in importance, particularly in universities.

In 1977 Wozniak and Jobs founded Apple Corp., produced Apple II, and launched the PC revolution. In 1984 Apple introduced the Macintosh PC, which was very easy to use (user friendly).

At first, most users wrote their own software, but eventually software was purchased from highly specialized companies. In 1975 W. Gates and P. Allen founded Microsoft Corp. which has been extremely successful in developing and marketing a variety of computer software.

In 1981 IBM entered the PC market that Apple had pioneered. Personal computers were widely accepted, and by 1995 one third of all US households had a PC.

The basic PC consists of three components:

- memory units

- central processing unit (CPU), the microprocessor
- input/output components

There are two types of memory—random access memory (RAM) and read only memory (ROM). The ROM stores the program to be executed and indicates where it is located. The RAM contains the data to be processed. The processor takes data from RAM, processes it according to the program stored in ROM and puts the results back into RAM.

Read only memory is usually stored magnetically on a spinning rigid platter inside a hard case (the hard disk). When a computer is turned off, information on RAM is lost. Therefore, any results residing there that are to be retained must be transferred to ROM by executing the SAVE command before shutting the computer down. It is important to periodically transfer important results from ROM to a compact disk (CD) or a floppy disk in case ROM is lost (crashes). A floppy disk is a magnetic storage unit contained in a rigid case. The present standard size is 3.5 inches and floppy disks are available as one and two sided models.

The input/output units handle data and convert it to digital form where necessary. Input devices include a keyboard and a mouse, which is a small device that moves a command cursor over the screen when the mouse is moved over a flat surface. Clicking the mouse selects an insertion point, a command to be used, or an Icon (small picture representing an action) to be activated.

Output devices include monitor screens (cathode ray tube or CRT as in television, or a liquid crystal display), a printer, or a storage device. Liquid crystal monitors are thin, flat, and consist of two very thin flat sheets of glass between which there is a substance having properties of both a liquid and a solid. It flows like a liquid, but like a solid its molecules are lined up. An electric field or temperature change can disrupt the alignment resulting in a change in color. Letters and numbers are created on a liquid crystal monitor screen by localized changes in voltage. The two types of printers in use today are the Laser and Inkjet types. A Laser printer activates the paper surface with a computer-controlled laser beam; a toner is applied that adheres to the activated regions and is secured in place by heat. An inkjet printer has a nozzle with many tiny holes through which ink is ejected to form patterns controlled by the computer.

The central processing unit (CPU) of the computer is on a circuit board (called the motherboard). It reacts to the program stored in ROM to carry out all the steps called for by the software. The CPU of a PC today is usually a very large scale integrated circuit (VLSI) on a silicon chip as discussed in Sec. 7.0 of this chapter.

An *algorithm* is a mathematical procedure that produces the solution to a problem in a finite number of steps. An algorithm that leads to a yes or no answer is a decision procedure. An algorithm that leads to a numerical answer is a computational procedure. Algorithms are located in ROM.

Software is a program of instructions to be followed written in terms of a language. A number of computer languages, which have been developed for specific purposes over time, are designated by the following acronyms:

- Fortran (Formula Translation) 1956—for mathematics problems—not for text
- Cobol (Common Business Oriented Language) 1959—for business and science
- Basic (Beginners All-purpose Symbolic Code) 1965—for circuit and mechanical analysis
- Pascal (named for a famous French scientist) 1970—for general purposes
- APT (Automatically Programmed Tools) 1957—for NC machine tools—discussed in Sec. 3.0
- C++, an object-oriented language
- Java, a popular language that is independent of the operating system

Personal computers are widely used for database management and process control. A database is a collection of data organized for rapid retrieval by a computer in data processing operations. The most popular platforms (PC computer systems) are Macintosh, IBM, Windows, and Linux. In addition to the basic PC components listed previously, a number of additional peripheral units may be involved:

- A modem (defined in Sec. 5.5 under e-mail)
- A speaker
- A CD-ROM and/or a floppy disk drive
- An additional hard disk drive to extend ROM
- A microphone

7.0 MICROELECTRONICS

Microelectronics involves digital solid state integrated circuits mounted on single crystal pieces of silicon called *chips*. The digital system used is binary which means that there are only two numbers involved (0 and 1). This is quite different from the conventional decimal number system where all members consist of combinations of 0 and 1 to 9. Figure 15.9 shows a few numbers in the decimal system together with their binary equivalents. Addition, subtraction, multiplication, and division are much simpler to perform in the binary number system than in the decimal system.

Decimal	Binary	Decimal	Binary
0	0000	8	1000
1	0001	9	1001
2	0010	10	1010
3	0011	11	1011
4	0100	••	••••
5	0101	••	••••
•	••••	14	1110
•	••••	15	1111

Figure 15.9. Comparison of numbers in decimal and binary codes.

A single 0 or 1 is called a *bit*. The number of bits required to characterize an equivalent analog signal is eight and this is called a *byte*. For example, the sequence of 8 bits corresponding to the byte representing the letter a is:

01000001

Other bytes correspond to other letters, numbers, punctuation, and mathematical symbols.

Modern microprocessor chips contain millions of switching elements (transistors, Ch. 10). A transistor used in a switching mode (going from 0 to 1 and vice versa) is called a *gate* and is analogous to an electrical relay, but very much faster and requiring far less energy. A relay employs an electromagnet as shown diagrammatically in Fig. 15.10 to close or open a circuit at 2 when energized (+ or –) at 1. Before the transistor was invented, vacuum tubes (diodes) were used in digital processors. Diodes are several orders of magnitude faster than electrical relays, but several orders of magnitude slower than transistor switching units.

The speed at which data is processed in a computer is extremely important, and this depends upon the length of a gate (distance between bits). Processor speed is measured in megahertz (1 MHz = 10^6 Hz). It is expected that soon chips having a speed of 1,000 MHz = 1 gigahertz = 1 GHz will be available having a gate length of 0.2 mm (about 1/500 the thickness of a human hair). To further increase the speed of a processor, pulses are transmitted by conductors in parallel. Use of 8 bit parallel transmission increases the speed by a factor of 8.

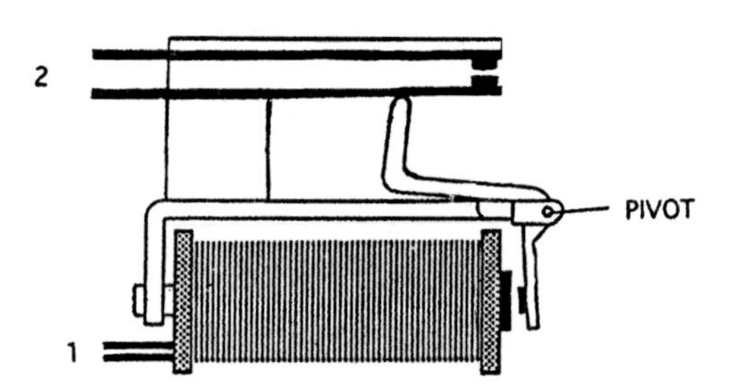

Figure 15.10. Diagrammatic sketch of an electrical relay.

In 1940 G. Stibitz of the Bell Laboratories used relays (Fig. 15.10) to demonstrate the possibility of using binary numbers to perform calculations. It soon became clear that relays were far too slow and energy intensive and they were replaced by diodes (Fig.10.19). The first NC machine (MIT 1952) used diodes. However, these were replaced by transistors as soon as they became available (late 1950s).

The starting point in microprocessor chip formation is the production of a cylinder (called a boule) of single crystal silicon dioxide by controlled crystallization. When integrated circuit technology was new, boules were only a few inches in diameter, but their diameter has grown with time, presently approaching a foot in diameter. A boule is next sliced into thin disks (wafers). The area of a wafer is sufficient to accommodate many chips, the size of a small fingernail, all of which are processed at the same time, employing a process called photolithography. This involves many steps of masking, deposition of other materials, heating, and removal of unwanted material by etching.

Masking is accomplished by applying a drop of photosensitive material (photoresist) to a rapidly spinning wafer which results in a thin layer of photoresist on the surface. This is exposed to light using a negative as in ordinary photography. The area exposed to light is converted to a polymer that is not removed by an acid washing solution. The remaining masking material determines the area that is not involved in further steps of doping (doping involves adding a small amount of material to the silicon to produce a p- or n-type semiconductor as described in Ch. 10). The final step is to evaporate a conducting material (usually Al) in a vacuum and remove unmasked areas to provide electrical connections between the many components of the chip. This is followed by packaging and testing.

Chip production is carried out in "clean rooms" where filtered air is maintained slightly above atmospheric pressure, since it is very important that dust particles not create problems in microprocessor chip production. Where possible, robots are employed, but where human intervention is required, workers are completely shrouded from head to foot in special lint-free material.

Integrated circuit chips are mainly produced at a few locations in the world:

- Near Palo Alto, CA (Silicon Valley)
- Texas (Silicon Plain)
- Phoenix, AZ area (Silicon Desert)
- Kyushu, Japan's southern island (Silicon Island)

Transistors were invented in 1947, but were not widely used in computers until about 1960. The integrated circuit was invented in 1958, but was not widely used in computers until well after 1970. In each case, it took somewhat over a decade for the new technology to be put to use. The integrated circuit chip initially involved only a few switching units, but once introduced the number increased rapidly. For example, in 1960 there were only about 10^2 components per chip, but by the end of the century (2000) this had increased to about 10^{10} components per chip.

8.0 COMPUTER SCIENCE

Computer technology was usually practiced in mathematics and electrical engineering departments of universities until about 1960 when computer science departments began to appear. Eventually, these departments gave courses associated with a variety of computer-related topics and some awarded degrees through the Ph.D. These departments covered computer architecture and design and many other computer details. An important detail is the design of digital logic circuits. A logic circuit is one that relates input to output in accordance with a set of rules. This involves knowledge of Boolean Algebra which consists of a collection of postulates and theorems associated with the mathematics of binary systems used in all modern PCs. It is interesting to note that Boolean Algebra had been developed by G. Boole in 1848 as an academic exercise long before there was any need for it in computer technology.

9.0 THE INTERNET

The internet had its origin in the 1960s. It is a huge network of computers that can send information by high-speed telephone and satellite links. It is expected to involve about 10^8 users soon after the beginning of the new millennium (2001). The internet is an open system with no central headquarters and it plays an ever-increasing role in society. It has been likened in importance to the invention of the printing press with movable type by Gutenberg at Mainz in the 1450s.

For information transfer, there is the World Wide Web (WWW) which consists of a large number of interconnected sites, (called web sites), each capable of providing information. These sites range in size from a small, personal single page to a huge volume of information such as the complete catalog of the Library of Congress, or the complete contents of the Encyclopedia Britannica. The software that enables one to navigate the web is called a browser. Two of the most popular browsers are Netscape Navigator and Microsoft Internet Explorer. Still another use of the internet is e-commerce. This involves buying and selling a wide variety of items ranging from books and compact disks (CDs) to stocks and bonds.

The internet's origin was the ARPANET, initiated in 1969, which linked military research contractors (including universities) doing work for the Advanced Research Projects Administration (ARPA). Beginning about 1980, there was a rapid increase in computer use in universities. The National Science Foundation (NSF) decided to fund a few very large and fast supercomputers and to use the ARPANET to enable universities to access these supercomputers. This swamped the ARPANET, and NSF replaced it by a more powerful NSFNET. The ARPANET was discontinued about 1990. The use of supercomputers proved to be a poor idea, but the NSFNET was so successful that beginning about 1994 it began to be taken over by a number of commercial companies (IBM, SPRINT, etc.). With this takeover, the internet was expanded to include other uses beyond research.

The public internet was so successful that large companies developed their own private networks of computers called intranets, which greatly reduced paperwork and speeded the design process.

A typical address for an internet web site would be: http//www.asu.edu. This is called a uniform resource locator (URL) and corresponds to the following English translation:

hypertext protocol//world wide web.ArizonaStateUniversity.education

The world wide web had its origin in the 1980s as a means of rapid communication between physicists to transmit new ideas and research results. This was initiated by T. Berners-Lee and a team of physicists at a laboratory in Switzerland (CERN). An important feature of this development was use of highlighted elements of the text called *links*. When these links are clicked, the computer brings up material concerning the topic of interest stored in other computers throughout the world. This makes it possible to obtain a very comprehensive international review of a wide variety of topics. Starting a search requires an index or director. This is the task of what is known as a search engine that brings up a large number of sites (addresses) related to a keyword (topic) of interest typed into the search box. There are a number of search engines available, two of the most popular being Yahoo and Alta Vista. The previously mentioned browser enables one to move from one subsite to another as links are activated on the WWW. The first browser for the Web was MOSAIC developed at the National Center for Supercomputing Applications (NCSA) in Illinois. The WWW is so effective in transmitting information throughout the world that it has been called the information superhighway.

10.0 ELECTRONIC MAIL

A very important aspect of the internet is the ability to send and receive electronic mail (e-mail). This involves composing a message on a computer, having the computer translate the message into base 2 digital format, and sending it over a telephone line to the local site of a server (an organization that provides access to the internet usually for a monthly fee). The coded message is transmitted to another server located almost anywhere in the world where it is stored until retrieved by the recipient. The recipient's computer translates the message back into conventional language, and the communication is complete.

The peripheral computer device that converts a message from normal language to digital form and vice versa is called a modem. This stands for modulation (conventional language to digital) and demodulation (the reverse conversion). Conversion to digital form before transmission is very important since data may be sent much more rapidly in digital form than in analog form. The speed of transmission is measured in bits per second. The speed of the modem should exceed 14,400 bits per second (abbreviated 14.4 K) since this is the maximum speed of digital transmission for an ordinary copper telephone line. To take advantage of a modem speed greater than 14.4 K requires a fiber optic telephone connection.

A fiber optic line transmits information by light waves rather than by electrical impulses. A *fiber optic* is a composite glass optical conductor having a refractive index from center to surface that prevents any light from escaping as the light moves down the fiber. Fiber optic conductors are superior to metal conductors since the speed of transmission is much greater. Several messages may be sent simultaneously along a fiber optic line by using light waves of different frequency instead of a single frequency as with a metal conductor. C. Kao, R. Maver, and J. MacChesney received the Draper prize for invention of the optical fiber for use in modern telecommunications systems.

An e-mail address is the cyberspace equivalent of a conventional mailing address. It must be unique and expressed in what is called hypertext protocol (abbreviated http://). A typical address in hypertext protocol would be:

jones.j@asu.edu
A B C

where:
A = a unique designation for the person at B
B = the location
C = the nature of the location (edu for educational institution in this case)

If the location is not in the United States, the country must also be indicated at the end of the address. For example:

For Japan: tyamasaka@u-Tokyo.ac.jp

For England: john.s@umist.ac.uk

In many countries, .ac is used as a designation for an educational institution instead of .edu.

Other organizational designations are:

- .com for a company or corporation
- .org for a nonprofit organization
- .gov for a government organization

Most software that provides use of the WWW also supports the use of e-mail. The items needed to use the internet (WWW and e-mail) include the following:

- A computer
- A modem
- An internet provider
- Appropriate software
- A browser with e-mail support

11.0 ENGINEERING APPLICATIONS

Probably the most important application of electronic computers in engineering is to the area of manufacturing. Development of NC and CAD discussed in Sec. 3.0 and Sec. 5.0 represent important breakthroughs. These had a slow start because the transistor invented in 1947 was not yet readily available, and the energy and space consuming vacuum diodes were still in use.

In 1959 there was another important manufacturing development. This was the machining center. Before its development, large items such as machine tool bases and engine blocks were moved along a transfer line from one machine to another where milling, boring, drilling, reaming, and other operations were carried out on machines that each performed a small part of the required processing. The machining center performed a large number of operations, one at a time, as the part remained stationary and secure. This was possible by having a number of tools in a storage unit that a computer could select, transfer to a spindle or chuck, and direct the machine to perform the necessary operation. Only after all the necessary procedures had been performed was the work moved. The tool changing unit of a machining center was a special form of robot that changed tools automatically, relieving the operator of the task.

There are several advantages to this method of production:

- A great savings of manufacturing space. This involved the space taken by the machining center vs the space occupied by several dedicated machine tools.
- In many cases, there is a large savings of capital associated with work in process. If 50 items were processed in a batch, a large percentage of these would be in a partial state of completion until the entire batch was processed.
- Greater flexibility of scheduling. Instead of batch production, it was possible to manufacture one unit at a time as needed. This was possible since changing the tools in the storage unit and changing the computer program takes a short time.
- A substantial reduction in the number of employees.

A major U.S. machine tool builder had just enlarged its factory to take care of an increased demand for milling machines. When the machining center was conceived and developed at this company, they suddenly found they had far more floor space than required and could find other uses for it. While the cost of a machining center was greater than any one machine, it was considerably less than the total cost of the machines it replaced.

By the early 1960s, practically every major machine tool builder had at least one NC machine in its program. Before about 1980, computers, in the form of workstations, were wired to central main frame computers. With the availability of PCs, the use of computers in all phases of manufacturing accelerated. Communication between design and manufacturing groups was greatly improved. In addition, production scheduling quickly became computer oriented. The need for many previously required drawings and paper work was quickly reduced, and there was talk of the possibility of a paperless factory.

A new computer-related development in industrial engineering was Computer-Aided Manufacturing which became closely linked with Computer-Aided Design (CAD/CAM).

In manufacturing, there are several activities that require careful coordination such as design, database control, production scheduling and planning of procurement, marketing, and scheduling to accommodate product delivery requirements. All of these were strengthened by the improved sharing of information a network of computers can provide.

Increased use of computers in manufacturing has led to many innovative changes. Most of these are covered by the term Computer-Integrated Manufacture (CIM). This covers all aspects of a manufacturing enterprise:

- Product design
- Process design
- Design for assembly
- Purchasing
- Concurrent engineering
- Product scheduling
- Quality control
- Marketing
- Customer support

All of these items must be integrated into a system, and PCs or workstations play an important role in doing this. A distinction is made in CIM between numerical control involving a computer for each machine tool, termed Computer Numerical Control (CNC), and where one computer services a group of NC machines. The latter is termed Distributed Numerical Control (DNC).

Automation is a term introduced in the automotive industry in the mid-1940s. This pertains to the use of controlled mechanisms to replace human operators. Robots represent an important form of automation. The first robots were used about 1961 to unload die cast parts. Robots are computer-controlled devices that perform tasks that might otherwise be done by humans. They are particularly useful in hazardous environments, and for repetitive tasks that require more energy or concentration than a person can be expected to provide over large periods of time. Robots are widely used in the automotive industry for spray painting, welding, and a variety of assembly operations. Robots have been termed steel collar workers as opposed to blue collar workers.

Servomechanisms are devices used to control motion. Two basic types involve hydraulic or electromechanical drives. The latter types involve stepping motors or DC variable speed motors. A stepping motor receives controlled pulses of energy that cause stepwise rotation in response to each pulse.

Computers are now employed in every branch of engineering, particularly in electrical engineering where most of the computer technology was originally developed. In electrical engineering, computers are used in circuit design and analysis and for power distribution and control.

In civil engineering, computers are used for stress analysis and the design of structures and also in the planning of highways and other transportation systems.

In chemical engineering, computers are used in the control of plants concerned with refinement operations, pharmaceutical production, and in a large number of operations involving chemical synthesis. Computers are particularly important in the control of plants which employ processes involving the continuous flow of material as opposed to batch processing.

In aeronautical engineering, computers are used in design, wind tunnel testing, and the production of complete aircraft. Also, embedded computers are used in many aeronautical instruments (radar, inertial guidance, flight simulation, etc.).

In all of the many applications of computers in engineering, the organization and management of databases are involved. A *database* is a collection of data organized for rapid retrieval.

Word processing is important to all areas of human endeavor. As early as 1936, the autotypist appeared which used punched tapes for storage. This was followed in 1964 by the electric typewriter produced by IBM which used magnetic tape and is considered to be the first word processor. Word processing is now an important feature of all modern PCs. This has made the conventional typewriter that uses an inked ribbon obsolete, just as the slide rule has been replaced by scientific electronic calculators.

A spreadsheet is a table consisting of rows and columns of data used to analyze data. In engineering, spreadsheets are used to evaluate equations when variables change through a range of values. Spreadsheets are also used in financial analysis, the development of budgets, and in statistical analysis. Special software (such as Excel) is available that makes it convenient to employ a spreadsheet.

12.0 THE BAR CODE STORY

In 1948 B. Silver, a graduate student at Drexel Institute of Technology (now Drexel University) in Philadelphia, overheard the president of a large food chain trying to convince a dean that there was a great need for a system which would automatically record the removal of items from inventory at the cash register. The dean alleged this was not an appropriate activity for a technical institution. Silver mentioned this to another graduate student, N. J. Woodland, who was fascinated enough with the problem to quit school temporarily to work on it. After some time he came up with an idea related to the Morse Code in which dots and dashes were extended downward to give thin and wide lines (a bar code). He planned to measure change in intensity of reflected light from the bar code using a vacuum tube that converted the changes in reflected light into an electrical signal. He returned to Drexel where he and Silver continued to develop the idea. A patent was filed by Woodland and Silver on October 20, 1949.

After graduation in 1951, Woodland took a job with IBM in Binghamton, New York, but continued to develop the idea. Based on some crude experiments, he and Silver found that the idea worked, but how to make this into a practical device for use in supermarkets remained a mystery. The patent was granted in 1952 and sold to Philco in 1962. Silver died at the early age of 38 in 1963. Philco later sold the patent to RCA.

R. J. Collins worked at Sylvania Corp. He was assigned to finding a use for a new computer that had been developed there. He had worked for the Pennsylvania Railroad during his undergraduate days and was aware of the difficulty of keeping tabs on freight cars as they wandered across the country. He suggested use of some sort of code that might be fed into a computer and presented this to Sylvania. When they rejected the idea, he left and founded the Computer Identics Corp. with another engineer. While his idea was successful, it proved to be too costly for the railroads.

In the late 1960s, lasers were becoming available, and Computer Identics successfully applied a helium-neon laser to read bar codes. In 1969 they developed a bar code system for General Motors in Pontiac, MI to monitor the production of automotive axles, and another system for the General Trading Company of Carlsbad, NM to monitor shipments of products. These were very simple systems, but were successful. In mid-1970 RCA, in conjunction with the Kroger Grocery Chain and an industry consortium, developed standards for bar code use in the grocery industry.

When IBM saw the potential for bar codes demonstrated by RCA and the consortium, they remembered that Woodland had been the co-inventor of the basic patent that had run out in 1969. He was transferred to an IBM plant in North Carolina where he helped develop the Universal Product Code (UPC) adopted in April 1973 which is now used throughout the world.

The two items that became available in the 1960s that made reliable scanners affordable were low cost lasers and integrated circuits. The first laser was produced by T. H. Maiman in 1960 (Ch. 10). Without the transistor, laser, and integrated circuit, the brilliant idea of Woodland and Silver would not have been possible. While Woodland never became wealthy for his contributions to the bar code revolution, he has the satisfaction of having been awarded the National Medal of Technology by President Bush in 1992. He was also elected to the prestigious Drexel 100, which is an association of the 100 most successful graduates of his Alma Mater.

This example is typical of new developments in engineering. Sometimes a good idea is ahead of its time, and to develop its full potential requires the invention and development of other essential technology. The development of NC and the bar code are excellent examples of this. In the case of NC, the new technology required was simplified programming, the transistor, and the integrated circuit. In the case of the bar code, the missing technology was fast, inexpensive computers and the laser. In all of these cases, the importance of patience, suggested by Dr. Inaba, is essential.

Epilogue

In closing, it is important to mention that there are two types of talented engineering students and faculty members. One type is mathematically oriented and mostly interested in analytical solutions of engineering problems. The major interest of the other type is in devising innovative practical solutions and in experimentally testing the feasibility of any new ideas that result. Both types play an important role in engineering. To maximize success in graduate studies, it is important that the student's interest be close to those of the thesis advisor with whom the student decides to work.

Today the tendency is for many talented graduates to enter the academic field immediately after receiving their Ph. D. or Sc. D. and before gaining experience in solving real world engineering problems. In the absence of broad consulting experience the result has been for an imbalance to exist between analytically-oriented professors and those interested in a more innovative and experimental approach.

Recently Dr. Dave Lineback has presented "A Short Discourse On Engineering Education" written in the manner Galileo might have expressed it in dialog form today:

Salviati: *"Ever noticed that, for the most part, engineers are educated by folks who, for the most part have had little–if any–opportunity to acquire practical engineering experience?"*

Simplicio: *"How's that?"*

Salviati: *"Seems to me that most of the engineering faculty I know are pretty deft at handling a differential equation, but only a few of them have actually been presented with a chance to apply their knowledge of science and mathematics in real engineering applications to find practical solutions to practical problems,"*

Sagredo: *"That's interesting. With the exception of those who go on to obtain advanced degrees and themselves become teachers, don't their former students experience just the opposite on the job?"*

Salviati: *"Yes, indeed, it's a safe bet that many–if not most–engineering students will never solve a single differential equation during their entire professional career. But, they do get a great deal of experience on the job working with others in designing and building such things as the roads, bridges and automobiles needed to get folks from Point A to Point B."*

Sagredo: *"Then I would think that learning to solve practical engineering problems is a very important part of their education."*

Salviati: *"Absolutely! And, while most folks involved in the practice of engineering would agree with you, we also know that design projects and laboratory exercises are difficult to do in the academic environment because of the time and effort they require to make them a meaningful part of an engineer's education."*

Simplicio: *"Not to mention the financial expense."*

Salviati: *"That, too, of course!"*

The above discourse appeared in the October 2000 issue of *Experimental Stress Analysis NOTEBOOK* published by the Measurements Group, Inc. of Raleigh, NC, and included here with permission of the Editor, Dr. L. D. Lineback.

Appendix A

A Historical Introduction to Calculus

Isaac Newton first introduced the mathematical subject we now call calculus in connection with his theory of universal gravity in the book *Philosophiae Naturalis Principia Mathmatica* first published in 1687. Calculus involves the mathematics of infinitesimals. For example, if a quantity (x) varies with time (t) as follows:

Eq. (A-1) $x = At^2$

where A is a constant. An infinitesimal change in x may be expressed as:

Eq. (A-2)

$$\Delta x = A(t + \Delta t)^2 - At^2 = A(t^2 + 2t\ \Delta t + [\Delta t]^2) - At^2$$

where Δt is a very small change in t.

Dividing both sides by Δt and letting Δt approach zero:

Eq. (A-3) $\Delta x/\Delta t = 2At$

Newton called $\Delta x/\Delta t$ the fluxion of x which he designated $\dot{x}$.

Similarly, if $x = At^3$, $\Delta x = A(t + \Delta t)^2 - At^3$, then:

Eq. (A-4) $$\lim_{\Delta t \to 0} \Delta x/\Delta t = 3At^2$$

In general, if $x = At^n$, $\dot{x} = nAt^{n-1}$. This also holds when n is negative. That is, if $x = A/t^2 = At^{-2}$, then $\dot{x} = -2At^{-3}$. It also follows that if $x = At^{-3}$, $\dot{x} = -3At^{-4}$.

1.0 DERIVATIVE CALCULUS

Soon after Newton invented calculus, Gottfried Leibnitz (a German mathematician) independently introduced a similar line of reasoning using different terminology. Instead of using $\dot{x}$ to denote the fluxion of x, he expressed it as dx/dt and called this the derivative of x. Despite the difference in terminology, the two systems are the same. Today, the derivative of a quantity (s) with respect to time is sometimes written $\dot{s}$, but when the independent variable is some other variable such as x, the Leibnitz notation is used (i.e., ds/dx).

If s is a function of x as in Fig. A-1, then physically the derivative ds/dx at a given value of x_1 is the slope of the curve at x_1. Similarly, if displacement (y) is a function of time (t) as in Fig. A-2, then the derivative dy/dt at time (t_1) is the velocity (v_1) at this point. Also, if velocity (v) is a function of time (t) as in Fig. A-3, then the derivative dv/dt at a given point t_2 is the acceleration at this point. This is also equal to $d(dy/dt)/dt$ and is called the second derivative of y with respect to t, and is written d^2y/dt^2, or $\ddot{y}$.

Figure A-4 shows the variation of a quantity (q) relative to a variable (x) having a maximum at (1) and a minimum at (2). The slopes of the q vs x curve at (1) and (2) will be zero (i.e., dq/dx at 1 and 2 will be zero).

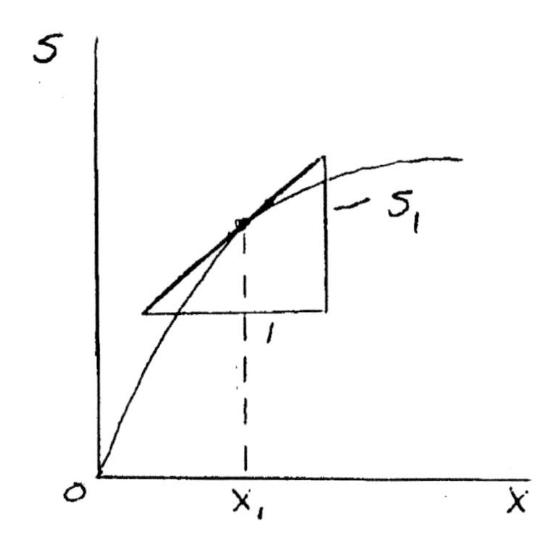

Figure A-1.

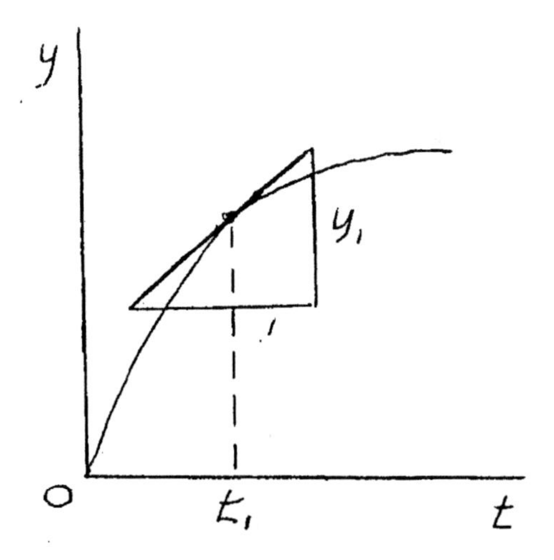

Figure A-2.

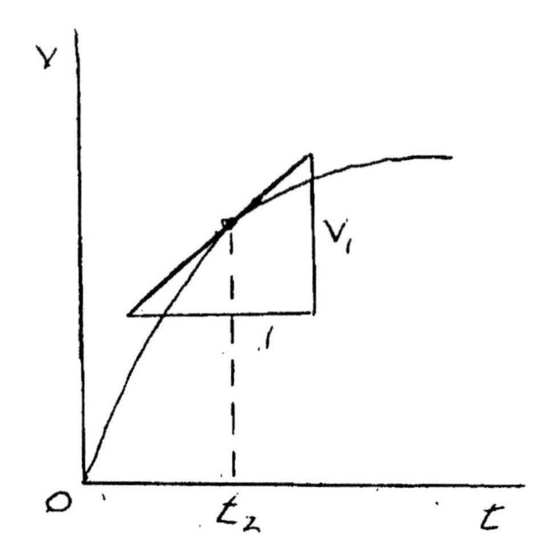

Figure A-3.

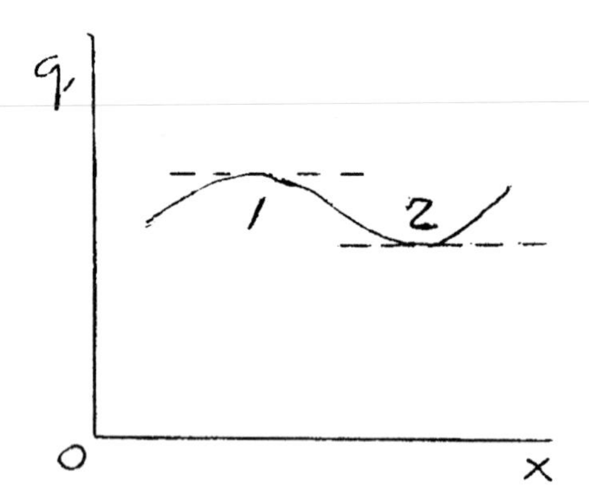

Figure A-4.

When derivatives other than those involving Aq^n or Aq^{-n} are required, they are not derived from scratch, but a table of derivatives found in many handbooks is consulted. A few of the more common derivatives are given below (where A is a constant, q represents some variable such as x, y, or t, and u and v are functions of some variable).

Eq. (A-5) $dA = 0$

Eq. (A-6) $d(Aq^n) = nAq^{n-1}dq$

Eq. (A-7) $d(Aq^{-n}) = -nAq^{-(n+1)}dq$

Eq. (A-8) $d(u + v) = du + dv$

Eq. (A-9) $d(A + u) = du$

Eq. (A-10) $d(uv) = udv + vdu$

Eq. (A-11) $d(u/v) = (vdu - udv)/v^2$

Eq. (A-12) $du^{0.5} = 0.5u^{-0.5}du$

Eq. (A-13) $d(\ln u) = du/u$

Eq. (A-14) $d(\sin u) = (\cos u)du$

Eq. (A-15) $d(\cos u) = -(\sin u)du$

At the end of the article on electronics in Ch.10, the derivative of $E^2R/(R + r)^2$ with respect to R was required (E and r being held constant). This is called a partial derivative and is designated $\partial[(E^2)/(R + r)^2]/\partial R$. This may be evaluated by use of Eq. (A-11) as follows:

$$\partial[(E^2R)/(R + r)^2]/\partial R = [(R + r)^2\partial(E^2R)/\partial R - (E^2R)\partial(R + r)^2/\partial R]/(R + r)^4$$
$$= [E^2(R + r)^2 - 2E^2R(R + r)]/(R + r)^4$$

(Eq. 10.30)

2.0 INTEGRAL CALCULUS

Integration is the inverse of differentiation. If we know the acceleration of a particle is a constant (a_0), and wish to know the change in velocity (Δv) between two times (t_1) and (t_2) (Fig. A-5), this will correspond to the sum of the incremental velocity differences ($a_0 \Delta t$) extending from $t = t_1$ to $t = t_2$. That is,

Eq. (A-16) $$\Delta v_{1-2} = \sum_1^2 (a_0 \Delta t)$$

where $\sum_1^2$ indicates that the sum of the incremental values ($a_0 \Delta t$) is to be taken from $t = t_1$ to $t = t_2$. This would be written in the terminology of calculus as follows:

Eq. (A-17)

$$\Delta v_{1-2} = \int_1^2 a_0 dt = a_0 (t_2 - t_1) \text{ [from the inverse of Eq. (A-6)]}$$

where $\int$ resembles a S for summation and is called the integral sign. When the limits over which the summation is to be taken are indicated below and above the integral sign, this is called a definite integral.

When limits are not specified, this is called an indefinite integral and a constant must be added since the derivative of a constant is zero [Eq. (A-5)]. Thus,

Eq. (A-18)

$$\Delta v_{1-2} = \left(\int a_0 dt + C\right)_{1-2} = a_0 t_2 + C - (a_0 t_1 + C) = a_0 (t_2 - t_1)$$

This is the same as the result from the definite integral [Eq. (A-17)].

Just as tables of derivatives are to be found in engineering handbooks, tables of integrals are also available. A few of the more common integrals are given below.

Eq. (A-19) $\int a\,du = a\int du$

Eq. (A-20) $\int (u + v)dx = \int u\,dx + \int v\,dx$

Eq. (A-21) $\int u\,dv = uv - \int v\,du$

Eq. (A-22) $\int dx/x = \ln x + C = \ln C'x$

Eq. (A-23) $\int e^x dx = e^x + C$

Eq. (A-24) $\int \sin x\,dx = -\cos x + C$

Eq. (A-25) $\int \cos x\,dx = \sin x + C$

Eq. (A-26) $\int dx/x^2 = -1/x + C$

3.0 EXAMPLES INVOLVING INTEGRATION

In Ch. 2, it was found that:

Eq. (2.14) $dT/T = f_s\,d\alpha$

where f_s is a static coefficient of friction.

Integrating both sides of this equation, using Eq. (A-24):

Eq. (2.15) $\ln T_1 - \ln T_2 = f_s(\alpha_2 - \alpha_1) = f_s\alpha$

and

Eq. (2.16) $T_1/T_2 = e^{fs\alpha}$

In Ch. 2, the motion of a projectile is considered. This will now be reconsidered in terms of definite integrals. Ignoring air drag, the projectile in Fig. 2.10 will be subjected to a constant deceleration due to gravity equal to $-g$, and:

Eq. (A-27) $d^2y/dt^2 = -g$

Its velocity will be:

Eq. (A-28) $dy/dt = \int -g dt = -gt$

where time is taken to be zero at launch and t is some later time of interest. The vertical displacement (y) at time (t) will be:

Eq. (A-29) $y = \int_C^t -gt dt = -(gt^2)/2 + Ct$

where C is the vertical velocity of the projectile at $t = 0$ ($= v_0 \sin\theta$), i.e.,

Eq. (2.24) $y = -(gt^2)/2 + v_0 \sin\theta$

Appendix B

Conversion Factors

Length: 1 in. = 2.54 cm = 0.0254 m
1 m = 39.37 in. = 3.28 ft
1 mile = 5280 ft = 1.609 km
1 μin = 25 nm = 250 Å
1 Å = 10^{-10} m = 0.1 μm
1 μm = 40 μin. = 10^{-6} m

Volume: 1 ft^3 = 0.0283 m^3
1 U.S. qt = 9.464×10^{-4} m^3
1 U.S. gal = 3.785×10^{-3} m^3 = 3.785 ℓ

Mass: 1 lb (mass) = 0.454 kg
1 dyne = 10^{-3} g (mass)

Density: 1 lb (mass)/$in.^3$ = 2.768×10^4 kg/m^3
1 g/cc = 10^3 kg/m^3

Speed: 1 mph = 0.447 m/s
1 m/s = 3.281 fps

Acceleration:	1 ft/s^2 = 0.3048 m/s^2
	1 m/s^2 = 3.281 ft/s^2
Force:	1 lb = 4.448 N = 454 g
	1 N = 0.2248 lb
	1 dyne = 10^{-5} N
Pressure:	1 Pa = 1 N/m^2
	10^3 psi = 6.895 MPa
	1 dyne/cm^2 = 0.1 Pa
Energy:	1 Btu = 1,065 J
	1 erg = 10^{-7} J
	1 J = 1 Nm
	1 kWh = 3.6×10^6 J
	1 cal, g = 0.0012 Wh = 4.185 J
Power:	1 Btu/s = 1.054×10^3 W
	1 W = 1 J/s
	1 hp = 746 W
Specific Heat:	1 cal/(gC) = 4.184×10^3 J/(kg.K) = 1.73 W/(mK)
Volume Specific Heat:	1 Btu/(in.^{3}F) = 2.77×10^4 J/(m^{3}K)
Thermal Conductivity:	1 Btu/(h.ft.F) = 1.73 W/(mK)
	1 cal/(s.cm.C) = 418.4 W/(mK)
Viscosity:	1 Reyn = 1 (lb.s)/in.^{2} = 68,950 p
	1 p = 1 dyne/cm^2
	1 cp = 10^{-3} (N.s)/m^2

Appendix C

Abbreviations

Å	=	Angstrom unit
dim. ind.	=	dimensionally independent
e	=	coefficient of linear expansion
h	=	hour
hp	=	horsepower
in.	=	inch
J	=	Joule
°K	=	degrees Kelvin
kg	=	kilogram
km	=	kilometer
kWh	=	kilowatthour
ℓ	=	liter
lb	=	pound
m	=	meter
mph	=	miles per hour
mi	=	mile
N	=	Newtons

nm	=	nanometer
Pa	=	Pascal
p	=	poise
psi	=	pounds per square inch
qt	=	quart
°R	=	degrees Rankin
s	=	second
W	=	Watts
μin.	=	microinch
μm	=	micrometer
Btu	=	British thermal unit
°C	=	degrees Celsius
cal	=	gram calorie
°F	=	degrees Fahrenheit
g	=	gram
gal	=	gallon

Answers to Problems

Where answers are not given they may be found in the text.

CHAPTER 2

2.1. a) 32.16 ft/s^2
b) 31.7 ft/s^2
c) 27.7 ft/s^2
d) 10.1 ft/s^2

2.2. $T_1 = 73.2$ lbs
$T_2 = 89.7$ lbs

2.3. a) 1,000 ft lbs
b) 100 lbs
c) 3,500 ft lbs
d) 600 lbs

2.4. $F_A = -100$ lbs
$F_B = 200$ lbs

2.5. $F_{CD} = 0$ lbs

2.6. $F_{DE} = +666$ lbs

2.7. $F_{BC} = -9$ lbs

2.8. $f_{\min} = 0.01$

2.9. a) 52.5 mph
b) 0.75 g

2.10. a) 279 m
b) 5.36 s
c) 56.7 m/s
d) 23.5^0

2.11. a) $P_{in} = Fy_1$, $P_{out} = 2Fy_2$
b) $y_1 = 2y_2$

2.12. $e = 4$

2.13. 17.32 ft

CHAPTER 3

3.2. a) $(\rho V^2 D^2)/F$
b) dim. ind.
c) dim. ind.

3.3. a) dim. ind.
b) $(Ba^3)/V^6$
c) dim. ind.
d) dim. ind.

3.4. a) $(\rho V^2 D^2)/T_e$
b) = dim. ind.
c) = dim. ind.

3.5. $h/d = \psi(T_e/\gamma d^2)$

3.6. $h/d = \text{const } (\gamma d^2)/T_e$

3.7. a) dim. ind.
b) 4
c) $F_D/(\rho S^2 V^2) = \psi[\mu/(\rho VS), (gS)/V, C/V]$

3.8. a) $V \sim (\lambda)^{0.5}$
b) if λ is larger, V will be larger

3.9. 417×10^{-6} lb/in.

3.10. a) $(P/d^{1.5})(T_e/\rho)^{0.5} = \text{const}$
b) freq inc as d decreases

3.11. $Te/(\gamma d^2) = \psi(B/d^3)$

3.12. a) $V/(gh)^{0.5} = \text{const}$
b) If h doubles, V increases by $(2)^{0.5}$

3.13. Same as Problem 3.7.

3.14. Same as Problem 3.12.

CHAPTER 4

4.1. $S_y = 10{,}000$ psi

4.2. $S = 1{,}273$ psi

4.3. a) $e = 42.4$ μin.
b) $\ell = 5.000212$ in.

4.4. 3,000 psi

4.5. 1,000 psi

4.6. 8.33 psi

4.7. a) 10,186 psi
b) 127 psi
c) 127 psi

4.8. 5,093 psi

4.9. S_m/τ at $A = 80$

4.10. 9,000 psi

4.11. 180 psi

4.12. $S_{mA}/S_{mB} = b/2c$

4.13. n

4.14. 1/8

4.15. 489 psi

4.17. 180 psi

4.18. S_m at $A = (2/3)(w\ell/bc^{2})$

4.19. $0.134\, d_0$

4.20. 0.0022 in.

4.21. 0.004 in.

4.22. $S_m = 339.5$ psi, $e = 11.3 \times 10^{-6}$

4.23. $S_m = 6{,}000$ psi, $e = 0.0006$

4.24. 145,000 lbs

4.25. 580,000 lbs

4.26. 1,667 lbs

4.27. 1.1 in^2 (steel), 92.3 in^2 (concrete)

4.28. a) $P(\sigma d/m)^{0.5} = \psi(D/d, H/d, \ell/d, R/d)$
b) 0.58

4.29. a) $\Delta\ell/d = \psi(D/d, Wd^{-2}/G, n)$
b) $\Delta\ell/d = Wd^{-2}/G\ \psi(D/d, n)$
c) W must be doubled
d) $W = 1/8$

4.30. a) not possible
b) yes, there is now no quantity including time in ψ_1 as given.
c) Must use mass/length (w/g) instead of weight/length (w);
d) If $\ell_1 = 5.05$ in, $f_1 = 2{,}000$ Hz. For $\ell_2 = 3.38$ in., $f_2 = 10^4$ Hz. For $\ell_3 = 2.84$ in., $f_3 = 20{,}000$ Hz.

CHAPTER 5

5.1. 1.12×10^{-3} lb.s.in^{-4}

5.2. 0.00092 psi

5.3. 2.4 psi

5.4. 0.2 psi

5.5. a) 0.0315 lb/in^3
b) below O_2

5.6. a) 1.05 oz
b) 0.89 g

5.7. a) 1.14 oz
b) 1.04 g

5.8. 6.696 lb/ft

5.9. 11.16 ft

5.10. 80 lbs

5.11. a) 14.88 psi
b) 0.52 hp

5.12. a) 32.68 psi
b) 4.63 hp

5.13. a) 148.8 psi
b) 5.21 hp

5.14. a) 0.00094 in.
b) 0.0071
c) 0.2 hp

5.15. a) 0.0008 in.
b) 0.0014
c) 0.2 hp

5.16. a) 0.0009 in.
b) 0.0048
c) 0.09 hp

5.17. a) 0.00061 in.
b) 0.001
c) 0.285 hp

5.18. a) $p_a = 0.462h$
b) 823 mm
c) 34.65 ft

5.19. 5.62 psi

5.20. a) 155,000 ft lbs
b) 15,500 ft lbs
c) 15,528 ft

5.21. a) larger
b) 17.6 psi

5.22. a) $(Q/d^2)(\rho/\Delta p)^{2.5} = \psi(D/d)$
b) Q increases by 2

5.23. b) $Q = \text{const } d^2(\Delta p/\rho)^{0.5}$

5.24. a) 34 ft
b) 35.9 fps
c) 3.13 ft^3/s

5.25. a) $Q/(g^{0.5}h^{2.5}) = \psi(\alpha)$
b) 565%

5.26. a) 5.78 psi
b) 2,721 hp

5.27. a) 0.0319 in.
b) 0.0161 in^2/s

CHAPTER 6

6.1. a) 85 tons
b) 3,196 lbs

6.2. a) 28.4 tons
b) 1,136 lbs

6.3. a) 21.3 tons
b) 799 lbs

6.4. a) 1,136 lbs
b) 1,212 hp

6.5. 28.4 tons

6.6. 33% greater force

6.7. 21% greater force

6.8. a) 246 Hz
b) 246 Hz

6.9. 0.21

6.10. Q increases by $2^{0.5}$

6.11. Sp. Gr. = 0.88

6.12. From 10.4 lbs to 41.6 lbs

6.13. yes

6.14. path *b*

6.15. a) Weight, buoyant, drag forces
b) 4.27×10^{-5} ips
c) $6.51h$

6.16. a) 869 ips
b) 81 ft

6.17. a) 15.62 lbs
b) 4.17 hp

6.18. a) 8.24 hp
b) 10.3 hp

6.19. a) 0.0005in.
b) 140 Hz
c) 72.9 fps (49.7 mph)

CHAPTER 7

7.1. 250 rpm

7.2. a) 0.002 in.
b) 0.002 in.
c) 360 lbs
d) 2,160 lbs
e) 150,000 lbs

7.3. a) $f = \psi(c/d, R/d, \mu N/P)$
b) 1

7.4. 88,900 lbs

7.5. 1,600 lbs

7.6. a) 65.2 hp
b) 261 hp

7.7. a) 2.25 ft^3/s
b) 0.97 in.
c) 8,230 hp

7.8. a) 159 ft
b) 1,320 hp

7.9. a) 1,581 rpm, 9.49 t^3/s
b) 30,682 hp
c) 90%

7.10. a) 468 ft
b) 15.85 ft^3/s
c) 819 hp

7.11. a) 112 ft^3/s
b) 64,000 hp

7.12. a) $P/(\rho N^3 d^5) = \psi(V/dN)$
b) $(\rho N^3 d^2)/\mu$

7.13. 66.4 hp

7.14. a) w, d, t, b, e,
b) $\delta/d = \psi(E/Wd^2, b/d, t/d)$
c) 6,667 lbs
d) 0.1 in.
e) $\delta \sim (Wd^7/Ebt^3)$

7.15. 0.0016 in.

7.16. a) For scale model $\delta/\ell = \psi(E/W\ell^2)$
b) 33,300 lbs

7.17. a) $\delta/\ell = \psi(I_n/\ell^4, \sigma_f\ell^3/U)$
b) $U_p = 20{,}000$ lbs, $\delta_p = 1$ in.

7.18. 1.188 lbs

CHAPTER 8

8.1. a) 0.0096 µin.
b) 0.00024 µm
c) 2.4 Å

8.4. *fcc*

8.5. Cu: 0.28 Pb: 0.62

8.6. a) cold worked
b) hot worked

8.7. a) 405 *C*
b) 27 *C*

8.8. $B = \psi_1(W, L, H)$ Eq. 8.8

$[L^3]$ $[F], [L], [FL^{-2}]$ Dimensions involved (F, L)

Take W and L as Dim. independent Quantities

$B/L^3 = \psi_2[(HL^2)/W]$ Eq. 8.9 By application of D.A.

8.9. Eq. 8.10 $B \sim L \rightarrow B = KL$ K= Constant

$$BW^aH^b = KL$$

$$[L^3]\,[F^a]\,[FL^{-2}]^b = [L]$$

For this to hold a = 1 and b = -1

$$[L^3]\,[F]/\,[FL^2] = [L]$$

8.10. a) Finer grain size
b) reduction in hardness

8.11. a) Increase
b) decrease
c) decrease

8.12. 167 mm^3

8.13. Elasticity

8.14. Elasticity

CHAPTER 9

9.4. Low—to prevent Cr depletion.

9.5. To retain austenite at room temp—prevents Cr depletion.

9.6. W and Mo

9.7. F

9.8. No

9.11. Roller bearings

9.15. Add S for cross linking.

9.16. Glass = amorphous, ceramic = crystalline structure.

9.17. SiO_2

9.18. Pure SiO_2

9.20. 1% portland cement, 2% sand, 4% crushed stone

9.21. Compression

CHAPTER 10

10.1. 89.9 N

10.2. a) 100 J
b) 100 W

10.5. a) 0.55 Ω
b) 0.55 W

10.6. a) 0.34 Ω
b) 0.56 Ω

10.7. 6.92 Ω

10.8. a) 5.805 Ω
b) 3.45 A
c) $I_2 = 0.67, I_3 = 2.78$

10.9. $I_1 = 3, 9A, I_2 = 0.867$ A, $I_3 = 3{,}033$ A

10.11. a) 2 Ω
b) $I_1 = 8$ A, $I_2 = 4$ A
c) 24 V

10.13. N pole

10.14. Counter clockwise

10.15. N to S

10.16. Opposite direction

10.17. Motor

10.18. 150 Ω

10.19. a) 1,833 turns
b) 13.6 mA

10.20. 10^5 turns

10.21. -1/2

10.22. a) 4 Ω
b) 0.57 Ω

10.23. a) 0.2 Ω
b) 1.8 Ω

10.24. a) 100 Ω
b) 10 Ω

CHAPTER 11

11.3. a) $\alpha = \psi(\ell_0/t, \Delta e \Delta\theta)$
b) $R/t = \psi(\ell_0/t, \Delta e \Delta\theta)$

11.4. a) $\alpha = (\ell_0/2t)(\Delta e \Delta\theta)$; $R = 2t/\Delta e \Delta\theta)$

11.5. 1/4

11.6. 8%

11.8. a) 13.3 ft^3
b) 2.49 ft^3
c) 1,917 lb/ft^3

11.9. 39%

11.10. a) $P = \psi_1$ (shape, ℓ, N, $\bar{p}$, e)

$P/\bar{p}N\ell^3 = \psi_2$ (shape, e)

b) $P/\bar{p}N\ell^3 = \psi_3$ (shape, e)

11.11. a) 13,335 W
b) 41,910 W
c) 191 W

11.12. a) 266,000 kW
b) 125,500 kW

11.13. 33 C

11.14. 2.37 ft

11.15. a) 17.6 C
b) 12 C

CHAPTER 12

12.3. a) 41.6 rpm
b) $L/D = 1.73$

12.4. $L = 5$ oz, $D = 3.54$ oz
b) $d = 25$ in.

12.5. a) 1.73 hp
b) 46.7 hp

12.6. a) 25 hp
b) 200 hp

12.7. 0.29 hp

12.8. a) 0.35 hp
b) 250 rpm

12.9. a) 118 rpm
b) 1.05 hp (in) 0.42 hp (out)

12.11. a) 6.68 g
b) 6.68 g

12.12. a) $V_p/V_m = 1$
b) $W_p/W_m = 1{,}000$

12.14. a) 3
b) net hp = 175
c) 180 hp OK
d) 51%

12.15. a) 21 hp
b) 0.424
c) 1,563 lbs
d) 41.7 hp

12.16. b) $\delta = \text{const}\ (Wbt^3)/(Er^5)$

12.17. $(Q\ell\mu)/(Gd^4) = \text{const}$
c) $Q/(G^{0.5}d^{2.5}) = \text{const}$

CHAPTER 13

13.1. $1,187

13.2. $1,480

13.3. $123.30

13.4. $265

13.5. *A* better

13.6. a) average yearly cost, a) $2,162 b) $2,002
b) *B* better

13.7. $1,935

13.8. *B* better

13.9. $813 saved next two yrs (plan *B*)

13.10. Different car costs more.

13.11. New car is better.

13.12. New car is better.

13.13. a) 0.381 in.
b) same

13.14. a) 7.67 ft
b) 95%

13.15. a) 10 in. pipe
b) 77%

13.16. \$0.93 difference

13.17. 6 yrs

CHAPTER 14

14.1. a) 0.3 m/s^2
b) 0.1 m/s^2
c) P most important

14.2. a) $\bar{x} = 1.108$
b) 1.11
c) 1.18

14.3. a) 1.01 to 1.20 in.
b) $\bar{d} = 0.068$ in.
c) $\sigma = 0.0704$ in.

14.4. a) 0.241
b) 0.243

14.5. a) 3.694×10^{-6}
b) 3.501×10^{-5}

14.6. 0.059

14.7. a) 1/20
b) $(0.05)^n$
c) $(0.95)^n$

14.8. 0.028

14.9. 6

14.10. 12

14.11. 5,040

14.12. 50,400

14.14. a) between $x = 10$ and 40
b) between $x = 20$ and 30

14.15. a) 81.5%
b) 17.1%
c) 1.35%
d) 0.05%
e) 0%

14.16. a) 88.6%
b) 10.7%
c) 0.6%
d) 0.0%

14.17. a) $V = \pm 1.95$, $v = \pm 5.95$
b) v is 3× as effective
c) for low roughness: high V, low v

Name Index

Subject Index